CRC Series in Chromatography

Editors-in-Chief

Gunter Zweig, Ph.D. and Joseph Sherma, Ph.D.

General Data and Principles
Gunter Zweig, Ph.D. and
Joseph Sherma, Ph.D.

Lipids
Helmut K. Mangold, Dr. rer. nat.

Hydrocarbons
Walter L. Zielinski, Jr., Ph.D.

Carbohydrates
Shirley C. Churms, Ph.D.

Inorganics
M. Qureshi, Ph.D.

Drugs
Ram Gupta, Ph.D.

Phenols and Organic Acids
Toshihiko Hanai, Ph.D.

Terpenoids
Carmine J. Coscia, Ph.D.

Amino Acids and Amines
S. Blackburn, Ph.D.

Steroids
Joseph C. Touchstone, Ph.D.

Polymers
Charles G. Smith,
Norman E. Skelly, Ph.D.,
Carl D. Chow, and Richard A. Solomon

**Pesticides and Related
Organic Chemicals**
Joseph Sherma, Ph.D. and
Joanne Follweiler, Ph.D.

Plant Pigments
Hans-Peter Köst, Ph.D.

**Nucleic Acids and
Related Compounds**
Ante M. Krstulovic, Ph.D.

CRC Handbook of Chromatography: Hydrocarbons

Volume I
Gas Chromatography

Author

Walter L. Zielinski, Jr.
Gas and Particulate Science Division
Center for Analytical Chemistry
National Bureau of Standards
Gaithersburg, Maryland

Editors-in-Chief

Gunter Zweig, Ph.D.
Deceased
School of Public Health
University of California
Berkeley, California

Joseph Sherma, Ph.D.
Professor of Chemistry
Chemistry Department
Lafayette College
Easton, Pennsylvania

CRC Press, Inc.
Boca Raton, Florida

Library of Congress Cataloging-in-Publication Data

Hydrocarbons.

(CRC handbook of chromatography)
Contents: v. 1. Gas chromatography / by Walter L.
Zielinski, Jr.
1. Hydrocarbons--Analysis. 2. Chromatographic
analysis. I. Zielinski, Walter L. II. Series.
QD305.H5H93   1987   547′.01046   87-9388
ISBN 0-8493-3041-6 (v.1)

71851

Direct all inquiries to CRC Press, Inc., 2000 Corporate Blvd., N.W., Boca Raton, Florida, 33431.

© 1987 by CRC Press, Inc.

International Standard Book Number 0-8493-3041-6 (v.1)

Library of Congress Card Number 87-9388
Printed in the United States

EDITOR'S PREFACE

This book represents the 14th volume of the *CRC Handbook of Chromatography* series published after the original two general volumes appeared in 1972. In it, Walter Zielinski of the National Bureau of Standards reviews methods and data related to the gas chromatography of various classes of hydrocarbons. My first exposure to Dr. Zielinski's work was through reading a prolific series of research papers and reviews on pesticide analysis that he wrote with Lawrence Fishbein in the 1960's, when I first started research in this area. Dr. Zweig and I were pleased when Dr. Zielinski agreed to produce the present *Handbook*.

We are very interested in extending the coverage to the liquid chromatography of hydrocarbons. I would be pleased to hear from anyone who can recommend an author for a companion volume on this topic. I would also appreciate communication with anyone who can offer corrections or comments concerning this volume, or suggestions for possible future volumes in the *CRC Handbook of Chromatography* series.

This is the first volume in the series that has been published since Dr. Gunter Zweig's death in January, 1987. I will do my best to carry on the series with the tradition and standards we established when beginning to write the original two volumes some 17 years ago.

Joseph Sherma, Ph.D.
Easton, PA
May, 1987

PREFACE

This volume provides a 10-year overview of research activities, methodological advances, experimental innovations, and a broad spectrum of applications reported in the World's literature in the use of gas chromatography for the analysis of hydrocarbons. As such, it represents the most comprehensive, unified volume to date on the gas chromatography of hydrocarbons.

The introductory sections of the book address the wide diversity of hydrocarbon structures, the basic operating components of the gas chromatographic system and their control for optimizing separations, and present a variety of chromatograms depicting the broad applications of gas chromatography in hydrocarbon analyses.

This is followed by three major sections on hydrocarbon analysis with respect to: (1) general applications (including geologic studies and process stream analysis); (2) specific studies (covering definitive sections on inlet systems, capillary columns, special GC systems and column arrangements, detectors, coupling with other instrumental systems, GC-mass spectrometry, and computers); and (3) special techniques (sampling, reaction GC, pyrolysis GC, hydrocarbon boiling point vs. GC retention [simulated distillation], preparative scale GC, and a detailed tabulation of thermodynamic studies on hydrocarbons). Each of these three sections categorize the referenced literature by general hydrocarbon applications and by individual hydrocarbon classes for easy reference.

The next major section covers definitive studies in air pollution (including polycyclic aromatic hydrocarbons) and in water pollution (including oil spills analysis and research and other hydrocarbons in aqueous systems).

The final two major sections of this test detail the properties and characteristics of and studies employing the broad spectrum of available liquid phases and adsorbents as separating media, and studies reporting retention data for the different hydrocarbon classes.

There are 45 comprehensive retention tables containing several thousand indices provided in a standardized retention index format, covering normal and isomeric alkanes, alkenes, cyclic hydrocarbons, alkyl benzenes, biphenyls and diphenylalkanes, 2-7 ring polycyclic aromatic hydrocarbons, alkylated naphthalenes, and terpenes.

Researchers and practitioners interested in the gas chromatography of hydrocarbons ranging from petrochemicals research to academic research in separation science will find this volume an indispensible reference text.

THE EDITORS-IN-CHIEF

Gunter Zweig, Ph.D., received his undergraduate training at the University of Maryland, College Park, where he was awarded the Ph.D. in biochemistry in 1952. Two years following his graduation, Dr. Zweig was affiliated with the late R. J. Block, pioneer in paper chromatography of amino acids. Zweig, Block, and Le Strange wrote one of the first books on paper chromatography which was published in 1952 by Academic Press and went into three editions, the last one authored by Gunter Zweig and Dr. Joe Sherma, the co-Editor-in-Chief of this series. *Paper Chromatography* (1952) was also translated into Russian.

From 1953 to 1957, Dr. Zweig was research biochemist at the C. F. Kettering Foundation, Antioch College, Yellow Springs, Ohio, where he pursued research on the path of carbon and sulfur in plants, using the then newly developed techniques of autoradiography and paper chromatography. From 1957 to 1965, Dr. Zweig served as lecturer and chemist, University of California, Davis and worked on analytical methods for pesticide residues, mainly by chromatographic techniques. In 1965, Dr. Zweig became Director of Life Sciences, Syracuse University Research Corporation, New York (research on environmental pollution), and in 1973 he became Chief, Environmental Fate Branch, Environmental Protection Agency (EPA) in Washington, D.C. From 1980 to 1984 Dr. Zweig was Visiting Research Chemist in the School of Public Health, University of California, Berkeley, where he was doing research on farmworker safety as related to pesticide exposure.

During his government career, Dr. Zweig continued his scientific writing and editing. Among his works are (many in collaboration with Dr. Sherma) the now 11-volume series on *Analytical Methods for Pesticides and Plant Growth Regulators* (published by Academic Press); the pesticide book series for CRC Press; co-editor of *Journal of Toxicology and Environmental Health;* co-author of basic review on paper and thin-layer chromatography for *Analytical Chemistry* from 1968 to 1980; co-author of applied chromatography review on pesticide analysis for *Analytical Chemistry,* beginning in 1981.

Among the scientific honors awarded to Dr. Zweig during his distinguished career are the Wiley Award in 1977, Rothschild Fellowship to the Weizmann Institute in 1963/64; the Bronze Medal by the EPA in 1980.

Dr. Zweig has authored or co-authored over 80 scientific papers on diverse subjects in chromatography and biochemistry, besides being the holder of three U.S. patents. In 1985, Dr. Zweig became president of Zweig Associates, Consultants in Arlington, Va.

Joseph Sherma, Ph.D., received a B.S. in Chemistry from Upsala College, East Orange, N.J. in 1955 and a Ph.D. in Analytical Chemistry from Rutgers University in 1958. His thesis research in ion exchange chromatography was under the direction of the late William Rieman III. Dr. Sherma joined the faculty of Lafayette College in Sepatember 1958, and is presently Charles A. Dana Professor and Head of the Chemistry Department. He is in charge of two courses in analytical chemistry, quantitative analysis and instrumental analysis. At Lafayette he has continued research in chromatography and had additionally worked a total of 14 summers in the field with Harold Strain at the Argonne National Laboratory, James Fritz at Iowa State University, Gunter Zweig at Syracuse University Research Corporation, Joseph Touchstone at the Hospital of the University of Pennsylvania, Brian Bidlingmeyer at Waters Associates, and Thomas Beesley at Whatman, Inc. and Advanced Separation Technologies, Inc.

Dr. Sherma and Dr. Zweig co-authored or co-edited the original Volumes I and II of the *CRC Handbook of Chromatography,* a book on paper chromatography, 10 volumes of theseries *Analytical Methods for Pesticides and Plant Growth Regulators,* and the *CRC Handbooks of Chromatography* of drugs, carbohydrates, polymers, phenols and organic acids, amino acids and amines, pesticides, terpenoids, lipids, steroids, peptides, and inorganics. Other books in the pesticide series and further volumes of the *CRC Handbook of Chromatography* are being edited, and Dr. Sherma has co-authored the handbook on pesticide chromatography. Books on quantitative TLC and advances in TLC were edited jointly with Dr. Touchstone. A general book on TLC was written with Dr. Bernard Fried, the second edition of which has been published. Dr. Sherma has been co-author of 10 biennial reviews of column and thin layer chromatography (1968—1986) and the 1981, 1983, 1985, and 1987 reviews of pesticide analysis for the ACS journal *Analytical Chemistry*.

Dr. Sherma has written major invited chapters and review papers on chromatography and pesticides in *Chromatographic Reviews* (analysis of fungicides), *Advances in Chromatography* (analysis of nonpesticide pollutants), Heftmann's *Chromatrography* (chromatography of pesticides), Race's *Laboratory Medicine* (chromatography in clinical analysis), *Food Analysis: Principles and Techniques* (TLC for food analysis), *Treatise on Analytical Chemistry* (paper and thin layer chromatography), *CRC Critical Reviews in Analytical Chemistry* (pesticide residue analysis), *Comprehensive Biochemistry* (flat bed techniques), *Inorganic Chromatographic Analysis* (thin layer chromatography), *Journal of Liquid Chromatography* (advances in quantitative pesticide TLC), and *Preparative Liquid Chromatography)* (strategy of preparative TLC). Dr. Sherma is editor for residues and elements of the Journal of the Association of Official Analytical Chemists (AOAC).

Dr. Sherma spent 6 months in 1972 on sabbatical leave at the EPA Perrine Primate Laboratory, Perrine, Fla., with Dr. T. M. Shafik, two summers (1975, 1976) at the USDA in Beltsville, Md., with Melvin Getz doing research on pesticide residue analysis methods development, and one summer (1984) in the food safety research laboratory of the Eastern Regional Research Center, Philadelphia, with Daniel Schwartz doing research on the isolation and analysis of mutagens in cooked meat. He spent 3 months in 1979 on sabbatical leave with Dr. Touchstone developing clinical analytical methods. A total of more than 300 papers, books, book chapters, and oral presentations concerned with column, paper, and thin layer chromatography of metal ions, plant pigments, and other organic and biological compounds; sthe chromatographic analysis of pesticides; and the history of chromatography have been authored by Dr. Sherma, many in collaboration with various co-workers and students. His major research area at Lafayette is currently quantitative TLC (densitometry), applied mainly to clinical analysis and pesticide residue and food additive determinations.

Dr. Sherma has written an analytical quality control manual for pesticide analysis under contract with the U.S. EPA and has revised this and the EPA Pesticide Analytical Methods Manual under a 4-year contract jointly with Dr. M. Beroza of the AOAC. Dr. Sherma has also written an instrumental analysis quality assurance manual and other analytical reports for the U.S. Consumer Product Safety Commission, and a Food Additives Analytical Manual and Animal Drug Analytical Manual for the FDA, these projects as scientific coordinator for the Association of Official Analytical Chemists. He is now editing an analytical field operations manual and preparing a second edition of the Food Additives Analytical Manual for the FDA.

Dr. Sherma taught the first, prototype short course on pesticide analysis, with Henry Enos of the EPA, for the Center for Professional Advancement. He was editor of the Kontes TLC quarterly newsletter for 6 years and also has taught short courses on TLC for Kontes and the Center for Professional Advancement. He is a consultant for numerous industrial companies and federal agencies on chemical analysis and chromatography and regularly referees papers for analytical journals and research proposals for government agencies. At Lafayette, Dr. Sherma, in addition to analytical chemistry, has taught general chemistry and a course in thin layer chromatography.

Dr. Sherma has received two awards for superior teaching at Lafayette College and the 1979 Distinguished Alumnus Award from Upsala College for outstanding achievements as an educator, researcher, author, and editor. He is a member of the ACS, Sigma Xi, Phi Lambda Upsilon, SAS, AIC, and AOAC.

THE EDITOR

Dr. Walter L. Zielinski's experience in gas chromatography dates from the 1950's, during his Master's Degree at North Carolina State University (1958, Biochemistry and Experimental Statistics). His Ph.D. thesis (Georgetown University, 1972, Analytical and Physical Chemistry) involved thermodynamic and analytical studies of the gas chromatographic retention behavior of diverse hydrocarbon structures, laying a basis for the predictability of retention from considerations of solute and stationary-phase chemical structures and thermodynamic principles.

During his professional career, he has held various chemistry research positions with industrial and private research firms, including Head of Analytical and Synthetic Chemistry at the Frederick Cancer Research Center. Dr. Zielinski's career at the National Bureau of Standards has advanced from the position of Air Program Manager in the NBS Office of Environmental Measurements to his present position as Leader of the Gas Metrology Research Group in the NBS Center for Analytical Chemistry, an area involved in the development of trace inorganic and organic gas measurements and NBS gaseous Standard Reference Materials.

Dr. Zielinski's research interests are in measurement science, statistical assessments of error, thermodynamic quantifications of molecular interactions, the broad field of analytical chemistry applications, and in the research areas of environmental pollutants and carcinogenesis. He has authored over 80 technical papers, received the Applied Analytical Chemistry Award from the Pittsburgh Conference, and currently is Editor of *CRC Critical Reviews in Analytical Chemistry*.

DEDICATION

This book is affectionately dedicated to my wife, Pat, and our daughters, Paige and Amy for their patience and loving support.

This work also is respectfully dedicated to the memory of Dr. Gunter Zweig who will be sorely missed by his many friends and colleagues.

TABLE OF CONTENTS

Chapter 1

INTRODUCTION

I. HYDROCARBONS — BACKGROUND

A. Hydrocarbon Classes

While a number of pure carbon-hydrogen compounds are known to exist in vivo (e.g., steranes) and numerous hydrocarbon structures have been synthesized, the term "hydrocarbon" is almost synonymous with carbon-hydrocarbon chemicals found geologically, their vast array of commercial products, or their combustion or waste products that contribute to air and water pollution.

The unique property of the carbon atoms of hydrocarbons for forming long carbon-carbon chains or carbon ring systems, coupled with the capability of the carbon atom of being sp^2 or sp^3 hybridized (resulting in saturated, unsaturated, cyclic, branched, or aromatic structures), offers the potential for theoretically postulating an almost unlimited number of possible chemical compound structures, ranging from the simple to the extremely complex, comprised solely of carbon and hydrogen.

The principal chemical elements in crude petroleum are carbon and hydrogen, usually in a carbon/hydrogen ratio ranging between 6 to 8. The hydrocarbons are predominantly liquids and gases, with some high-molecular-weight solids in dispersion or solution. Many other materials present include heteroatom (oxygen, sulfur, nitrogen) derivatives of hydrocarbons, trace elements, and water.

From a standpoint of petroleum chemistry, hydrocarbons are often considered as existing in one of three major structural classes: paraffins (normal and branched, noncyclic, nonfused-ring hydrocarbons), naphthenes (cyclic, nonaromatic hydrocarbons), and aromatics, with the three classes often designated by the acronym, "PNA".

In crude oil, the simpler molecular structures are more abundant than highly branched hydrocarbons in the lower boiling fractions, evidencing simpler PNA classes with only fractions of isomeric hydrocarbon species. In contrast, higher-boiling fractions contain increased levels of polycyclic aromatic hydrocarbons and polycycloparaffins and diminished levels of normal, branched, and monocyclic paraffins. Natural gas (the gaseous component of petroleum crude) may be found at some geological distance from a petroleum oil deposit and mainly consists of light, volatile hydrocarbons (methane to butane), with some traces of higher boiling paraffins (up to hexanes), along with nonorganic gases. The wide variety of structural classes comprising hydrocarbons is illustrated in Section II.

B. Significance of the Hydrocarbon Class

While the exact mechanisms for the conversion of decayed organisms and vegetation to geological hydrocarbon deposits is unknown, it is believed that degrading, sedimented life forms were initially changed to the asphaltic material known as kerogen. Heat, pressure, and possibly geo-catalysis subsequently converted the kerogen over millions of years into crude oil and gaseous hydrocarbons. These raw, naturally occurring materials are industrially processed today into a wide variety of products, ranging from energy fuels to plastics, solvents, and intermediates for other organic chemicals, without which, life as we know it would not be possible.

1. Industrial Products Dimension

Crude oil is processed by refineries into a wide variety of products using a variety of industrial processes and chemical reactions. The products may include liquefied petroleum gases, gasolines, fuel oil, diesel fuel, kerosine, lubricating oils, greases, waxes, coke, asphaltic materials, etc. Processes may include thermal cracking, hydrocracking, continuous catalytic cracking procedures, catalytic reforming, fluidized coking, distillation, vacuum distillation, solvent extraction, dewaxing, deasphalting, desulfurization, and processes based on a variety of chemical reactions. A number of chemical reactions are employed for the development of various hydrocarbons (hydrogenation and dehydrogenation, aromatization, polymerization, alkylation, cyclization) and hydrocarbon products (partial oxidation, chlorination, hydration, dehydration, hydrolysis).

The distribution of products produced by a refinery will depend on the characteristics of the crude oil being processed and the characteristics of the processes used and how they are controlled. The mean U.S. distribution pattern in 1965 was 45% gasolines, 23% gas oils and distillates, 8% residual fuel oils, 6% kerosine, and 18% of other products, with about 4% of the crude oil processed into petrochemicals.[1]

2. Health Concerns

A number of hydrocarbons have been identified as health hazards. These range from benzene (having a NIOSH/OSHA limit of 1 ppm (as of 2/11/78) and a level "immediately dangerous to life or health" of 2000 ppm[2]), to "volatile organic compounds" (hydrocarbons) as monitored by EPA, to polycyclic aromatic hydrocarbons (PAH; certain members of which have been shown to be carcinogenic (e.g., benzo[*a*]pyrene)).

Information on the potential health hazard of certain types of hydrocarbons may be found in a number of texts, such the excellent one edited by Stern.[3] This includes observations that (1) sufficient exposure to benzene can produce interferences in the formation of red blood cells in the bone marrow and has been linked with leukemia in some individuals having long-term occupational exposures; (2) the mouse skin test has been shown to be useful in assessing the tumorigenic potential of several fractions of catalytically cracked petroleum oils and coal tar; (3) nonvolatile hydrocarbons (C-10 to C-16), found in the organic fraction of air particulates, in high concentrations may increase the carcinogenic activity of PAH; (4) nonvolatile olefins may photooxidatively decompose to tumorigenic epoxides and peroxides; and (5) some studies have correlated PAH (especially, benzo[*a*]pyrene) in polluted air with the urban factor for lung cancer, with the major potentially carcinogenic PAH having 4, 5, and 6 rings and some carcinogenic alkylpolycyclics. A special chapter by Dipple[4] in a text on chemical carcinogens published by the American Chemical Society detailed the carcinogenicity of members of the PAH class, and a recent text edited by Pucknat[5] specifically focused on health effects of PAH. The health hazards offered by benzene and toluene are reviewed by Fishbein.[6] A recent report by the Council on Environmental Quality[7] cites that little overall change has occurred in the U.S. national emission estimates for volatile organic compounds (precursors of total oxidants, including ozone) over the 1970 to 1978 period; emission rates increased during the last 3 years of this period. The primary national ambient air quality standard for health for these emissions (defined as nonmethane hydrocarbons) is set at 0.24 ppm averaged over 3 hr (6 to 9 a.m.). Monographs developed by the International Agency for Research on Cancer provide epidemiological and experimental information on the carcinogenic activity and potential health hazards of PAH and industrial hydrocarbon processes containing them.[8]

II. CHEMICAL STRUCTURES

Representative structures for eight structural classes of hydrocarbons are given under this section. These include: saturated and unsaturated normal and branched chain acyclic hydrocarbons; saturated and unsaturated cyclic hydrocarbons; polycyclic aromatic hydrocarbons; bridged (multicyclic) hydrocarbons; spiro hydrocarbons; hydrocarbon ring assemblies; substituted (side-chain) cyclic hydrocarbons; and terpene hydrocarbons.

For further details on hydrocarbon structures and IUPAC rules concerning their nomenclature, the reader is referred to alternative reference sources (e.g., *CRC Handbook of Chemistry and Physics,* 68th Ed. (1987 to 1988), pp. C-1 to C-24).

A. Acyclic Hydrocarbons
1. Normal Chain Alkanes

Following the first four normal chain alkanes (methane, ethane, propane, butane) the names of longer normal chain hydrocarbons consist of a numerical prefix and the suffix "ane", as follows, where n is the number of carbon atoms in the normal chain:

n		n	
5	Pentane		
6	Hexane	32	Dotriacontane
7	Heptane	33	Tritriacontane
8	Octane	34	Tetratriacontane
9	Nonane		.
10	Decane		.
11	Undecane		.
12	Dodecane	40	Tetracontane
13	Tridecane	50	Pentacontane
14	Tetradecane	60	Hexacontane
15	Pentadecane		.
16	Hexadecane		.
17	Heptadecane		.
18	Octadecane	100	Hectane
19	Nonadecane		.
20	Eicosane		.
21	Heneicosane	132	Dotriacontahectane
22	Docosane		
23	Tricosane		
24	Tetracosane		
	.		
	.		
	.		
30	Triacontane		
31	Hentriacontane		

2. Branched Alkanes

(1) Branch Radicals

Examples of common saturated alkyl branches attached to a normal alkane to produce a branched alkane are as follows:

Radical name	Radical structure
methyl	CH_3-
ethyl	CH_3CH_2-
propyl	$CH_3CH_2CH_2-$
isopropyl	$(CH_3)_2CH-$
butyl	$CH_3(CH_2)_2CH_2-$
isobutyl	$(CH_3)_2CHCH_2-$
sec-butyl	$CH_3CH_2CH(CH_3)-$
tert-butyl	$(CH_3)_3C-$
pentyl	$CH_3(CH_2)_3CH_2-$
isopentyl	$(CH_3)_2CHCH_2CH_2-$
neopentyl	$(CH_3)_3CCH_2-$
tert-pentyl	$CH_3CH_2C(CH_3)_2-$
hexyl	$CH_3(CH_2)_4CH_2-$
isohexyl	$(CH_3)_2CH(CH_2)_2CH_2-$
....etc.	

(2) Nomenclature rule for point of attachment:
Number the longest normal carbon chain, such that radical substituents are assigned the lowest number possible. For example,

7   6   5   4   3   2   1
$CH_3CH_2CH_2C(CH(CH_3)_2CH(CH_3)CH_2CH_3$
3-Methyl-4-isopropyl-heptane

3. Unsaturated Acyclic Hydrocarbons

(1) Common Unsaturated Acyclics

- ethylene $CH_2 = CH_2$
- acetylene $CH \equiv CH$
- allene $CH_2 = C = CH_2$

(2) Nomenclature rule for numbering the position of double- and triple-bonds in unsaturated acyclics that contain either a double bond(s) or a triple bond(s):
Number the double bond(s) in the acyclic alkene with the lowest number possible. For example,

1   2 3 4   5 6
$CH_2{=}CHCH_2CH{=}CHCH_3$
1,4-Hexadiene

Number the triple bond(s) in the acyclic alkyne with the lowest number possible. For example,

1   2 34   5
$CH_3C{\equiv}CCH_2CH_3$
2-Pentyne

(3) Nomenclature rule for numbering positions of attachment in unsaturated acyclics containing both double and triple bonds:
When possible, assign the lowest number to the double bond(s). For example,

1   2 3 4 5
$CH_2{=}CHCH_2C{\equiv}CH$
1-Penten-4-yne

(4) Examples of unsaturated radicals attached to a normal chain saturated or unsaturated acyclic:

- vinyl $CH_2{=}CH-$
- ethynyl $CH{\equiv}C-$
- allyl $CH_2{=}CHCH_2-$
- 1-propenyl $CH_3CH{=}CH-$
- 2-propynyl $CH{\equiv}CCH_2-$
- isopropenyl $CH_2{=}C(CH_3)-$
- 2-butenyl $CH_3CH{=}CHCH_2-$
- 1,3-butadienyl $CH_2{=}CHCH{=}CH-$

- 2-pentenyl $CH_3CH_2CH=CHCH_2-$
- 2-penten-4-ynyl $CH\equiv CCH=CHCH_2-$

Example of branched unsaturated acyclic:

$$\underset{1}{C}\equiv \underset{23}{CCH}=\underset{4}{CHCH}(\underset{5}{CH_2CH_2CH}=CHCH_3)CH=\underset{6\quad 7\ 8\quad 9\ 10}{CHCH}=CHCH_3$$

5-(3-Pentenyl)-3,6,8-decatrien-1-ynyl

B. Saturated and Unsaturated Cyclic Hydrocarbons

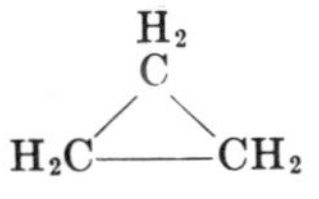

Cyclopropane

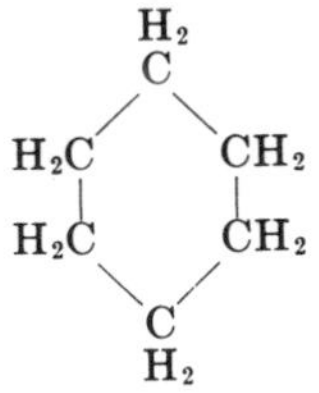

Cyclohexane

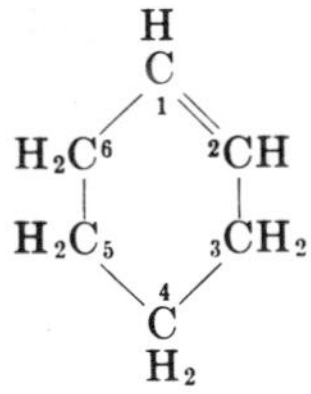

Cyclohexene

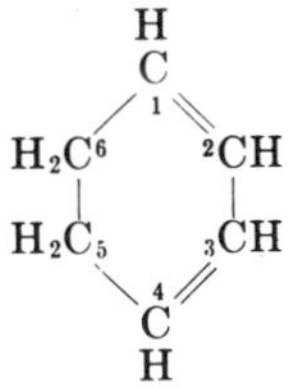

1,3-Cyclohexadiene

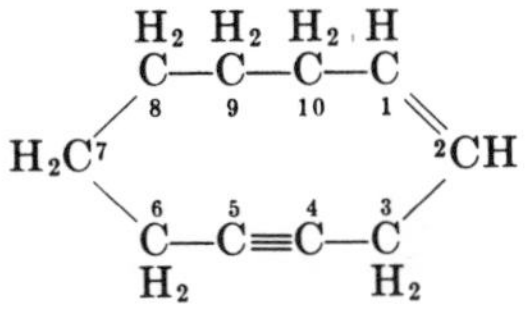

1-Cyclodecen-4-yne

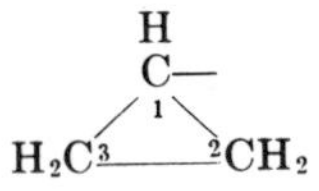

Cyclopropyl

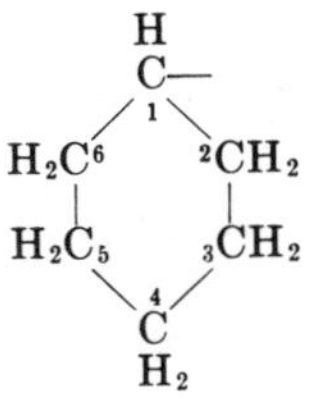

Cyclohexyl

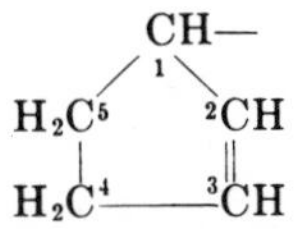

2-Cyclopenten-1-yl

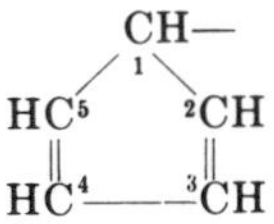

2,4-Cyclopentadien-1-yl

Phenylene (*p*-shown)

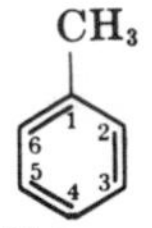

Toluene

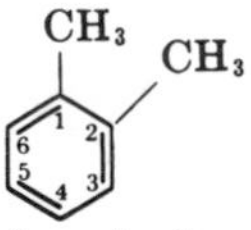

Xylene (*o*-shown)

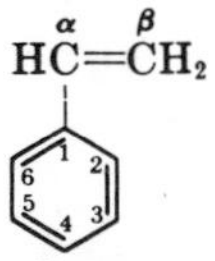

Styrene

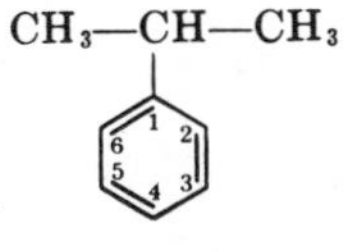

Cumene

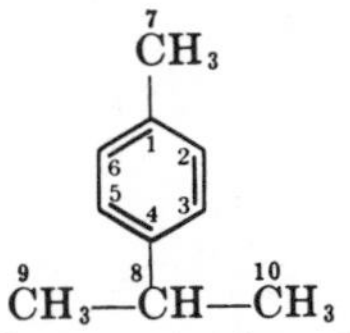

Cymene (*p*-shown)

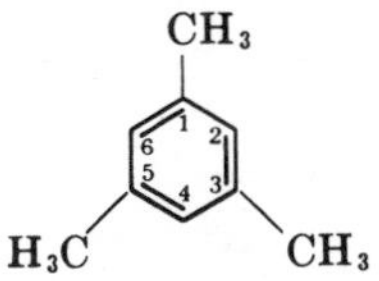

Mesitylene

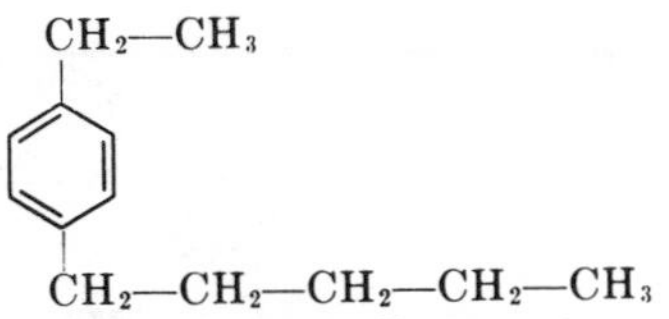

1-Ethyl-4-pentylbenzene or *p*-Ethylpentylbenzene

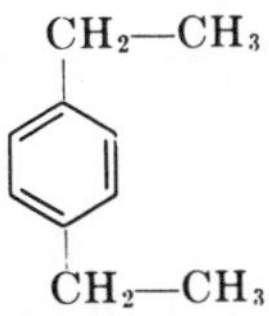

1,4-Diethyl-benzene or *p*-Diethyl-benzene

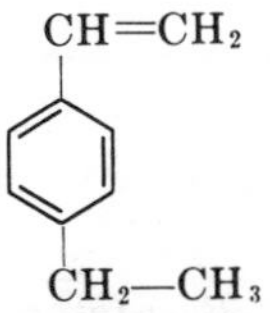

4-Ethylsty-rene or *p*-Ethylstyrene

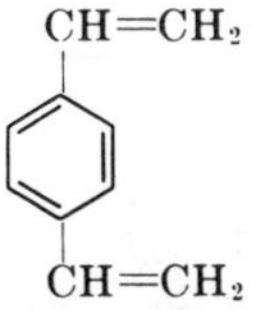

1,4-Divinylben-zene or *p*-Divinyl-benzene, not *p*-Vinylstyrene

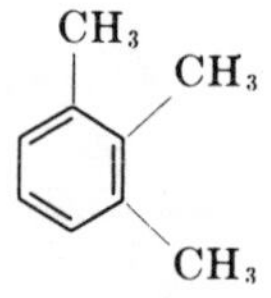

1,2,3-Trimethyl-
benzene, not
Methylxylene
nor Dimethyl-
toluene

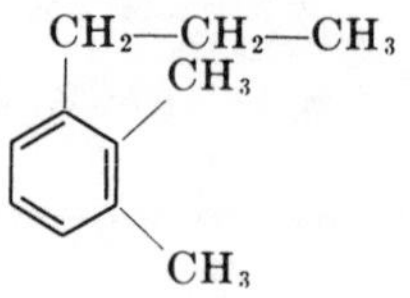

1,2-Dimethyl-3-
propylbenzene
or 3-Propyl-
o-xylene

Phenyl	$C_6H_5—$
Benzyl	$C_6H_5—\overset{\alpha}{CH_2}—$
Benzylidyne	$C_6H_5—C\!\!=$
Phenethyl	$C_6H_5—\overset{\beta}{CH_2}—\overset{\alpha}{CH_2}—$
Styryl	$C_6H_5—\overset{\beta}{CH}\!\!=\!\!\overset{\alpha}{CH}—$
Trityl	$(C_6H_5)_3C—$
Cinnamyl	$C_6H_5—\overset{\gamma}{CH}\!\!=\!\!\overset{\beta}{CH}—\overset{\alpha}{CH_2}—$
Cinnamylidene	$C_6H_5—\overset{\gamma}{CH}\!\!=\!\!\overset{\beta}{CH}—\overset{\alpha}{CH}\!\!=$

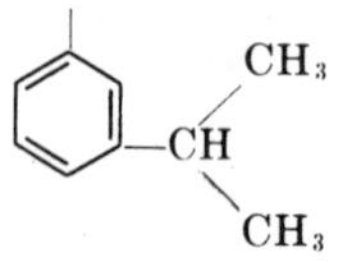

Cumenyl (*m*-shown) 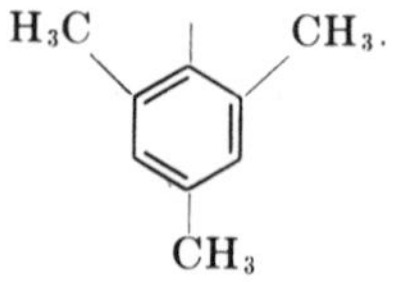   Mesityl 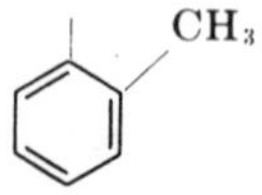   Tolyl (*o*-shown) 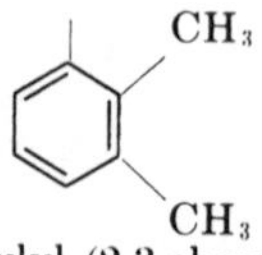   Xylyl (2,3-shown)

C. Fused Unsubstituted Polycyclic Aromatic Hydrocarbons

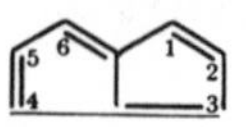

(1) Pentalene

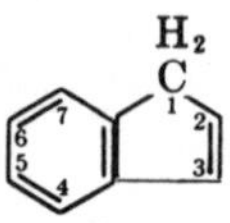

(2) Indene

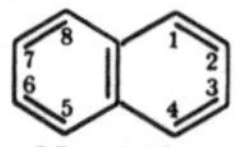

(3) Naphthalene

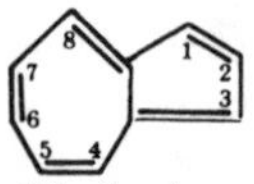

(4) Azulene

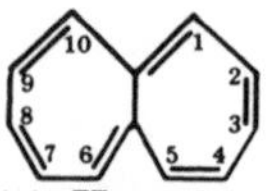

(5) Heptalene

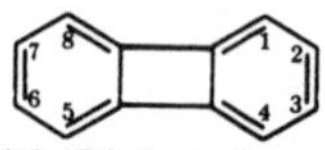

(6) Biphenylene

(7) *as*-Indacene

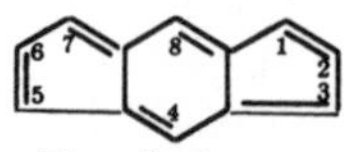

(8) *s*-Indacene

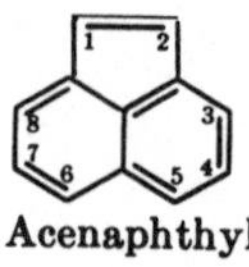

(9) Acenaphthylene

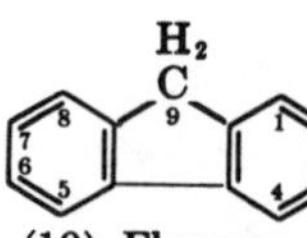

(10) Fluorene

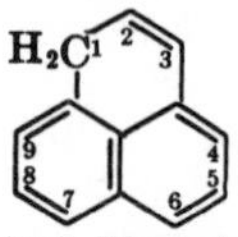

(11) Phenalene

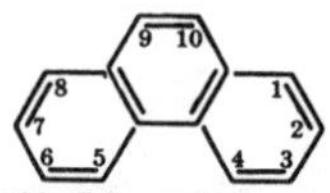

(12) Phenanthrene

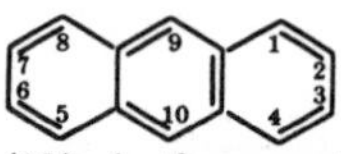

(13) Anthracene[1]

(14) Fluoranthene

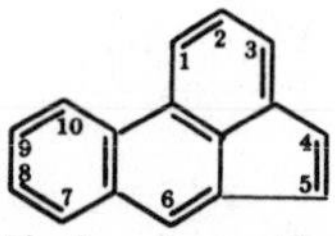

(15) Acephenanthrylene

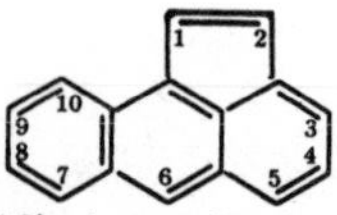

(16) Aceanthrylene

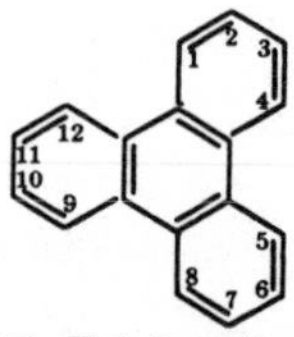

(17) Triphenylene

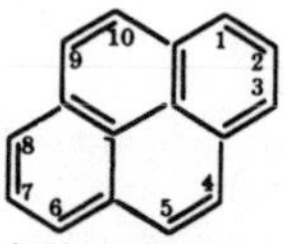

(18) Pyrene

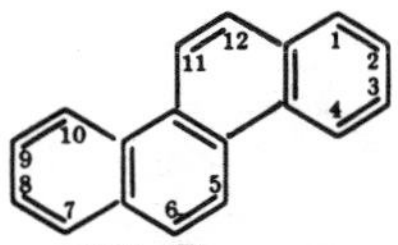

(19) Chrysene

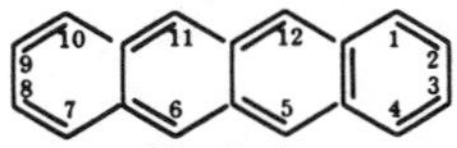

(20) Naphthacene

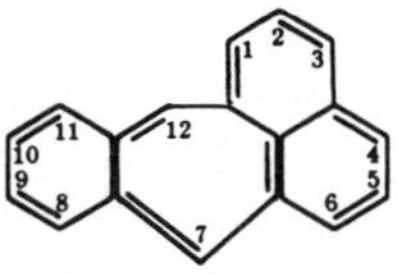

(21) Pleiadene

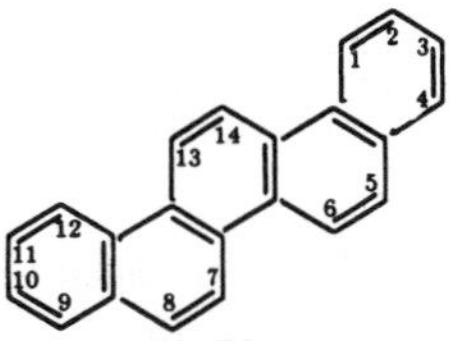

(22) Picene

(23) Perylene

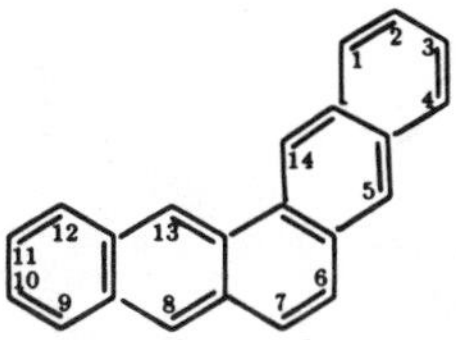

(24) Pentaphene

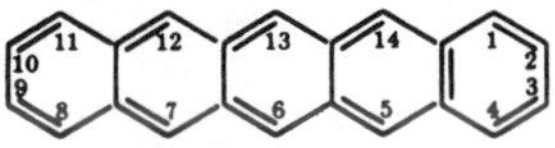

(25) Pentacene

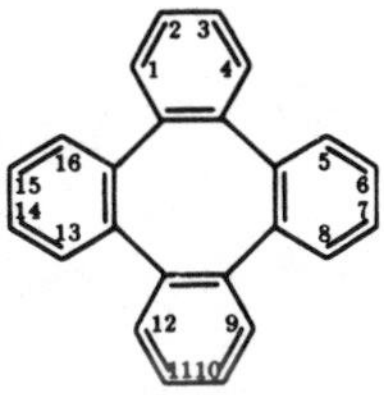

(26) Tetraphenylene

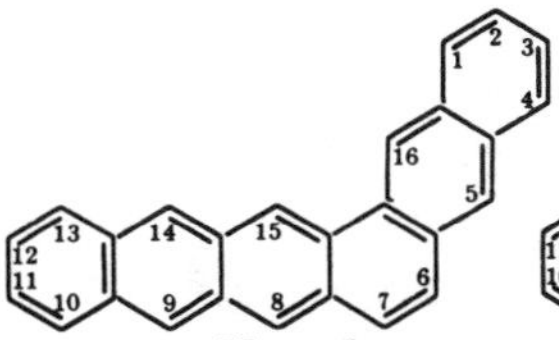

(27) Hexaphene

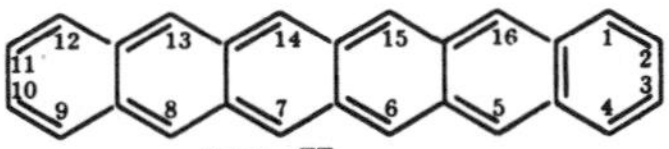

(28) Hexacene

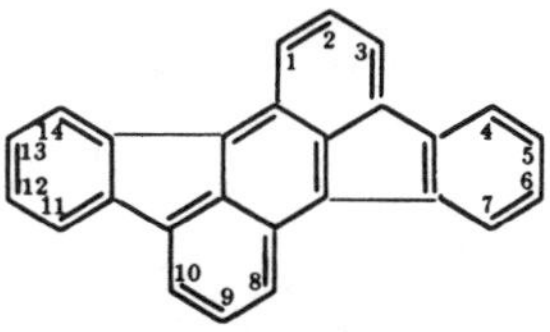

(29) Rubicene

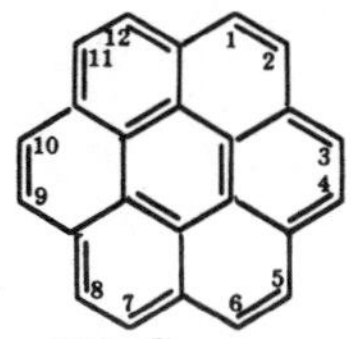

(30) Coronene

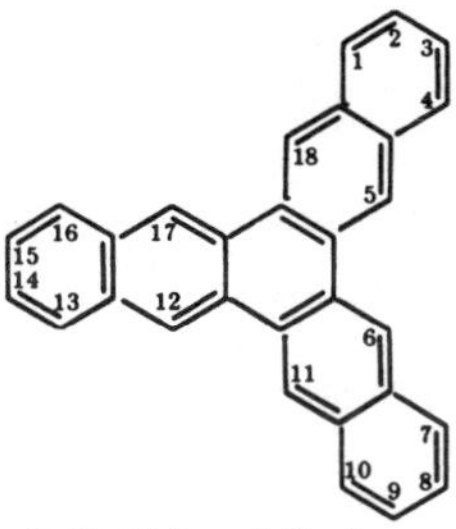

(31) Trinaphthylene

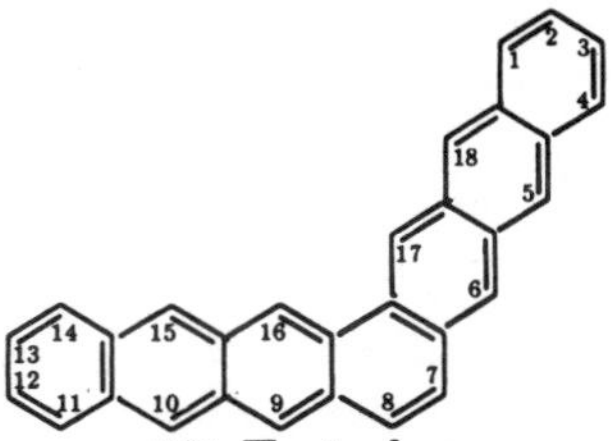

(32) Heptaphene

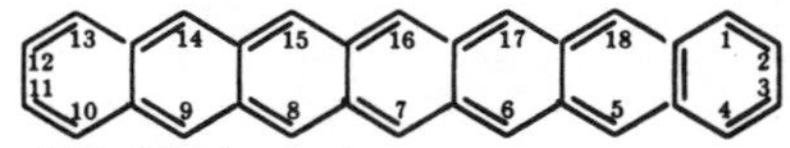

(33) Heptacene

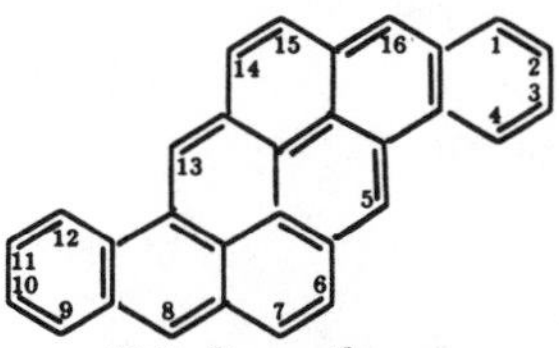

(34) Pyranthrene

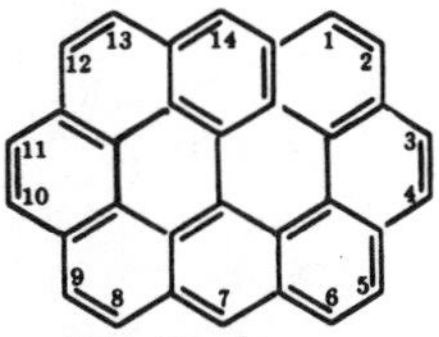

(35) Ovalene

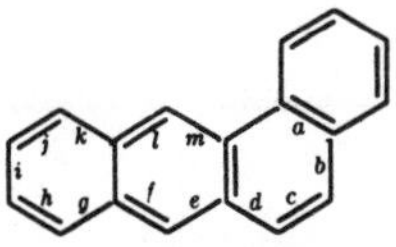

Benz[*a*]anthracene

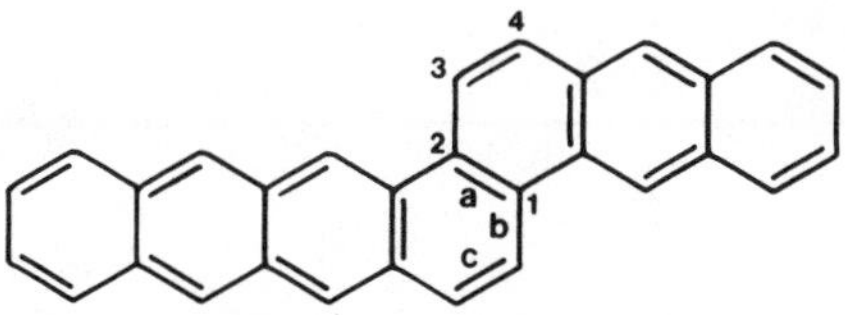

Anthra[2,1-*a*]naphthacene

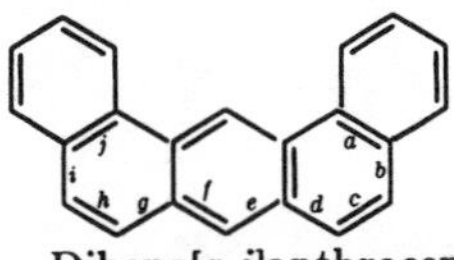

Dibenz[*a,j*]anthracene
(not Naphtho[2,1-*b*]phenanthrene)

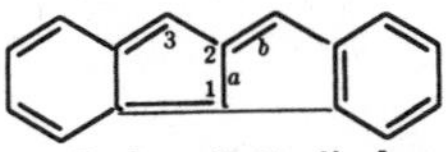

Indeno[1,2-*a*]indene

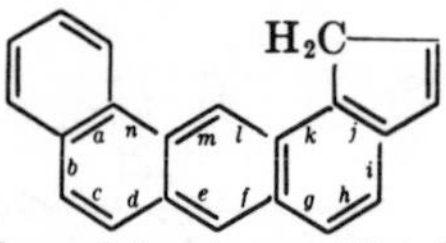

1*H*-Benzo[*a*]cyclopent[*j*]anthracene

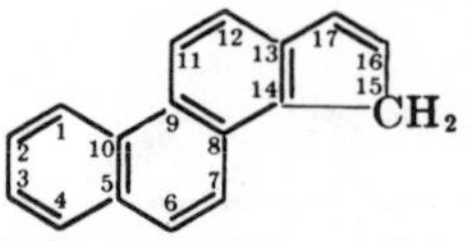

Cyclopenta[*a*]phenanthrene
(15*H*- shown)

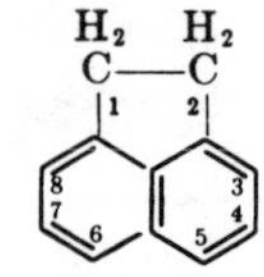

Acenaphthene

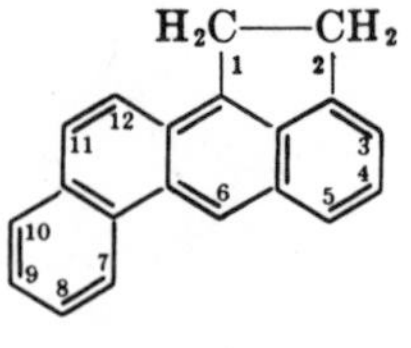

Cholanthrene

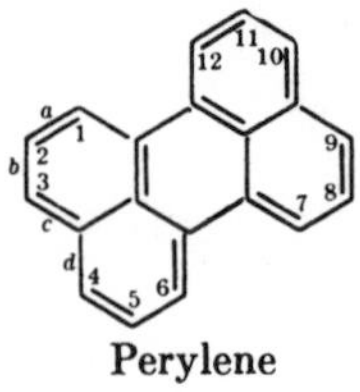

Perylene

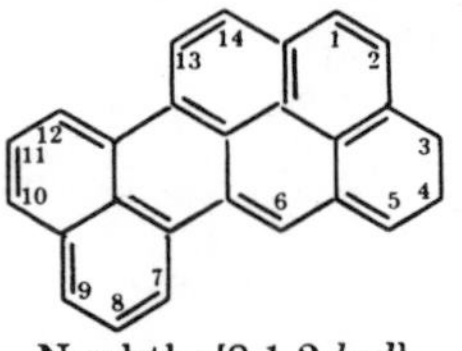

Naphtho[8,1,2-*bcd*]-
perylene

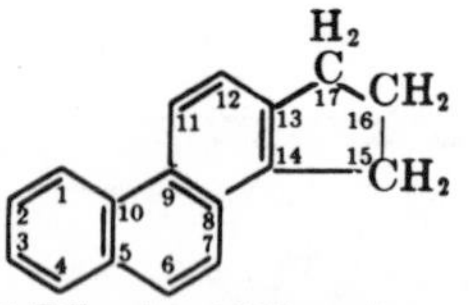

16,17-Dihydro-15*H*-cyclopenta[*a*]
phenanthrene

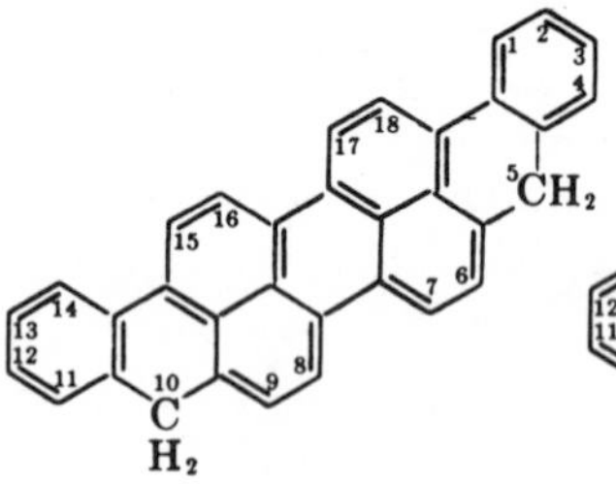

Violanthrene

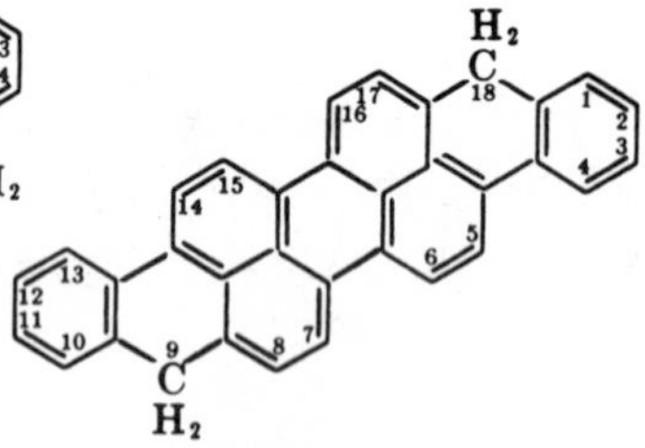

Isoviolanthrene

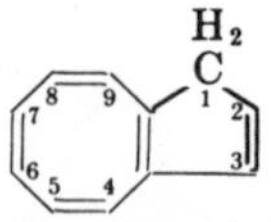

1*H*-Cyclopentacyclooctene

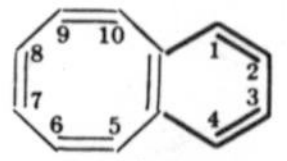

Benzocyclooctene

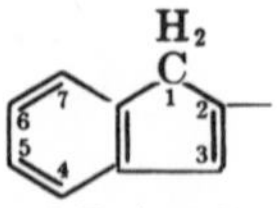

2-Indenyl

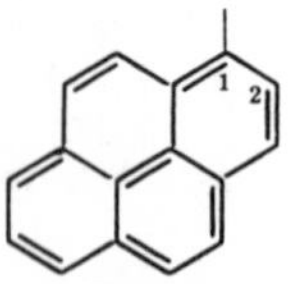

1-Pyrenyl

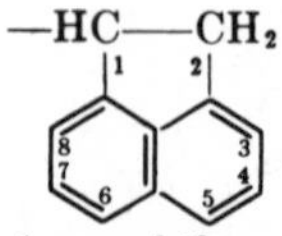

1-Acenaphthenyl

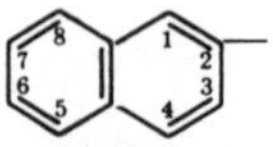

Naphthyl
(2-shown)

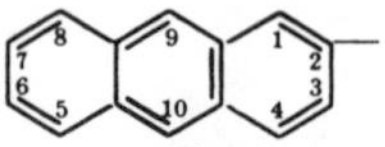

Anthryl
(2-shown)

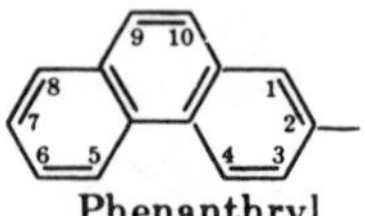

Phenanthryl
(2-shown)

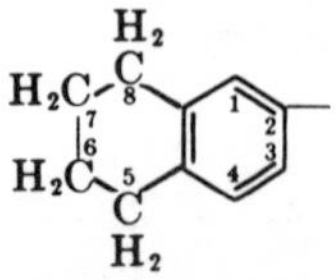

5,6,7,8-Tetrahydro-
2-naphthyl

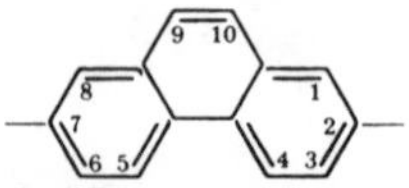

2,7-Phenanthrylene
or 2,7-Phenanthrenediyl

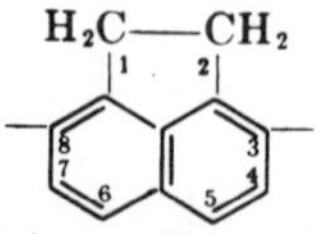

3,8-Acenaphthenylene

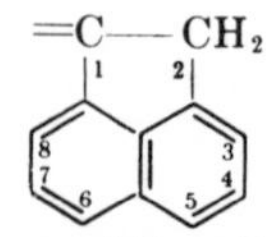

1-Acenaphthenylidene

D. Bridged (Multicyclic) Hydrocarbons

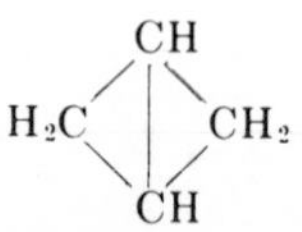

Bicyclo [1.1.0]-
butane

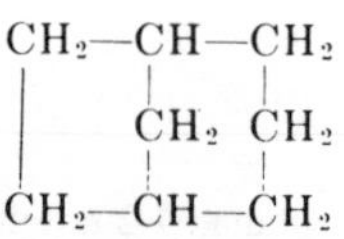

Bicyclo [3.2.1]-
octane

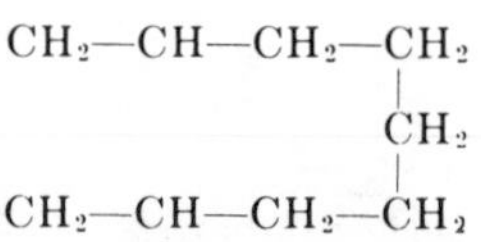

Bicyclo [5.2.0]nonane

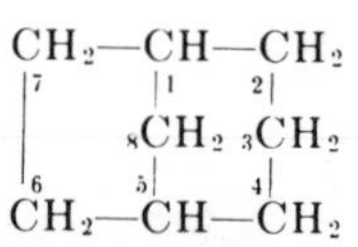

Bicyclo [3.2.1]octane

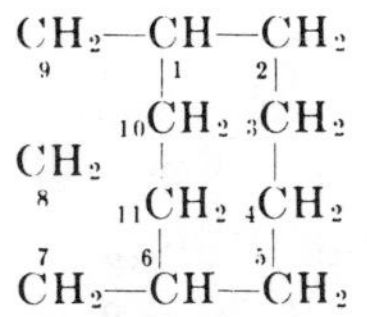

Bicyclo [4.3.2]undecane

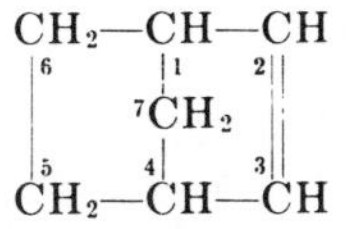

Bicyclo [2.2.1]hept-2-ene

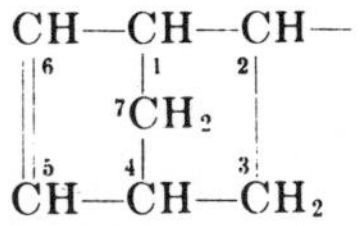

Bicyclo[2.2.1]hept-5-en-2-yl

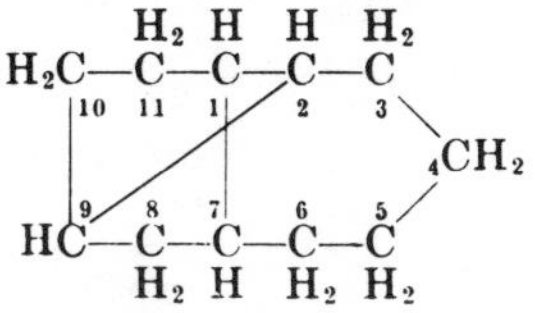

Tricyclo[5.4.0.0^{2,9}]undecane

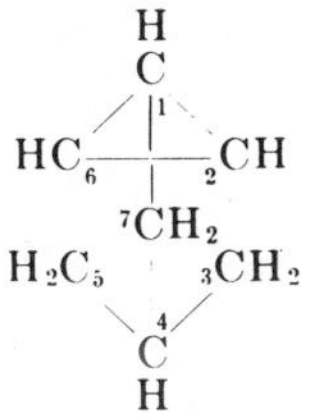

Tricyclo[2.2.1.0^1]heptane

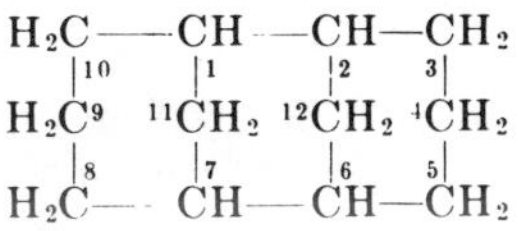

Tricyclo[5.3.1.1^1]dodecane

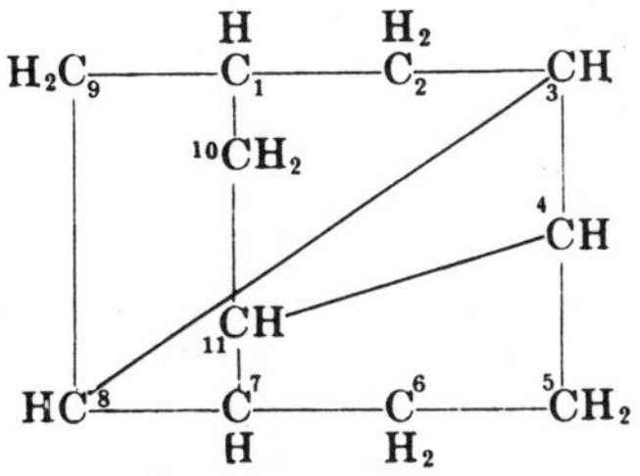

Tetracyclo[5.2.2.0^{3,8}.0^{4,11}]-undecane

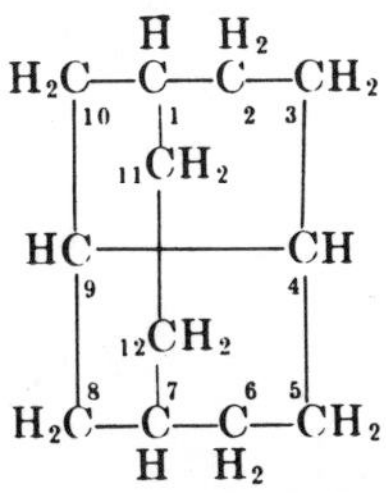

Tricyclo[5.3.2.0^{4,9}]dodecane

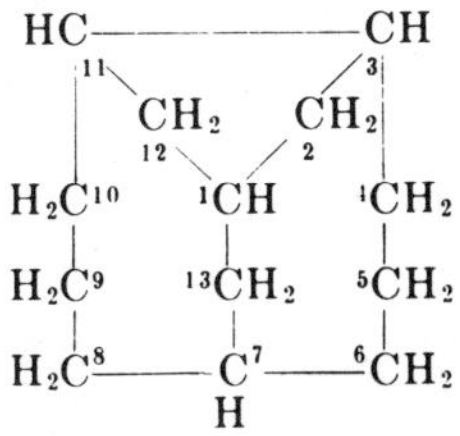

Tricyclo[5.5.1.0^{3,11}]tridecane

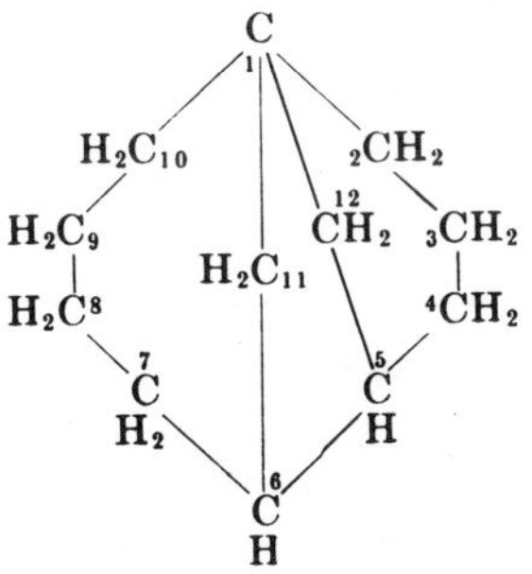

Tricyclo[4.4.1.1^{1,5}]dodecane:

Butano	—CH_2—CH_2—CH_2—CH_2—
Benzeno (*o-, m-, p-*)	—C_6H_4—
Ethano	—CH_2—CH_2—
Etheno	—$CH=CH$—
Methano	—CH_2—
Propano	—CH_2—CH_2—CH_2—

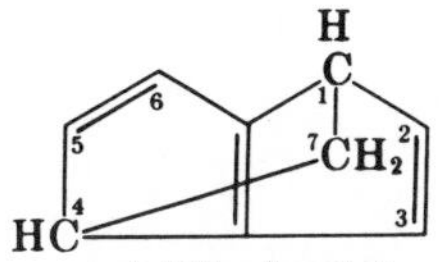

1,4-Dihydro-1,4-methanopentalene

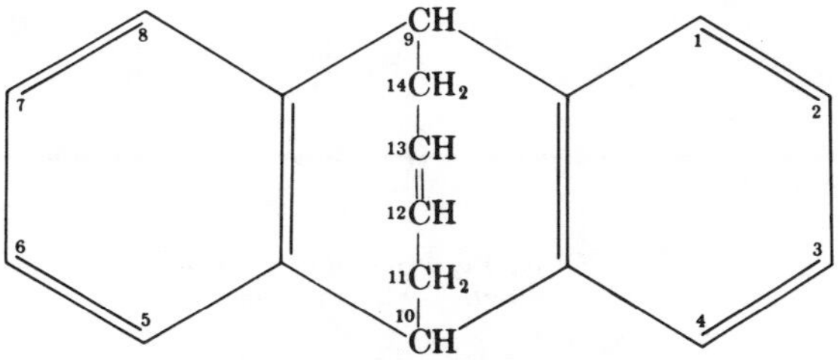

9,10-Dihydro-9,10-(2-buteno)-anthracene

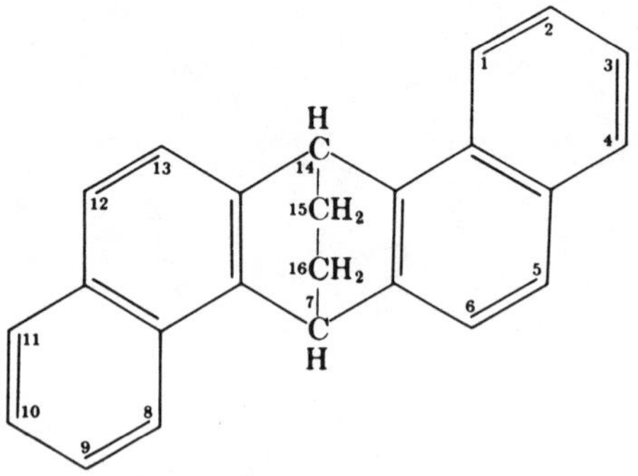

7,14-Dihydro-7,14-ethano-
dibenz[a,h]anthracene

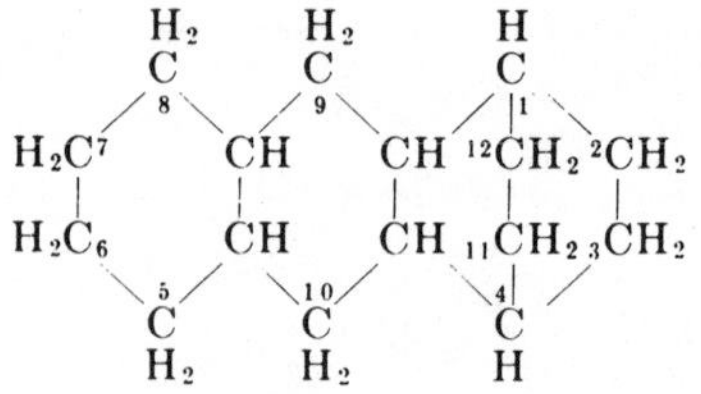

Perhydro-1,4-ethanoanthracene

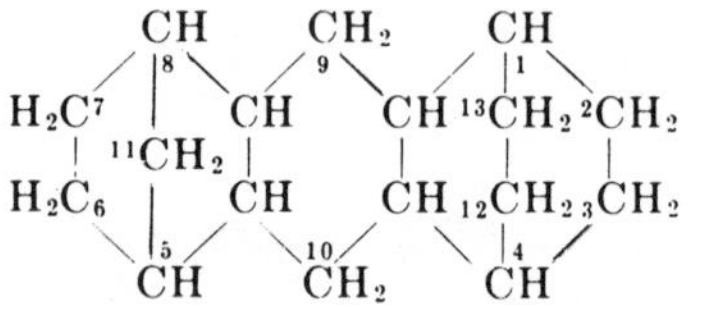

Perhydro-1,4-ethano-5,8-methanoanthracene

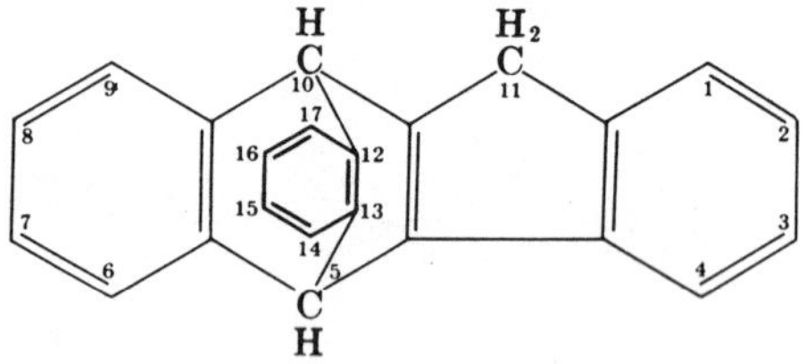

10,11-Dihydro-5,10-*o*-benzeno-5*H*-benzo[*b*]fluorene

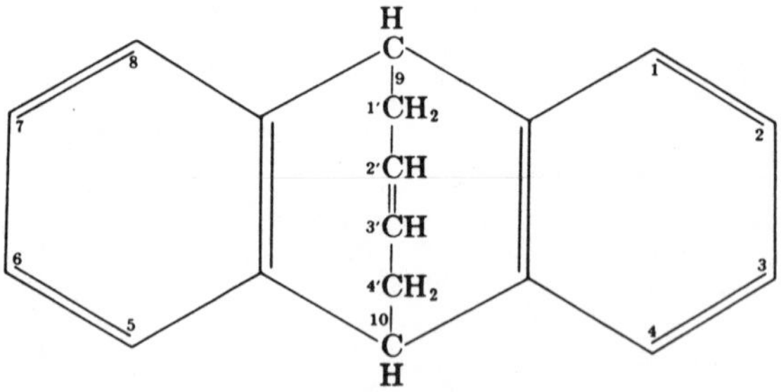

9,10-Dihydro-9,10-(2′-
butylene)anthracene

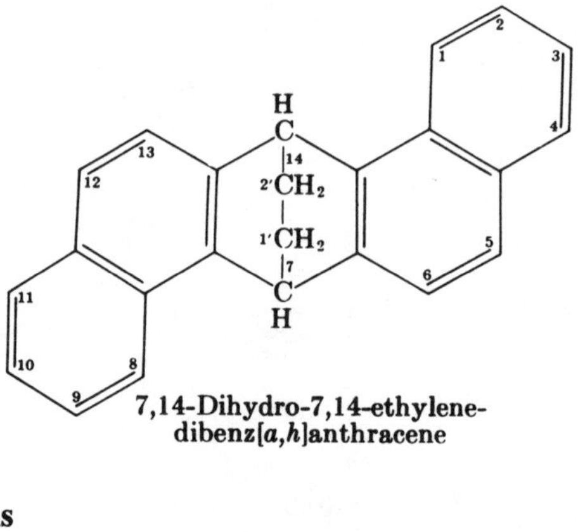

7,14-Dihydro-7,14-ethylene-
dibenz[a,h]anthracene

E. Spiro Hydrocarbons

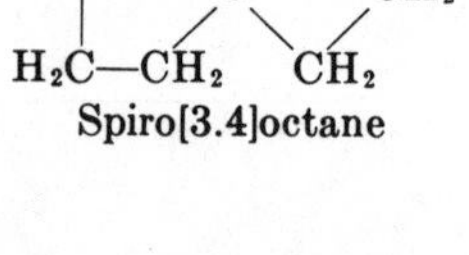

Spiro[3.4]octane

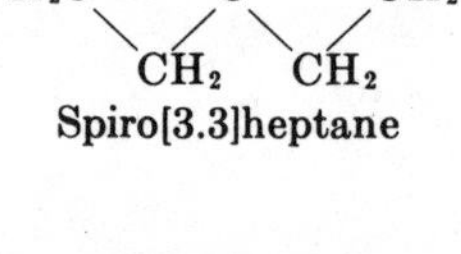

Spiro[3.3]heptane

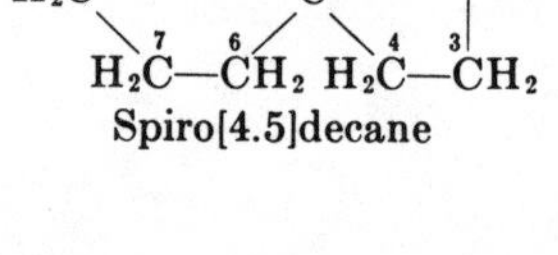

Spiro[4.5]decane

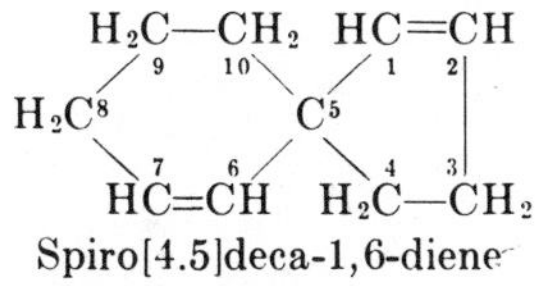

H₂C—CH₂ HC=CH

H₂C⁸   C⁵

HC=CH   H₂C—CH₂

Spiro[4.5]deca-1,6-diene

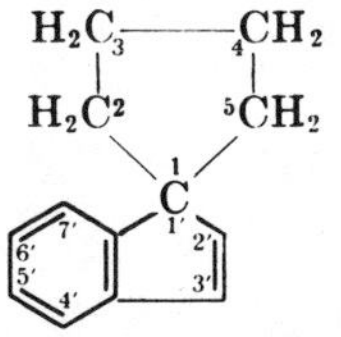

Spiro[cyclopentane-1,1'-indene]

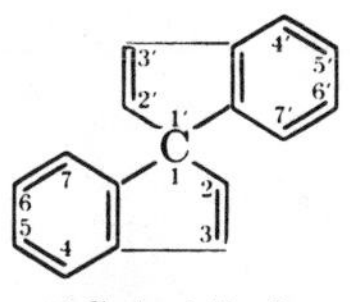

1,1'-Spirobiindene

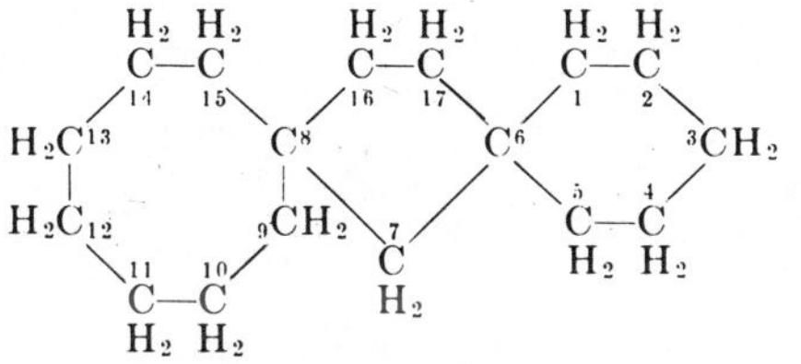

Dispiro[5.1.7.2]heptadecane

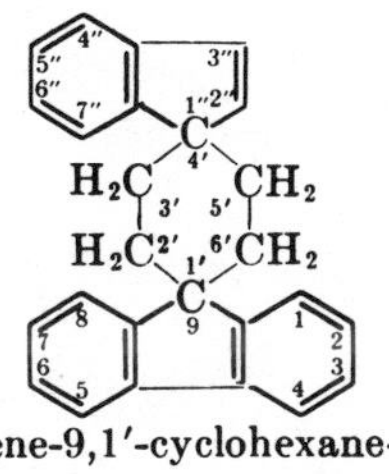

Dispiro[fluorene-9,1'-cyclohexane-4',1''-indene]

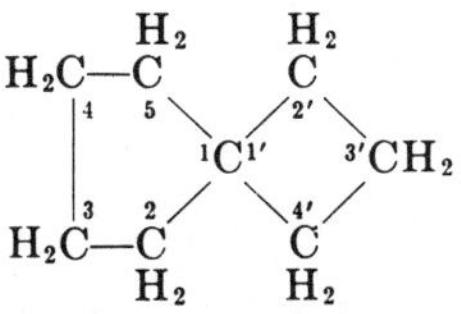

Cyclopentanespiro-
cyclobutane

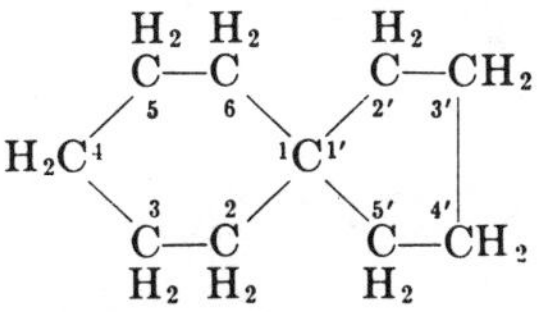

Cyclohexanespirocyclo-
pentane

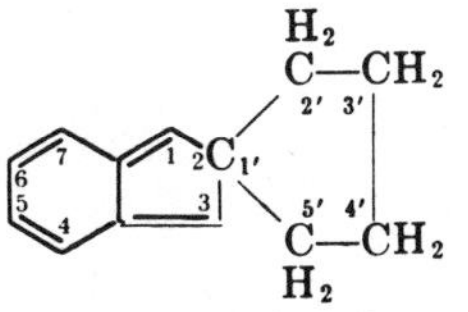

2H-Indene-2-spiro-1'-cyclopentane

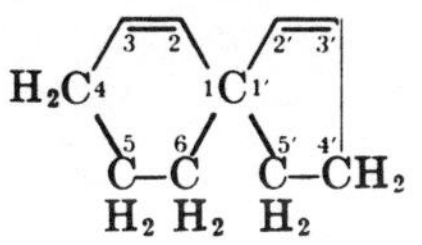

2-Cyclohexenespiro-(2'-cyclopentene)

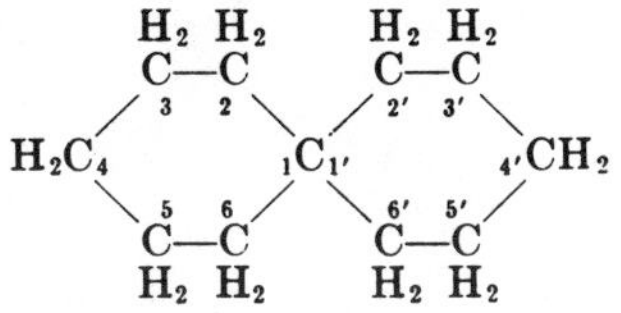

Spirobicyclohexane   but 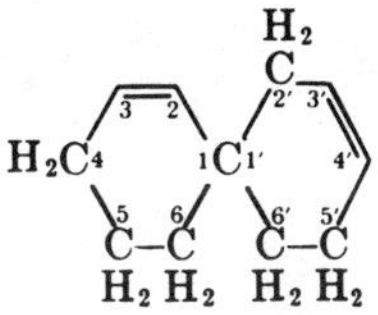   2-Cyclohexenespiro-
(3'-cyclohexene)

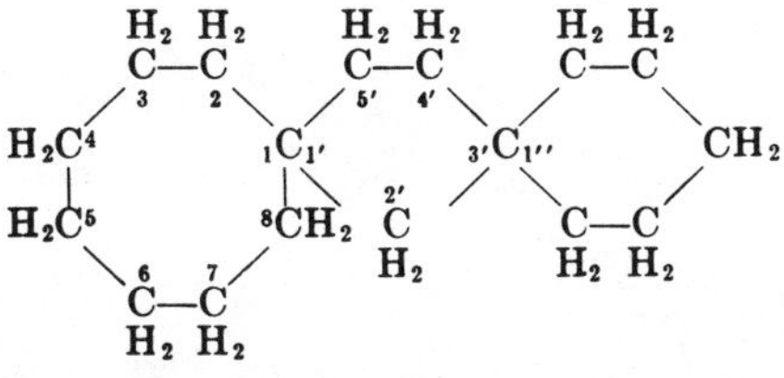

Cyclooctanespirocyclopentane-3'-spirocyclohexane

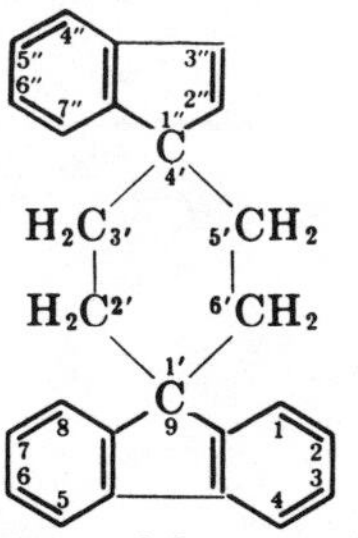

Fluorene-9-spiro-1'-cyclohexane-4'-spiro-1''-indene

F. Hydrocarbon Ring Assemblies

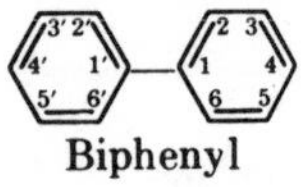

Biphenyl

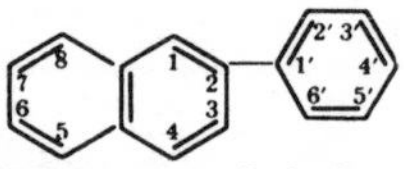

2-Phenylnaphthalene

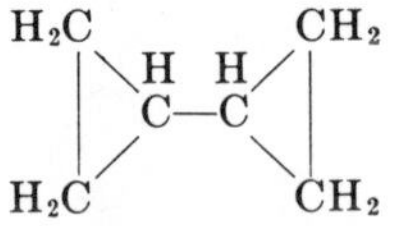

1,1'-Bicyclopropyl
or 1,1'-Bicyclopropane

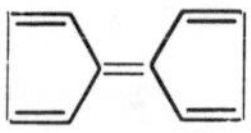

1,1'-Bicyclopentadienylidene

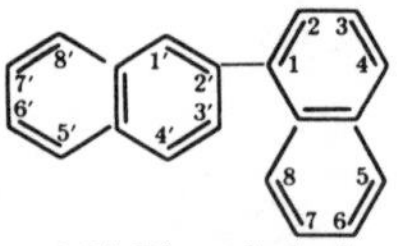

1,2'-Binaphthyl
or 1,2'-Binaphthalene

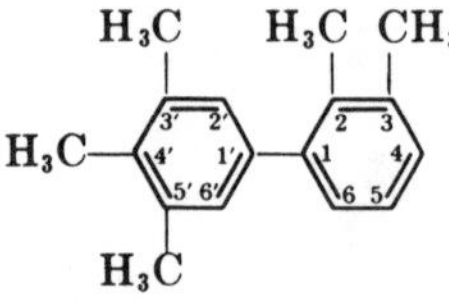

2,3,3',4',5'-
Pentamethylbiphenyl

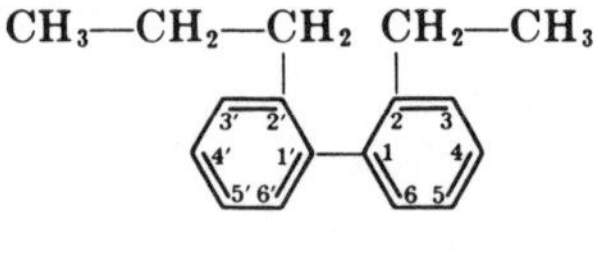

2-Ethyl-2'-
propylbiphenyl

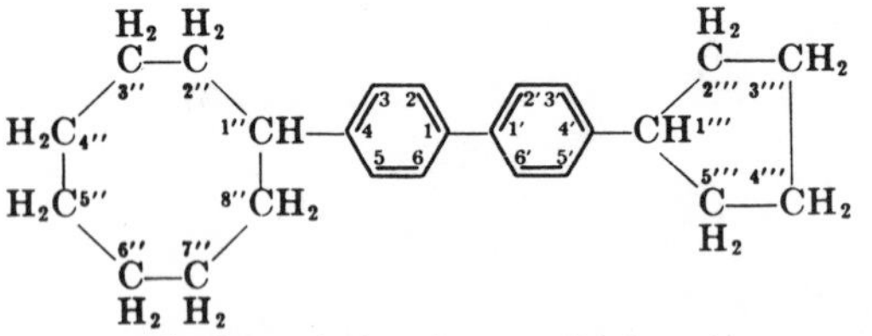

4-Cyclooctyl-4'-cyclopentylbiphenyl

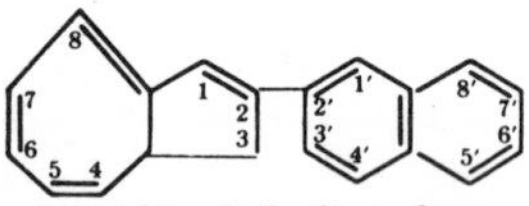

2-(2'-Naphthyl)azulene

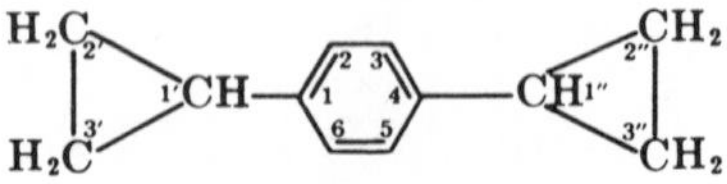

1,4-Dicyclopropylbenzene
or *p*-Dicyclopropylbenzene

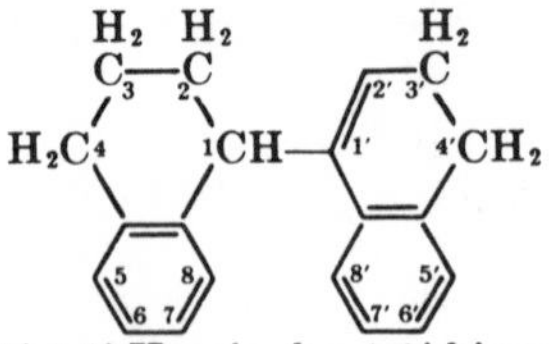

1,2,3,3',4,4'-Hexahydro-1,1'-binaphthyl

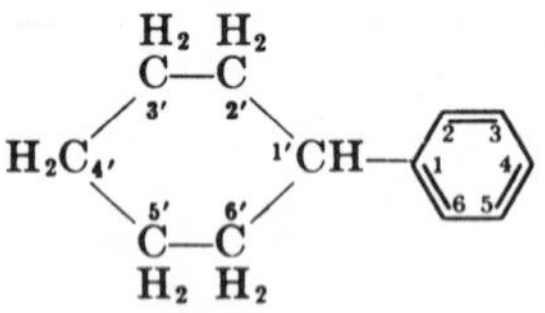

Cyclohexylbenzene

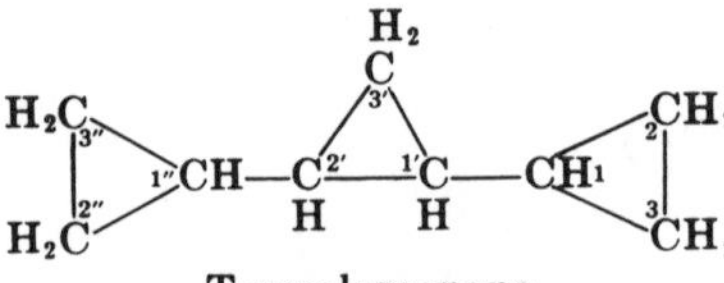

Tercyclopropane

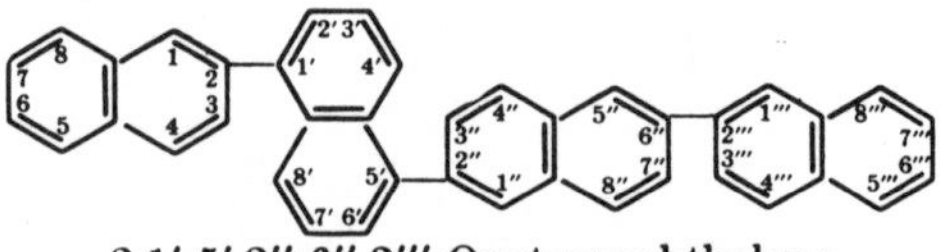

2,1':5',2'':6'',2'''-Quaternaphthalene

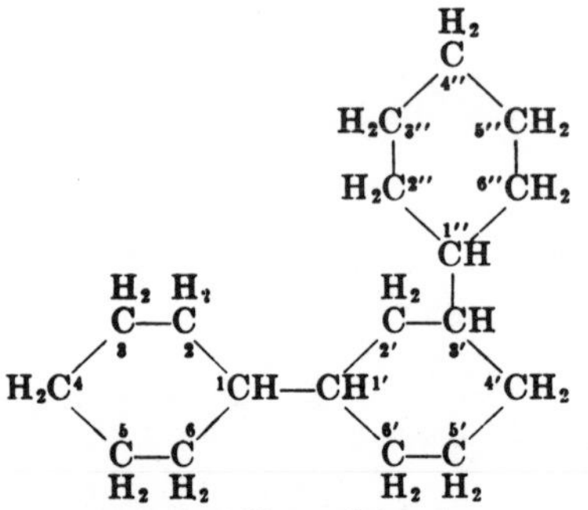

1,1':3',1''-Tercyclohexane

p-Terphenyl
or 1,1':4',1''-Terphenyl

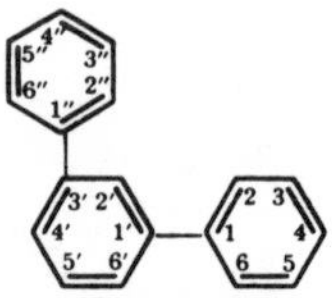

m-Terphenyl
or 1,1':3',1''-Terphenyl

G. Substituted (Side-Chain) Cyclic Hydrocarbons

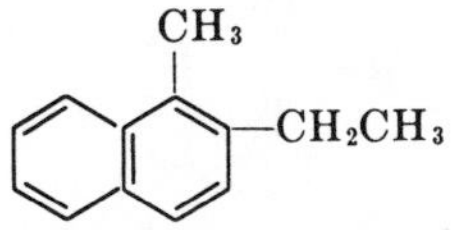
CH3
CH2CH3

2-Ethyl-1-methylnaphthalene

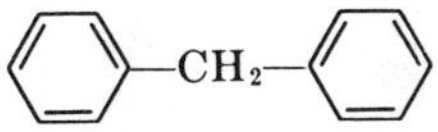
CH2

Diphenylmethane

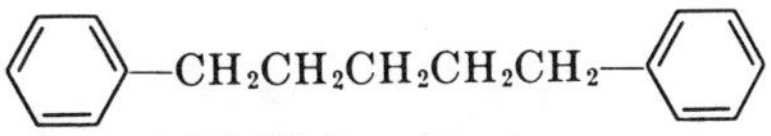
CH2CH2CH2CH2CH2

1,5-Diphenylpentane

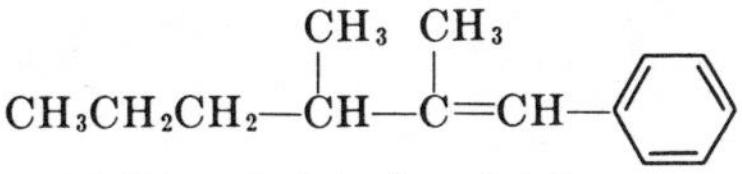
CH3 CH3
CH3CH2CH2—CH—C=CH—

2,3-Dimethyl-1-phenyl-1-hexene

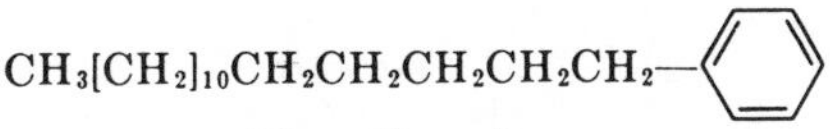
CH3[CH2]10CH2CH2CH2CH2CH2—

1-Phenylhexadecane

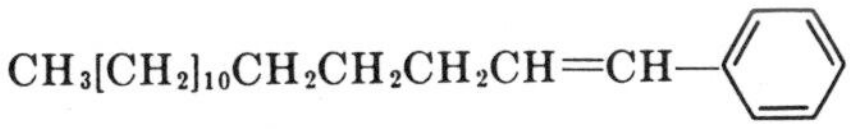
CH3[CH2]10CH2CH2CH2CH=CH—

1-Phenyl-1-hexadecene

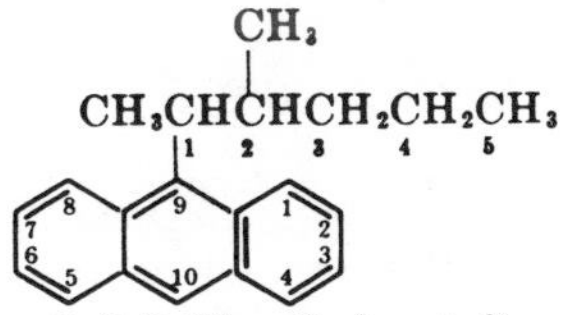
CH3
CH3CHCHCH2CH2CH3
1 2 3 4 5

**9-(1,2-Dimethylpentyl)-
anthracene**

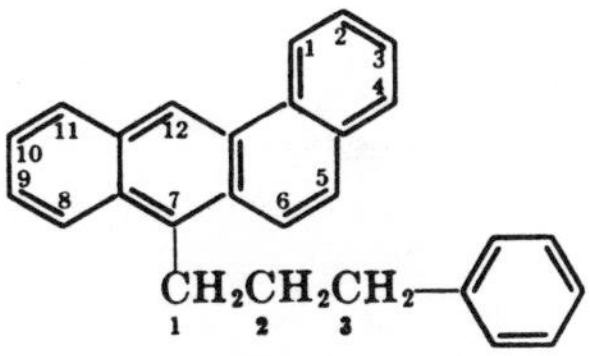
CH2CH2CH2
1 2 3

**7-(3-Phenylpropyl)-
benz[*a*]anthracene**

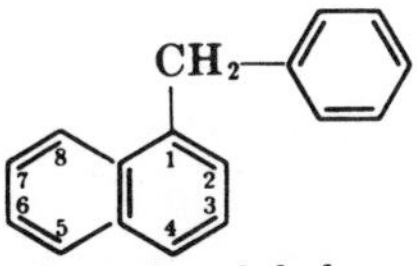
CH2

1-Benzylnaphthalene

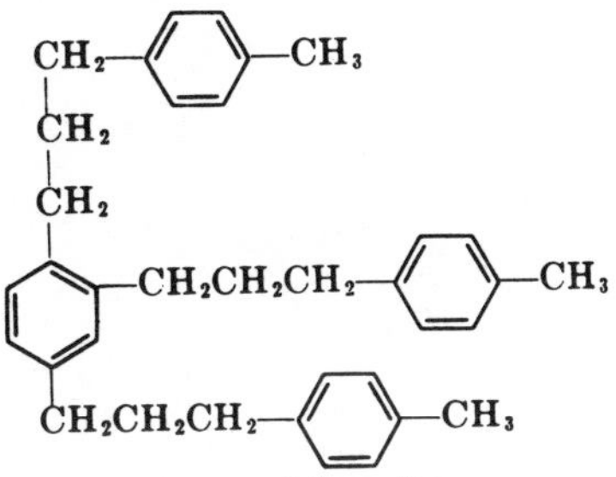
CH2
CH2
CH2
CH3
CH2CH2CH2
CH3
CH2CH2CH2
CH3

1,2,4-Tris(3-*p*-tolylropyl)benzene

H. Terpene Hydrocarbons

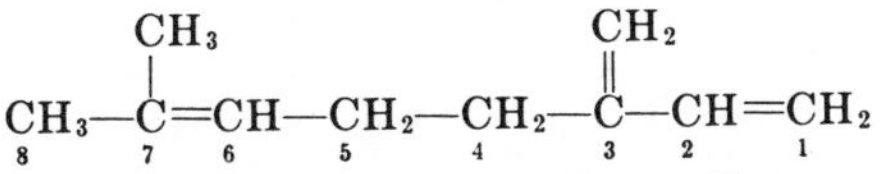
CH3 CH2
CH3—C=CH—CH2—CH2—C—CH=CH2
8 7 6 5 4 3 2 1

7-Methyl-3-methylene-1,6-octadiene

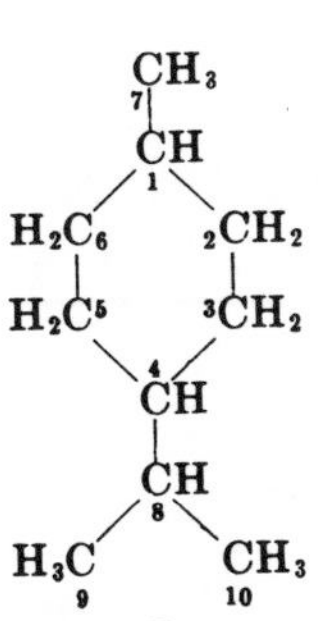

I
Menthane (*p*-form)

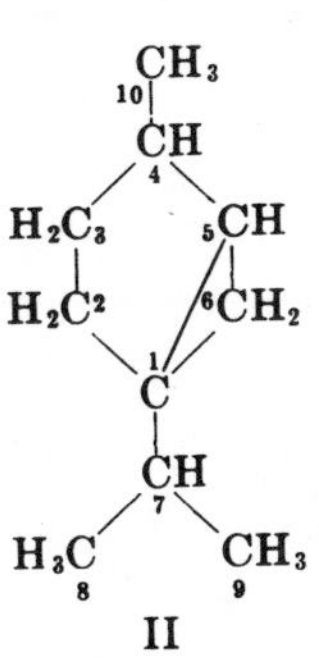

II
Thujane

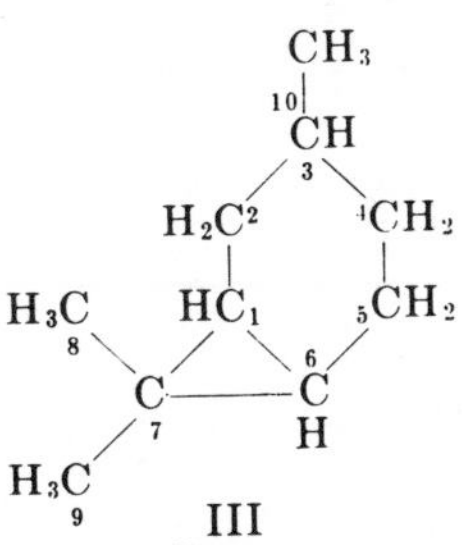

III
Carane

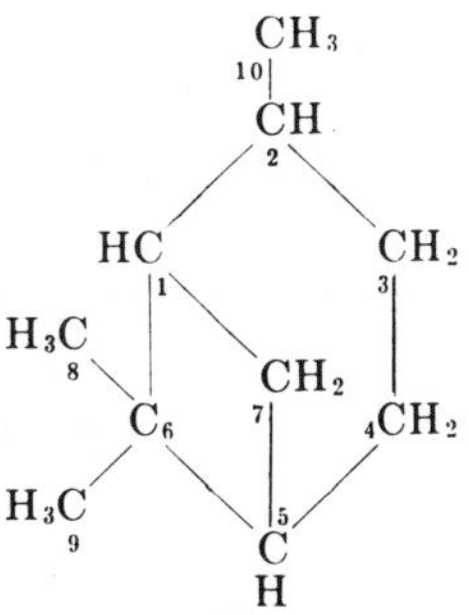

IV
Pinane

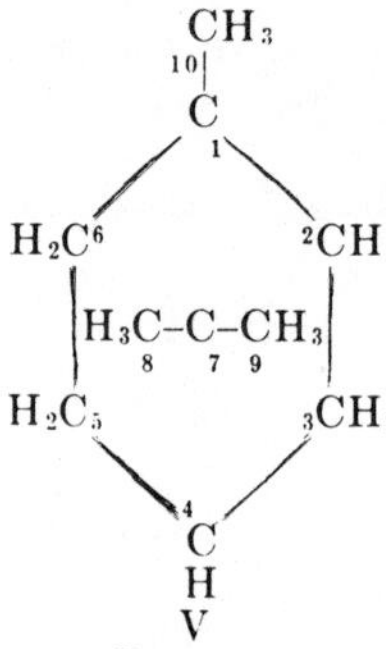

Nor-structures:

V
Bornane

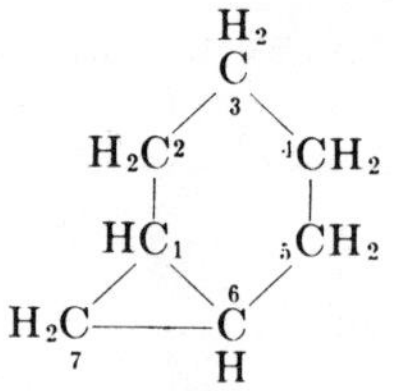

VI
Norcarane

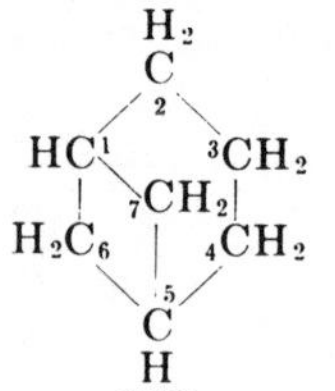

VII
Norpinane

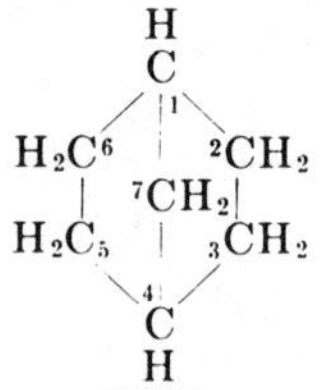

VIII
Norbornane

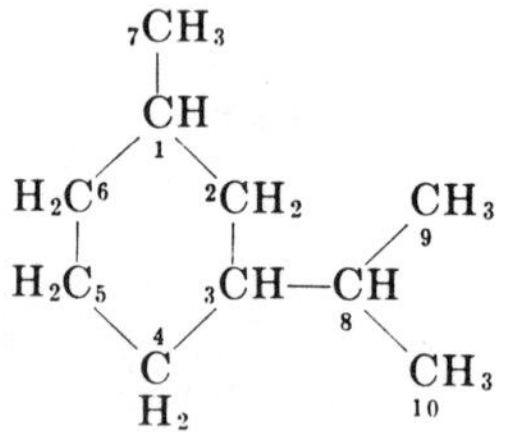

m-Menthane

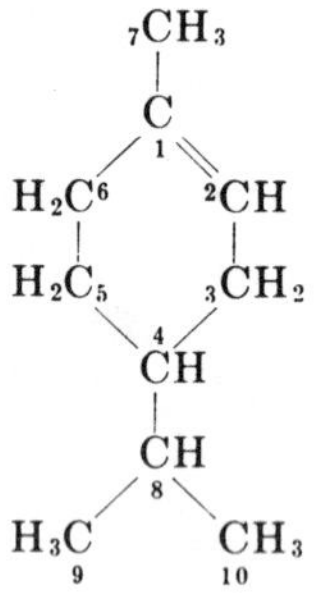

1-*p*-Menthene

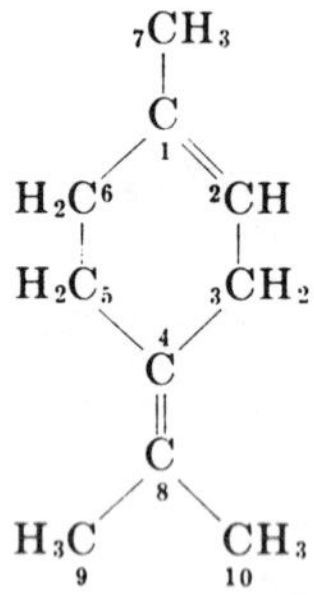

1,4(8)-*p*-Menthadiene

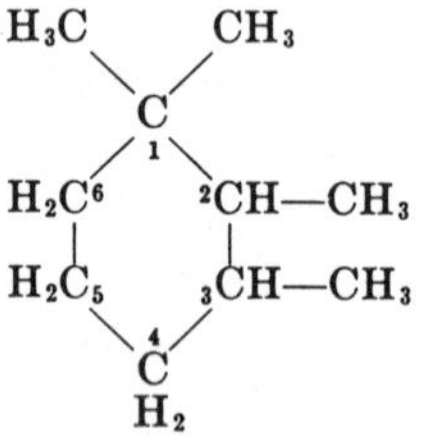

1,1,2,3-
Tetramethyl-
cyclohexane

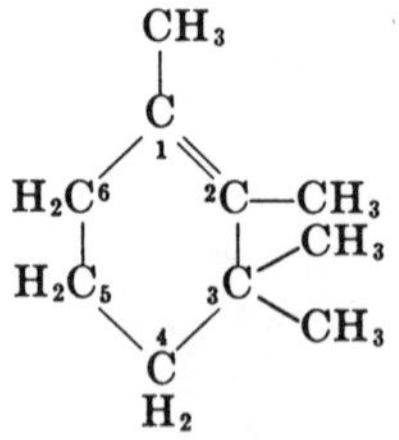

1,2,3,3-
Tetramethyl-
cyclohexene

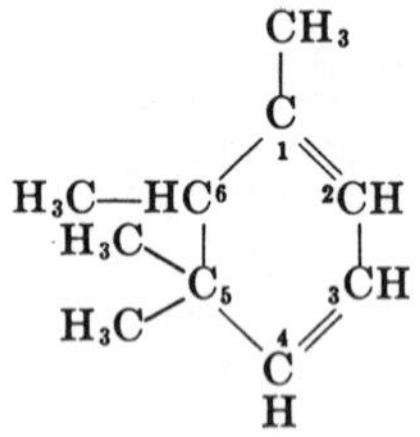

1,5,5,6-
Tetramethyl-
1,3-cyclohexadiene

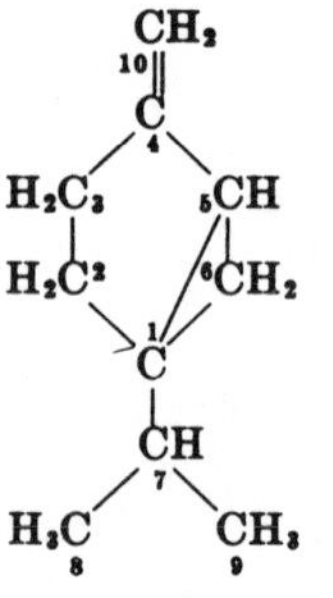

4(10)-Thujene

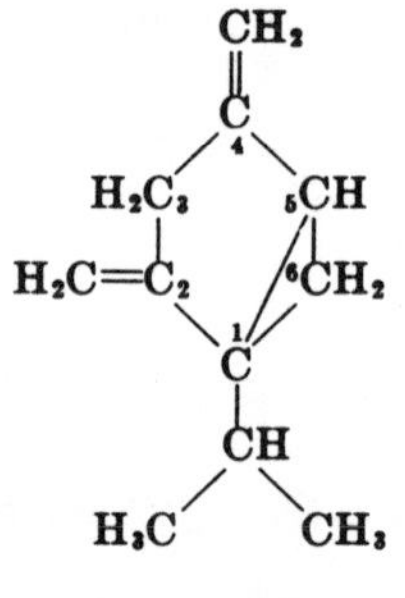

1-Isopropyl-2,4-
dimethylenebicyclo-
[3.1.0]hexane

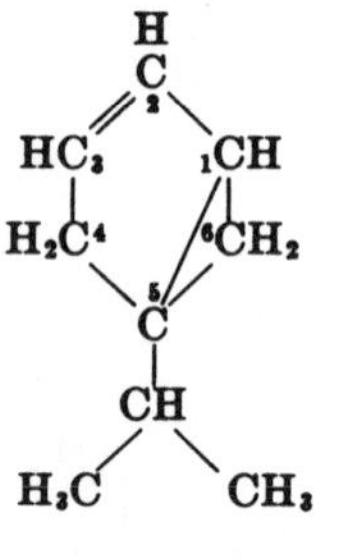

5-Isopropyl-
bicyclo[3.1.0]hex-
2-ene

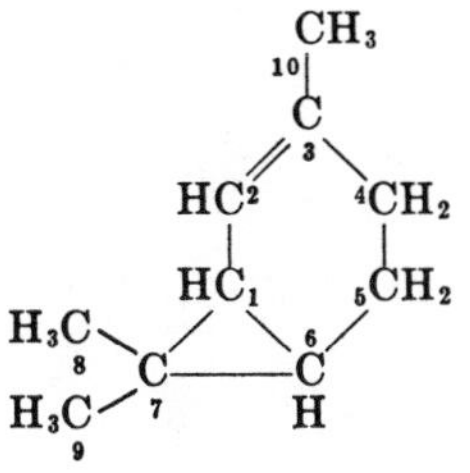

2-Carene

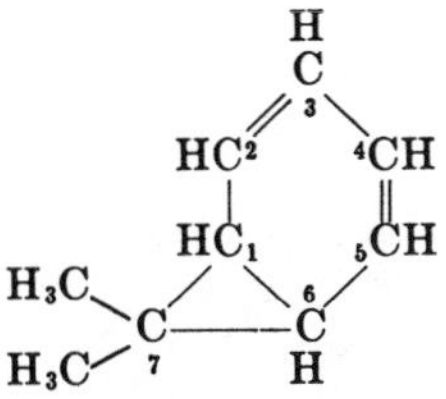

7,7-Dimethyl-2,4-norcaradiene

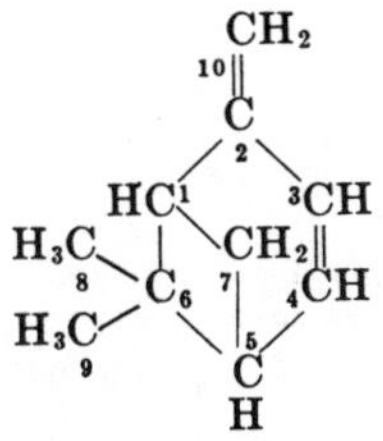

2(10),3-Pinadiene

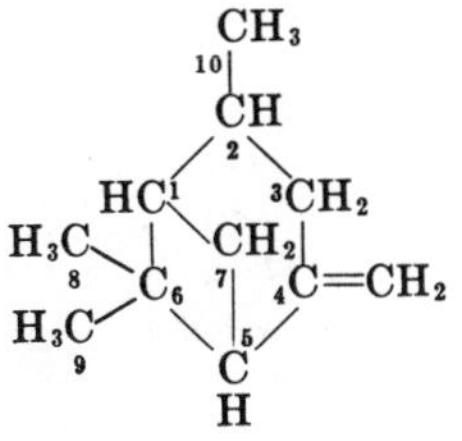

4-Methylenepinane

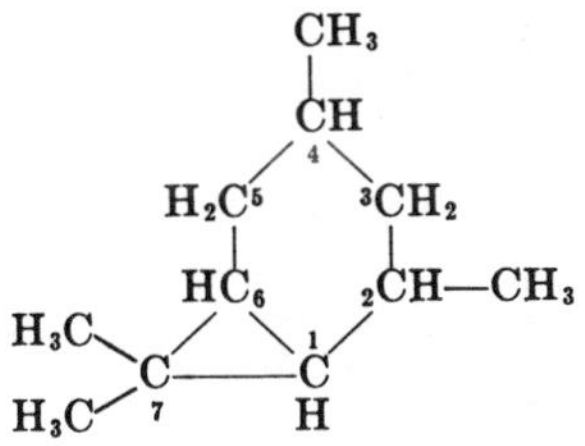

2,4,7,7-Tetramethylnorcarane

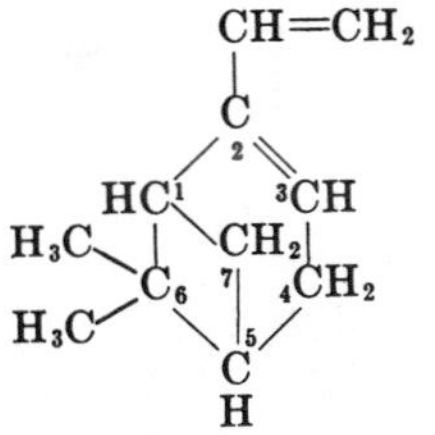

6,6-Dimethyl-2-vinyl-2-norpinene

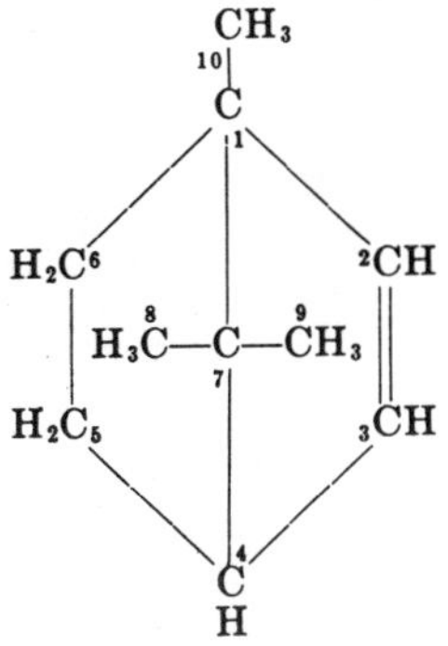

2-Bornene

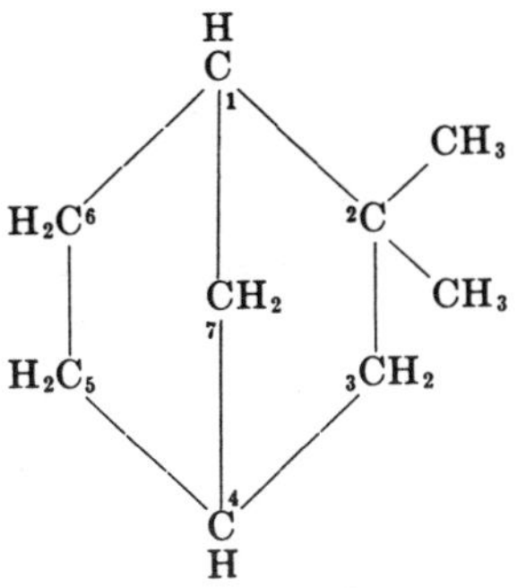

2,2-Dimethylnorbornane

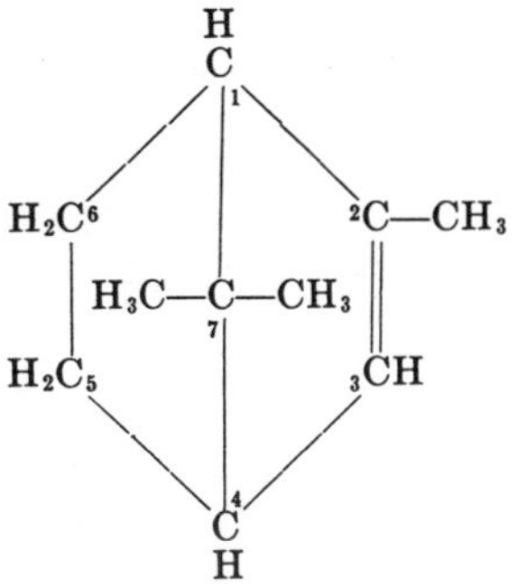

2,7,7-Trimethyl-2-norbornene

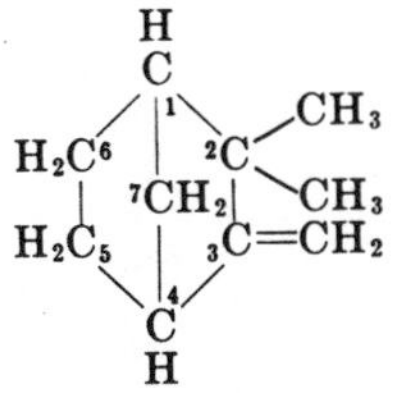

Camphene

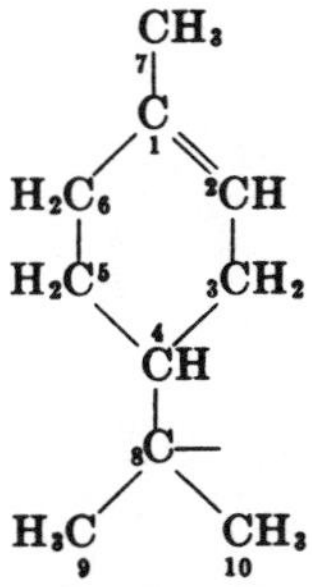

-*p*-Menthen-8-yl

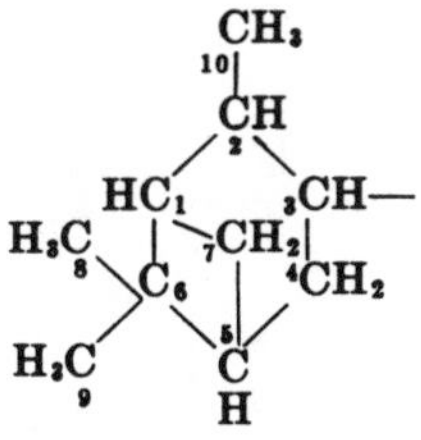

3-Pinanyl

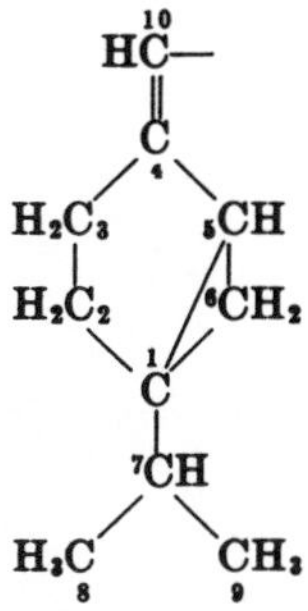

4(10)-Thujen-10-yl

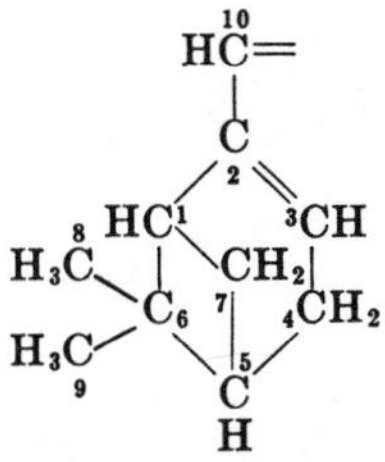

2-Pinen-10-ylidene

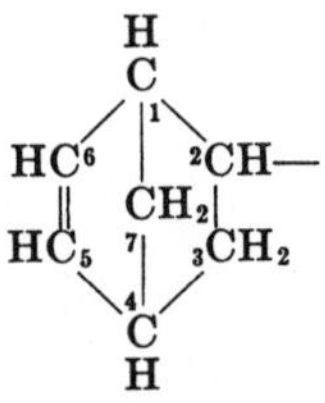

5-Norbornen-2-yl

REFERENCES

1. **Bland, W. F. and Davidson, R. L., Eds.,** *Petroleum Processing Handbook,* McGraw-Hill, New York, 1967.
2. **Mackison, F. W., Stricoff, R. S., and Partridge, L. J., Jr., Eds.,** NIOSH/OSHA Pocket Guide to Chemical Hazards, U.S. Government Printing Office, Washington, D.C., 1980.
3. **Stern, A. C., Ed.,** *Air Pollution,* Vol. II, Academic Press, New York, 1977.
4. **Dipple, A.,** in *Chemical Carcinogens,* Searle, C. E., Ed., ACS Monograph Series No. 173, American Chemical Society, Washington, D.C., 1976, 245.
5. **Pucknat, A. W., Ed.,** Health Impacts of Polynuclear Aromatic Hydrocarbons, Noyes Data Corporation, Park Ridge, N.J., 1981.
6. **Fishbein, L.,** *Potential Industrial Carcinogens and Mutagens,* Elsevier, New York, 1979.
7. Environmental Quality — 1980: The 11th Annual Report of the Council on Environmental Quality, U.S. Government Printing Office, Washington, D.C., 1980.
8. Certain Polycyclic Aromatic Hydrocarbons and Heterocyclic Compounds, IARC Monographs on the Evaluation of Carcinogenic Risk of the Chemical to Man, Vol. 3, International Agency for Research on Cancer, Lyon, France, 1973.
9. **Weast, R. C. and Astle, M. J., Eds.,** *Handbook of Chemistry and Physics,* 68th ed., CRC Press, Boca Raton, Fla., 1987, C-1—C-24.

Chapter 2

GAS CHROMATOGRAPHY

There are numerous excellent texts detailing the theory and practice of gas chromatography (GC). This chapter highlights some of the more important principles and techniques of GC and discusses the underlying theoretical and practical aspects of solute band broadening and the thermodynamic basis for GC separations.

Recent research and applications in the use of GC for the analysis and characterization of hydrocarbons for a wide variety of activities are described and cited in Chapters 3 to 6. Information on liquid and solid stationary phases used and on retention data reported in these activities is summarized in detail in Chapters 7 and 8, respectively.

I. THE BASICS OF GAS CHROMATOGRAPHY

A. Principles

In GC, a sample is introduced into an essentially inert gas stream (carrier gas) moving through a tube (the GC "column"), usually metal or glass, that is thermostatically controlled. The sample generally contains one to several hundred analytes dissolved in a solvent, or one to tens of analytes in a matrix gas sample. The sample is injected into the GC inlet system generally using a microliter syringe for liquid solutions or pure liquids, or using a gas sampling/switching multiport valve as the inlet system for gaseous samples. The temperature of the inlet system of the GC may range from ambient for very volatile samples to over 200°C for the vaporization of samples of low volatility. Hence, the sample analytes enter the GC column in the gaseous (or vapor) state.

Once the analytes enter the moving gas in the column, they move through the column with the carrier gas. In the absence of any substrate (packing material or coating) in the column, the analytes elute from the column as unretained components. Packed columns contain a premeshed particulate material known as the column packing. Column packings are generally either: (1) a very low volatility substrate known as the "stationary phase" (e.g., a high-boiling silicone polymer) coated from less than 1% to over 20% by weight on an "inert" particulate support material (e.g., diatomaceous earth) covering a particle size range of anywhere from coarse (40 to 60 mesh) to fine (100 to 120 mesh) particles; or (2) a solid sorbent (e.g., porous polymers or treated graphite). Analyses of samples on columns containing liquid-coated support packings are known as gas-liquid chromatographic (GLC) analyses; analyses of samples on columns containing solid sorbent packings are known as gas-solid chromatographic (GSC) analyses. Coated columns differ from packed columns in that there is no solid support or solid sorbent in the column; instead, the inner wall of the column is coated with a thin film of a very high boiling liquid stationary phase. Such columns generally have a very narrow bore diameter ("capillary columns") and are known for their markedly enhanced power for separating analytes of similar volatility (i.e., "resolution") relative to packed columns. The trade-off in the use of capillary columns is (1) that they have a much lower sample size capacity than packed columns (which are generally 2 to 4 mm or more in internal diameter and contain a much greater amount of stationary phase), and (2) that they generally require high sensitivity detection capabilities owing to the sub-resolving power of such columns for very complex samples have made these types of columns even more popular than packed columns in a number of cases. An alternative semipopular type of "capillary column" is one in which a solid support material has been bonded to the inner wall of the column, which is then coated with the liquid stationary phase ("support-coated open-tubular "SCOT" columns). The newest type of capillary columns are the flexible fused-silica quartz capillaries, some of which contain either polymerized or bonded "liquid" stationary phases. The stationary phase represented by the column packing (GLC or GSC) or coating provides the mechanism of separation of the sample analytes. In the ideal case, the analytes in the samples are retained to different degrees by the stationary phase, emerging as separate, discrete zones (or "peaks") from the end of the column. In the case of GLC-packed or -coated columns, the mechanism involves the "partitioning" of the sample analytes between the stationary phase and the moving carrier gas ("mobile phase"). In essence, this differential partitioning is due to the differences in the partial molar free energy of mixing (or the free energy of adsorption in the case of GSC) of the sample analytes in or with the stationary liquid phase. That is, the stronger the attractive forces between the sample analyte and the stationary phase (i.e., the more negative the free energy of mixing or association), the longer it will take to elute from the column (i.e., its "retention" will be prolonged). The observed total retention of an analyte is a function of (1) the magnitude of the interaction between the analyte and the stationary phase, (2) the vapor pressure of the analyte, and (3) the column temperature. Other than the stationary phase, the column temperature is the most important controllable parameter influencing analyte retention in the column. Increasing the column temperature will both decrease the strength of interaction between the analyte and the stationary phase (for isotropic stationary phases) and will increase the vapor pressure of the analyte; hence, increasing the column temperature will almost always result in a decrease in analyte retention in the column. To a first approximation, the use of "nonpolar" stationary phases (e.g., methyl silicone or hydrocarbon phases) will result in the elution of hydrocarbon analytes in the order of their vapor pressures. The petrochemical industry has used this observation for the development of GLC methods for the determination of boiling points of analytes in complex hydrocarbon mixtures. Owing to the high dependence of analyte retention on column temperature, the use of linear and multilinear temperature programing of the column has proven to be a powerful technique for the analysis of very complex mixtures that contain analytes having wide boiling point ranges.

As the analytes exit the end of the GC column, they enter a detector that produces an electrical signal, the strength and duration of which is directly related to the amount of mass and type of the analyte and the width of its eluting zone (or "peak"). The signal is subsequently amplified and passed to an electronic integrator, an integrating computer, and/or a strip chart potentiometric recorder by which the amount of analyte can be quantified and/or observed. Such signal sensing systems also provide information on the retention time (i.e., the amount of time an analyte is retained by the column) of the eluting analytes, which is a quasiqualitative parameter related to its identity and is, under well-controlled conditions of column temperature and carrier gas flow rate, highly reproducible.

Additional aspects of some of the more important areas of GC analysis are given in the following sections on band broadening theory, the GC column, column temperature, GC detectors, qualitative analysis, quantitative analysis, applications, instruments and column packings, computers, and the thermodynamic basis for separations in Sections II to IX, respectively.

B. Basic Apparatus

The basic components of a GC system are illustrated in Figure 1. These components include the following.

1. Carrier Gas Source

A high pressure cylinder containing the carrier gas is used as the carrier gas source. The gas must be suitable for the detector being used and should be an inert, high-purity gas. A higher molecular weight carrier gas, such as nitrogen, will produce less band broadening of sample components than a light carrier gas, such as helium. High-purity nitrogen is useful

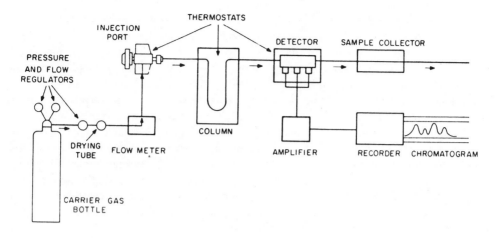

FIGURE 1.   Basic Components of a GC System

for capillary GC applications, but a carrier gas having a high thermal conductivity, such as helium or hydrogen, is required for adequate sensitivity using a thermal conductivity detector. Hydrogen has been used as a carrier gas for both thermal conductivity and flame ionization detectors in Europe since it is much less expensive than helium, but extreme care must be exercised in its use and leaks must be carefully excluded due to its explosive nature. The purity of the carrier gas is of utmost importance for high quality GC analysis. To ensure this, a drying tube and/or sorbent trap is recommended between the gas cylinder pressure regulator and the GC injection port to remove trace levels of water vapor, hydrocarbons, fine particulate matter, and other impurities.

2. Pressure and Flow Controllers

A two-stage regulator is used at the outlet of the carrier gas cylinder to (1) monitor the residual pressure in the cylinder and (2) to set an appropriate cylinder outlet pressure to the GC system. Most GC instruments are equipped with a flow controller and a flow meter. The flow controller contains a diaphram that ensures a constant downstream flow despite changes in pressure and pressure drops throughout the GC system. The flow meter is generally used to set a carrier flow rate to an acceptable level and to monitor the stability of the carrier gas flow. In practice, the flow meter may be calibrated against actual measurements of the column exit flow rate using a soap bubble flow meter at the end of the column at ambient temperature (or at the detector outlet when the column is operated at an elevated temperature). The location of particle and impurity traps for water vapor and trace hydrocarbons will aid in preventing the ball float in the GC flow meter from sticking.

3. Injection Port

Samples are generally injected rapidly onto the GC column through a self-sealing septum in the injection port. Liquid samples and solutions are injected with a liquid syringe. Gas samples may be injected with a gas-tight syringe or a gas sampling/switching valve containing a gas sampling loop. Sample injection sizes will depend on the concentration of the sample components being analyzed, the capacity of the column, and the sensitivity of the detector. Sample sizes for capillary GC analysis may range from 0.1 to 10 $\mu\ell$ for gases and 0.004 to 0.5 $\mu\ell$ for liquid and solution samples; sample sizes for packed analytical GC columns may range from 0.1 to 50 mℓ for gases and 0.05 to 10 $\mu\ell$ for liquids and solutions. Sample sizes for preparative-scale GC may range from 0.05 to 5 ℓ for gases and from 0.02 to 2 mℓ for liquids and solutions. The injection port is generally heated at a temperature that is sufficient to allow instantaneous volatization of the sample without thermally decomposing the sample components. The internal volume of the injection port should be negligible to avoid unnecessary band broadening prior to the sample entering the column. To minimize this, injection ports are often designed to allow the tip of the sample syringe to enter the head of the column packing. When using capillary GC columns, the injection port often contains a heated low dead-volume splitter to exhaust most of the sample to prevent column overload. A recommended technique for syringe injection of liquids and solutions is using the solvent-flush method. This method involves drawing about 1 $\mu\ell$ of solvent into the syringe, followed by about 1 $\mu\ell$ of air, followed by the sample. The syringe plunger is then further withdrawn and the volume of sample is read from the syringe barrel. The entire contents of the syringe is then introduced into the injection port, with the solvent plug flushing the syringe needle and the injection port to ensure total transfer of the recorded amount of sample into the GC column. This technique works well with packed GC columns, but is not especially appropriate with capillary columns.

4. The Column

GC columns are hollow tubes of copper, stainless steel, glass, or other materials, and are usually coiled in a helix for internal packing or coating.

Capillary columns are typically coated by a dynamic or static method. In the dynamic method a dilute coating solution is passed slowly through the column at a controlled rate, followed by dry nitrogen. In the static method the column is filled with the coating solution which is then evaporated in a laminar fashion using a special oven, leaving a thin film deposition of the coating on the internal wall of the column. Special types of alternative methods of producing coatings also are used. Two of these include *in situ* thermal polymerization of a coating material and coating a film on fibral alumina which had been previously affixed to the internal wall by a special procedure.

Packed columns are prepared by introducing a previously prepared packing material into one end of the column with gentle tapping while pulling a vacuum on the other end that also contains a small silanized glass wool plug. The column packing can be prepared in several ways. Two of the most popular include: (1) the use of a rotary evaporator to remove the solvent from a mixture of a known amount of the support material and a solution containing a known amount of the stationary phase while gently heating under vacuum; and (2) evaporating the solvent from a mixture of the support and coating solution in a large evaporating dish resting on a large beaker of warm water, using a gentle stream of dry nitrogen.

Packed analytical GC columns are generally 6 to 20 ft in length and 1/8- or 1/4-in. O.D.; standard capillary GC columns are 0.25 up to 0.75 mm I.D. ("wide-bore"), and typically vary in length from 30 to 300 m.

5. Detector

The GC detector senses the emergence of a compound as it exits from the GC column, producing an electrical signal that is proportional in intensity and lasting in duration with the amplitude and width of the eluting compound band. A useful detector is highly sensitive to the sample components being analyzed, possesses a linear response over a wide concentration range, and is insensitive to carrier flow and temperature variations.

6. Sample Collection

Isolated sample components may be collected by passing the GC carrier gas outlet stream through tubing immersed in a cryogenic bath. Sample collection is rarely used in analytical GC, but is routinely practiced in preparative GC applications. In preparative-scale GC a very small fraction of the effluent from the GC column is passed to a GC detector to monitor the emergence of sample components, the collection of which can be readily automated. Direct sample component collection from the outlet stream of a nondestructive detector (e.g., thermal conductivity detector) also is possible. This is obviously not feasible with a destructive detector such as the flame ionization or electron-capture detectors, or in GC-mass spectrometry.

7. Signal Amplification and Recording

The detector signal is passed to an amplifier where the signal is amplified and electronically massaged. The amplifier output signal is then attenuated to a potentiometric strip-chart recorder to monitor the progress of the chromatography and produce a record of qualitative (retention time) and quantitative (peak area or height) information concerning the GC analysis. Typical GC systems today employ electronic integrators and computers for data storage, acquisition, and various data treatments. Computerized GC data acquisition systems can generate reports of the GC analysis, identify peaks on the basis of a retention data library, and store raw data for future reference or manipulation.

8. Thermostating

The temperatures in a GC system must be controlled in three locations: the injection port; the column oven; and the detector.

The injection port temperature is usually maintained 10 to 50° above that of the column oven for liquid samples, but the injection port temperature need not be heated for gaseous samples.

The temperature in the column oven is selected to provide adequate resolution of the sample components in a reasonable analysis time. Linear temperature programing of the column oven is used to obtain optimum information from samples that contain wide boiling components. The temperature program is selected in a manner that resolves early peaks and elutes later peaks in a reasonable time. Isothermal operation is appropriate for samples that contain components that cover a relatively narrow to moderate boiling point range and is preferred over temperature programing for highly accurate quantitative analysis. Temperature control of the column oven to within 2° is usually adequate, but it is not unusual for modern GC systems to control column oven temperatures to less than 1°. Column oven temperature control will usually decrease with increasing temperature. Temperature readings on a GC temperature meter are typically inaccurate, substantially worsen at elevated temperatures, and generally do not reflect the actual excellent temperature control designed into the GC oven.

The detector temperature must be higher than the column oven temperature to avoid possible condensation of sample components (or their decomposition products in the case of destructive detectors) in the internal detector flow lines. Detectors may occasionally require cleaning due to deposition of column stationary phase bleed. Silane deposition typically occurs over a period of time when using silicone stationary phases; this can sometimes be reduced *in situ* by injecting a solution into the GC that converts the silane deposit to SF_4 gas. Temperature control is critical to ensure the stability of some detectors (e.g., thermal conductivity), but not as critical for others (e.g., flame ionization).

II. BAND BROADENING

In principle, once an analyte has been injected into the GC and has been vaporized in the inlet system, it enters the GC column as a narrow-width "band" of its composite molecules. In the absence of any column packing or stationary phase coating, this band of molecules will broaden somewhat due to some degree of molecular diffusion in the moving carrier gas stream. In the presence of a stationary phase wall coating (capillary columns), the band will be further somewhat broadened due to the free energy of its interaction with the stationary phase that will retain the band in the column longer than a nonretained analyte. The magnitude of this free energy of interaction will decrease with increasing column temperature, hence reducing the "retention time" of the band. In the presence of a column packing (packed columns), the band will undergo a much enhanced broadening due to two factors: (1) the molecules comprising the band can take a multitude of paths through the packing bed, some of which will be longer or shorter than the average path, and (2) packed columns contain a much greater liquid depth than wall-coated capillary columns, resulting in different depths of penetration of the molecules into the liquid phase and different diffusion times back through the stationary phase into the carrier gas stream.

A. Estimating Column Efficiency

Capillary columns are said to provide much greater column "efficiencies" and "resolution" of analytes than packed columns, simply meaning that such columns produce substantially less band broadening, resulting in narrow bands of consecutively eluting peaks that are well separated or resolved from each other.

Efficiency in GC analysis is defined in terms of "theoretical plates", using an analogy to classical distillation columns. A "theoretical plate" in GC is conceived of as being equivalent to the linear column distance between equilibrations that take place between the gaseous analyte and the stationary phase as the analyte passes through the column. The number of theoretical plates in a capillary column are much greater than in a packed column. This is due to the rapid transfer of analyte molcules between the thin stationary phase in capillary columns and the carrier gas phase, that allows for numerous equilibrations of the analyte molecules with the stationary phase as they move down the column towards its exit. The number of theoretical plates (N) in any GC column is estimated by the thinness of the band that emerges from the column, and can be estimated by the use of an equation such as:

$$N = 5.545[X/w]^2 \tag{1}$$

where X is the time (or distance), on a strip-chart recorder, from the peak maximum of an unretained component (e.g., the "air" peak) to the peak maximum of the analyte of interest, and w is the time (or distance) width of the analyte peak at half-height. Measurement of X, hence, is a measurement of the duration that the analyte has spent in contact with the column substrate, since subtracting the time (or distance) for the unretained component corrects for the total void volume of the entire GC system (inlet plus column plus column-detector connection). "N", hence is calculated as a ratio of times or of distances, and is dimensionless. The greater the value of N, the sharper the eluting peak (i.e., the narrower and less spread is the analyte eluting band), and the more "efficient" is the column performance.

The efficiency of the column also can be estimated in terms of "plate height" (H). This is ideally conceived as being the average column distance between consecutive equilibrations between the analyte molecules and the stationary phase, asnd is generally estimated by

$$H = L/N \tag{2}$$

where L is the length of the column in millimeters.

B. Band Broadening Theory

In GC band broadening theory, the magnitude of H is controlled by a number of parameters. A simplified version of the association of these parameters is given by

$$H = A + [B/u] + Cu \tag{3}$$

The "A" term is a measure of the analyte band broadening due to the variability of paths that the analyte molecules can take through the bed of the column packing. The magnitude of A is controlled by the uniformity of the particles comprising the column packing and their particle diameter; hence, the use of small particles of essentially equal diameter will decrease the contribution of the A term to H. The B term in Equation 3 is a measure of the longitudinal diffusion of the analyte molecules in the carrier gas stream; it is directly proportional to uniformity of the column packing and the diffusion coefficient of the analyte molecules and inversely proportional to carrier gas flow rate (u). Given a uniform column packing, the use of very low carrier gas flow rates will increase the time with which diffusion in the gas phase can occur, while the use of high flow rates will increase the mixing of the analyte molcules with the carrier gas. Either of these factors will increase the B term and, correspondingly, contribute to an increase in H. Hence, in packed GC columns there is an "optimum" flow rate that is intermediate to either of these extremes at which the contribution of the B term to H is a minimum. Diffusion in the gas phase also is slightly enhanced when the molecular weight of the carrier gas is increased; hence, for the same set of conditions, the use of nitrogen as a carrier gas will have a somewhat lower effect on longitudinal band broadening than using helium. The C term in Equation 3 is a coupled function of the resistance to mass transfer in the liquid phase and in the gas phase (the latter of which is increased with increasing column diameter). This term is directly proportional to carrier gas flow rate and the thickness of the liquid phase (i.e., its percent loading) on the packing support material and is inversely proportional to the diffusion coefficient of the analyte molecules in the liquid phase. Hence, H can be notably reduced by using low loaded column packings, liquid phases that have low viscosities and do not have strong affinities for the analyte molecules, and moderate carrier gas flow rates.

In summary, solute zone broadening with packed GC columns can be reduced by using uniform packings having a small particle diameter and thin low-viscosity films, narrow internal diameter columns, and neither excessively low or high carrier gas flow rates. In wall-coated capillary columns, the A term is zero due to the absence of a column packing bed, thereby eliminating one of the major contributions to band broadening. Resistance to mass transfer (the C term) also is reduced in the case of capillary columns in both the liquid phase (since a very thin stationary phase film is used) and in the gas phase (since very narrow internal diameter columns are used, usually in the order of 0.25 mm I.D.). In the balance then, capillary GC columns have much lower values of H than packed GC columns, thereby affording substantially more theoretical plates per foot of column length (from Equation 2), sharper peaks, and higher efficiency columns. It also should be obvious from the above that in the case of high performance liquid chromatography, the B term is essentially negligible due to the low diffusivity of solutes in a liquid mobile phase (roughly 10^4 lower than in a gas phase).

C. Definitions of Separations

There are three popular parameters that are conventionally used to define retention behavior and separations on a given GC system (i.e., GC stationary phase and operating conditions): (1) the "resolution" (R) between two consecutively eluting peaks; (2) the separation factor (α); and (3) the retention index (I). The resolution (R) of two consecutively eluting peaks 1 and 2 is estimated by

$$R = 2d/[W_1 + W_2] \qquad (4)$$

where d is the distance between the band centers (i.e., the distance between the peak maxima for symmetrical peaks), and W_1 and W_2 are the respective widths of the two peaks at their base. A value of R as unity indicates a separation between the two peaks that is close to baseline (i.e., 100% resolution).

The separation factor, α, is a measure of the selectivity of the column and the operating conditions for separating two peaks, x and y, and is defined as

$$\alpha = t'(y)/t'(x) = V'(y)/V'(x) \qquad (5)$$

where t' and V' are the corrected retention time and the corrected retention volume (retention volume = retention time times carrier gas flow rate) of peaks x and y, and where y elutes at some point after x. The retention times and the retention volumes in Equation 5 are corrected for system "dead volume" by subtracting out the retention time and retention volume of an "unretained" component (usually estimated by the retention time or retention volume of air or methane); hence, the "corrected" values represent a measure of the net duration of solute-stationary phase contact during the passage of a solute zone through the column. The separation factor determined in Equation 5 may have a slight dependence on column temperature, but nowhere near the magnitude of temperature dependence exhibited by retention time or retention volumes, themselves. Hence, separation factors have been extensively used in reporting GC data, since they are far more reproducible than their components, and they represent one conventional way of reporting normalized data. A major disadvantage in using separation factors for reporting retention data lies in the fact that separation factors decrease in precision when comparing retention of solutes that elute much earlier or much later than the solute chosen as the reference peak (i.e., having a separation factor of unity).

A more acceptable way of reporting retention data is to use a number of reference solutes, such as used in the retention index system. The conventional index system uses a homologous series of normal-chain hydrocarbons as the reference solutes. In this system, the retention index (I) of a solute is given by

$$I = 100\{\log[t'(x)/t'(P_z)]\}/\log[t'(P_{z+1})/t'(P_z)] + 100z \qquad (6)$$

where I is the retention index of solute x on a given stationary phase at a given temperature, P_z is an *n*-alkane eluting before x, P_{z+1} is an *n*-alkane containing one more methylene group than P_z, and z is the number of carbon atoms in P_z. Hence, solutes in this system are reported as having GC retentions on a normal hydrocarbon scale. To illustrate, a solute having a retention index of 847 is described as having a retention analogous to a normal hydrocarbon having a chain length of 8.47 carbon atoms. The beauty of this system is that retention indices of solutes are quite reproducible between laboratories using the same GC column temperature and stationary phase. In addition, since the retention index is on a log scale, absolute differences between retention indices of two solutes that differ by a single structural moiety represent, to a first approximation, a measure of the retention contribution of the structural moiety to the retention of the solute containing it. The major criticism of retention indices based on a hydrocarbon scale is that such indices lose some meaning when applied to other classes of organic compounds. For this reason, various workers have employed other homologous series (e.g., fatty acid methyl esters) or sets of other reference compounds (e.g., polycyclic aromatic hydrocarbons for indices of same). Nonetheless, the utility of retention indices remains a popular mechanism for reporting retention data in the GC literature.

III. THE GC COLUMN IN HYDROCARBON ANALYSIS

Conventionally, one of the most critical steps in establishing a GC method for the analysis of solute mixtures is the selection of the GC column and its stationary phase. This is particularly the case when one is dealing with solute mixtures containing a wide range of polarities. While this is usually not the case in hydrocarbon analyses, nonetheless, different column types and stationary phases are more appropriate for certain hydrocarbon mixtures than for others. The hydrocarbon class, per se, is fairly nonpolar in nature, but the chemical compounds in this class can range from low- to high-molecular-weight paraffins and include unsaturated, cyclic, aromatic, polycyclic aromatic, and terpene hydrocarbons. The choice of column for these different applications can vary considerably. To illustrate: spherical carbosieves can be used to separate low-molecular-weight hydrocarbons; graphitized carbons spiked with picric acid have shown utility for separating low-molecular-weight unsaturated hydrocarbons; tris[cyanoethoxy]propane has been used for the separation of mixtures of aromatic and aliphatic paraffins; Dexsil 300 can be used for hydrocarbon waxes; silicone phases (e.g., SE-30, SE-52, SE-54) have gained considerable popularity for separations of high-molecular-weight hydrocarbons; and nematic liquid crystals afford unique isomeric separations of substituted benzenes and polycyclic aromatic hydrocarbons. As a guide to the selection of columns and stationary phases, the reader is referred to applications in the 32 chromatograms in Chapter 3, and to the reference citations and tabulated retention data given in Chapter 8, Sections I and II respectively.

Packed columns generally contain a support material that has been specially treated to enhance its inertness, coated with a thin layer of a stationary phase. As has been pointed out in the previous section, small, uniform particle sizes (e.g., 100 to 120 mesh) of the support material (e.g., diatomaceous earths) are recommended for enhancing column efficiency and decreasing zone broadening.

The stationary liquid phase should have a low viscosity and afford some differences in solute solubilities for hydrocarbon mixtures that contain various classes of hydrocarbons. Liquid phase loadings of 1 to 10% by weight of support material are generally used, although lower loadings will produce less band broadening and sharper peaks as long as the stationary phase capacity is not exceeded by using samples that are too large or too concentrated. Since solute retention is a function of liquid phase loading, retention decreases with decreasing phase loading. For this reason, solute retention data sometimes are expressed in terms of the specific retention volume (i.e., retention volume per gram of stationary phase). In addition, liquid phases should be chosen such that phase "bleed" does not occur at the column temperature(s) needed for the analysis. Commercial GC supply catalogs generally indicate an upper recommended column temperature for the various stationary phases. Finally, it should be noted that some stationary phases have unique selectivity characteristics for certain classes of hydrocarbons and provide large differences in retention times between solute peaks (e.g., the use of a nematic liquid crystal as a stationary phase for separating disubstituted benzene isomers); such columns generally require fewer theoretical plates to effect a useful separation.

Capillary GC columns still command the lead for separating highly complex mixtures such as are found in petroleum fractions. Such columns contain a thin uniform film of the stationary liquid phase in the order of 0.5 μm, but the state-of-the-art in open-tubular GC columns has advanced such that the nature of the film, its thickness, the inner diameter of the column, and the column material itself have been varied considerably to afford a broad spectrum of applications. For example, wider-bore, thicker film columns offer the advantage of direct (splitless) injection of samples, while the new fused silica capillaries provide a flexability not possible with conventional glass capillary columns. The reader is referred to the literature for broad applications of capillary columns (see Chapter 4, Section II) to hydrocarbon analysis.

IV. COLUMN TEMPERATURE

The choice of column temperature and the use of linear column temperature programing offer powerful advantages to GC analysis, which, on the one hand allows the selection of an optimum column temperature for separating the sample components of interest, and, on the other, permitting the analysis of complex mixtures containing low, intermediate, and high-boiling hydrocarbons. Generally, within some restriction, lowering the liquid phase loading decreases the column temperature required for an analysis. Use of a higher temperature will reduce the analysis time. An exception to this is given by stationary phases that, within a finite temperature range, become better solvents for the sample components with increasing temperature.

For mixtures containing a wide boiling point range of components, improved separations can be obtained for early eluting components and sharper peaks and faster elutions can be obtained for late eluting components using linear temperature programing of the column oven. An adjunct to this type of analysis involves cryogenically concentrating dilute samples at the head of the column, followed by column temperature programing at a low to moderate programing rate (e.g., 2 to 5°/min). Column "bleed" of the stationary phase is a common occurrence of temperature programing, particularly if the recommended upper temperature limit of the stationary phase is exceeded. The use of matched columns in the oven and a dual detector can correct some of this. Column bleed, however, does occur to some finite extent at most temperatures. This is sometimes used to advantage when using a mass spectrometer as the GC detector, resulting in reference m/e lines in the mass spectrometer. Column bleed can be detected by periodically monitoring the retention time of a reference compound over a prolonged period (days or weeks, depending on the bleed rate). Flow programing of the carrier gas also has been used by some investigators to advantage; however, since elution times of sample components are a far greater function of column temperature than flow rate, flow programing has not attained the popularity that temperature programing has enjoyed.

V. DETECTORS

A. Detector Requirements

There are three principal factors that determine the utility of a GC detector: stability, sensitivity, and linear dynamic range.

Detector stability is a general, but essential characteristic of any detector. Stability is related to electronic drift which directly affects the reproducibility of results. If detector instability is encountered and is unpredictable, reproducible analytical results may be difficult to impossible to obtain. Major advances in solid-state electronics have greatly enhanced the stability of FI and TC detectors. Nonetheless, investigators should routinely evaluate and calibrate their detector to ensure that detection stability exists.

Detector sensitivity is a direct function of the amount and type of compound being detected, and is really comprised of two aspects: mean sensitivity and absolute sensitivity. The mean sensitivity of a detector for a given compound can be estimated from the slope of a plot of signal vs. concentration (e.g., microcoulombs per gram for FI detectors). The absolute sensitivity of a detector for a given compound is defined as the lower practical detection limit for the compound, usually estimated as the amount of compound that will produce a signal that is twice that of the random electronic noise level of the detector.

The linear dynamic range of a detector can be estimated by the analysis of a reference compound over a wide concentration range. Departure from linearity (and, hence, the linear dynamic range) can be determined by plotting either the detector response (e.g., peak area or peak height) or the detector response factor vs. concentration. Clearly, analyses conducted above or below the linear dynamic range of a detector will result in unreliable and misleading results and false conclusions.

B. Common Detectors for Hydrocarbon Analysis

A wide variety of detectors have been developed for GC analysis. Of these, however, only a handful have gained popularity and have exhibited a high degree of utility. Many of those that have failed the test of time have lacked adequate stability, durability, linearity, sensitivity, and/or possessed too narrow a selectivity. Three specific detectors have gained a high popularity in the GC analysis of hydrocarbons: flame ionization (FI), thermal conductivity (TC), and mass spectrometry (MS).

In using the FI detector, when a hydrocarbon elutes from the GC column it is mixed with hydrogen at the base of a hydrogen flame jet and is subsequently burned in the flame to produce a small ion current. This small current is collected by the collector electrode in the detector housing and is passed through a megohm resistor to produce a measureable voltage ($E = IR$) which is then electronically amplified by an electrometer. The magnitude of ion current that is produced in the flame by the eluting compound is a function of the structure and the amount of the compound. The FI detector has been called a carbon detector since its response is roughly proportional to the weight percent of carbon in the compound eluting from the GC column. This varies, to be sure, with the nature of the compound structure, but is essentially true for hydrocarbons. The FI detector is much less sensitive to variations in carrier gas flow rate than the TC detector. In addition, the detection limit of the FI detector is roughly three orders of magnitude greater than the TC detector, and a linear dynamic range of 10,000 or more is typically attainable for the FI detector.

The TC detector was designed as a Wheatstone bridge network in which a resistance wire or thermistor bead is located in each of two legs of the four legs of the bridge network. One of these is exposed to carrier gas only; the other is exposed to carrier gas emerging from the GC column. When no sample components are eluting from the column, the bridge is electrically balanced and a baseline trace on the strip-chart recorder is generated. When a sample component elutes from the column, its thermal conductivity (i.e., cooling capacity) is less than the carrier gas itself, and the sensing leg of the bridge network becomes hotter, producing a bridge imbalance that results in a displacement of the pen on the strip-chart recorder. As the sample component band clears the detector, the bridge returns to balance, resulting in a GC peak on the recorder trace. The greater the difference between the thermal conductivity of the carrier gas and the sample component, the greater will be the response of the detector. For this reason, carrier gases of high thermal conductivity are used. The gas with the highest thermal conductivity is hydrogen, with helium as next highest. Owing to the potential explosive nature of hydrogen, helium typically is used as the carrier gas. Due to advances in materials and miniaturization design in TC detectors over the past 15 years, the sensitivity and response characteristics of TC detectors have been markedly improved.

MS coupled with a computerized data system and a capillary GC is unique in that it is the only GC detector that affords direct structural information (except, possibly for specific hydrocarbon isomers) on hydrocarbons eluting from a GC column. When a sample component eluting from the GC column enters the source of the MS, it is exposed to electronic bombardment that breaks the component into positively charged fragments. The fragments then migrate at different rates down a sweeping magnetic or rf field, depending on their mass, and are collected by an electron multiplier. Analytical GC-MS systems typically scan a mass fragment range of 10 to 800 AMU, resulting in a mass spectrum of the component in which the occurrence and amplitude of the AMU signals for the various fragments serve to identify the component itself. While a number of quantitative techniques have been developed for GC-MS detectors (e.g., isotope dilution, single and multiple ion monitoring), the MS detector is principally used as a qualitative detector for the identification of sample components emerging from a GC column. In contrast, FI and TC detectors have been well established for quantitative analysis, and remain as the work-horses for the quantitative analysis of hydrocarbon mixtures.

Specific applications employing FI, TC, and other detectors in the GC studies of hydrocarbons are described in Chapter 4, Section IV. A and B; applications of the MS detector for hydrocarbon analyses are reported in Chapter 4, Section V.

VI. QUALITATIVE ANALYSIS

The retention time, t, of an eluting compound under well-controlled GC conditions of temperature and flow rate is generally highly reproducible and a unique characteristic of the compound in question. The retention time of a compound is defined as the time required to elute one half of the GC peak (i.e., the peak maximum) of the compound. To properly describe a compound's retention by a GC stationary phase, however, it should be adjusted by subtracting out the time spent by the compound in the carrier gas phase from the point of its injection to the point of its sensing by the detector. This "noninteractive" time (i.e., the time that the compound is not interacting with the stationary phase) may be estimated from the retention time of a noninteractive ("unretained") component, t_u. In practical GC analysis, the retention time of air or methane is often used for this purpose. Thus,

$$t' = t - t_u \qquad (7)$$

where t' is the "corrected" retention time of the component of interest. The corrected retention volume, V', of the component can be determined from

$$V' = t'F \qquad (8)$$

where F is the measured carrier gas flow rate. Hence, V' is independent of flow rate at a given column temperature. The carrier gas flow rate, however, is more rapid at the exit of the column where the pressure is least and slowest at the inlet of the column where the pressure is greatest. Therefore, the corrected retention volume, V', should properly be corrected for carrier gas compressibility by multiplying it by the gas compressibility factor, j, to give the adjusted retention volume (V°), as

$$V^\circ = jV' \qquad (9)$$

where j is calculated from the column inlet and outlet pressures, p_i and p_o, respectively, as

$$j = 1.5([(p_i/p_o)^2 - 1]/[(p_i/p_o)^3 - 1]) \qquad (10)$$

The basis of qualitative analysis is that the corrected retention volume of a compound reflects the partition coefficient, K, of the compound between the stationary phase and carrier gas phase. An expression can be derived relating K to V°, as

$$K = V^\circ/V_L \qquad (11)$$

where V_L is the volume of the stationary liquid phase. Tabulations of K values for various reference compounds for a given stationary phase at a given temperature can be used to tentatively identify unknown GC peaks.

Alternative qualitative methods for the identification of unknown GC peaks or for the characterization of known GC peaks employ the use of relative retention data (i.e., the use of separation factors) or retention indices, both of which have been discussed previously. For positive identification, GC peaks usually must either be collected and characterized using other analytical techniques, or be characterized by coupling the GC with detectors that provide structural information, such as MS or infrared absorption spectrometry (IR).

Great strides have been made in the characterization of GC peaks in the development of computerized Fourier-transform GC-IR systems. However, GC-MS remains by far the most popular technique for qualitative analysis of GC peaks.

VII. QUANTITATIVE ANALYSIS

Reliable quantitative GC analysis is a function of a number of factors, including sample preparation and introduction, detector reproducibility and linearity of response, and data collection and treatment. Samples must be accurately prepared, with or without an accurately measured amount of an internal standard. The prepared sample must be reliably introduced into the GC such that repeated introductions are reproducible within a small but known error. The signal obtained from the detector must be reproducible in replicate analyses of the same sample. If the detector is calibrated with reference mixtures, the response of the detector must be either linear or predictable over the concentration range in which samples are to be analyzed. For accurate quantitative analysis, it is best to calibrate the detector with reference mixtures that cover as small a concentration range as possible, since the use of wide calibration ranges can introduce a greater uncertainty in the analyzed result. Finally, sufficient (but not excessive) data must be obtained that allow a valid estimate of the analytical uncertainty.

The analyst should be aware that GC analysis represents a "relative" method, not an "absolute" method, an "absolute" method being defined as a method that is directly traceable to one of the standard absolute units of measurement (e.g., the kilogram, the meter, or the coulomb). The analyst also should be aware that all analytical methods possess an inherent total uncertainty that is comprised of individual uncertainties associated with each step in the method.

The two most popular methods used for quantitative analysis are the external standard method and the internal standard method. The external standard method relates to the development of a standard calibration curve covering the concentration range of interest, using accurately prepared standard solutions of the analyte in question. The internal standard method involves the quantitative addition of a structurally related reference compound that is readily separated from the analyte peak in the GC chromatogram. The external standard method requires that the analyst must accurately know the amount of sample injected into the GC; the amount of sample injected into the GC using the internal standard method does not have this requirement since measurements are made on a relative basis and quantification is based on the amount of internal standard added to the sample prior to its analysis.

One final comment regarding quantitative analysis must be made. The value reported for the concentration of a sample component is analytically incomplete without specifying an estimate of the uncertainty of the value. The uncertainty of the value is determined from two sources of errors: (1) errors involved in the preparation of the sample for analysis and (2) errors involved in the analysis itself. Errors in the preparation of the sample (U_p) relate to the uncertainty of the purity of the sample and the external or internal standards used and to the imprecision associated with the preparation of the samples and standards for analysis. Errors in the analysis (U_a) relate to the analytical imprecision estimated from the reproducibility of replicate analyses of the analyte of interest (which includes the imprecision of sample injection (when using the external standard method) and the imprecision of the response of the detector). The variances of these two major sources of errors (i.e., preparative and analytical) are pooled in quadrature and multiplied by two to obtain an estimate of the total uncertainty (U_T) of the measured concentration at 95% confidence, by the equation

$$U_T = 2([U_a^2 + U_p^2]^{1/2}) \tag{12}$$

VIII. COMPUTERS

The predominance of analyses of hydrocarbons involving GC today employ automated data systems for the injection of samples, real-time measurement of retention times and peak areas, normalization, control of the GC, and the issuance of reports of analysis. For a more comprehensive treatment of this topic, the reader is referred to the numerous texts now available on this topic. An example of the use of computers in analysis is provided in the recent text by Barker.[1] Applications to the use of computers in the GC analysis of hydrocarbons are provided in Chapter 4, Section VI.

IX. THERMODYNAMIC BASIS FOR SEPARATIONS

GC separations (i.e., the separation of one component from another component in a mixture) are controlled by the differential degree of interactions between the sample components and the column substrate (i.e., the stationary liquid phase in GLC and the sorbent in GSC). A number of thermodynamic relationships have been developed to characterize such interactions. Some of the more fundamental relationships are identified in this section. Zielinski[2] reported the application of thermodynamic relationships to the treatment and prediction of retention behavior for hydrocarbon separations in GLC. Criteria for the development of reliable thermodynamic data from GLC have been discussed by Martire.[3,4] Several texts have explored the use of thermodynamic treatments in both GLC and GSC (e.g., Reference 5).

A. Gas-Liquid Chromatography

A solute specific retention volume (V_g^o) is calculated from the retention time (t') of the solute (corrected for dead volume; see Equations 7 and 8), using the expression:

$$V_g^o = (t'F)(P_o - P_w)T_o j/P_o T_a W_L \tag{13}$$

where V_g^o is in $m\ell/g$ ($m\ell$ carrier gas needed to elute one half of the solute peak per gram stationary phase), t' is the corrected retention time at peak height (minutes), F is the carrier gas flow rate, P_o is the column outlet pressure (atmospheric; mmHg), P_w is the water vapor pressure at T_a (mmHg; using a soap-bubble flowmeter for measuring carrier gas flow rate), T_o is 273.16 K, T_a is ambient temperature (K), W_L is the weight of stationary phase in the column (in grams), and j is the gas compressibility factor (see Equation 10) which corrects for the column pressure drop.

The solute activity coefficient (γ) may be related to the specific retention volume (V_g^o) through:

$$\gamma = 1.704 \times 10^7/p^o V_g^o M_L \tag{14}$$

where p^o is the saturated vapor pressure of the pure solute (mmHg) at the column temperature at which the specific retention volume is determined, and M_L is the molecular weight of the stationary phase. The specific retention volume is a true solution property which should only be a function of the column temperature. The solute activity coefficient is associated with a state of infinite dilution and static equilibrium and can be corrected for any vapor phase nonideality through the use of a second virial coefficient expression.[6]

Partial molar excess free energies of mixing (G) in calories per mole can be obtained from the solute activity coefficients through the expression

$$G = RT \ln\gamma \qquad (15)$$

where R is the gas constant (1.9872 cal/deg-mole), and T is the column temperature (K).

Solute partial molar excess enthalpies (H, in cal/mole) and entropies (S, in cal/deg-mole) can be estimated from

$$G = H - ST \qquad (16)$$

where T is the column temperature (K), using a linear regression of values of G determined at various column temperatures vs. column temperature (with −S as the slope, and H as the "y"-intercept). For other thermodynamic treatments of GLC retention data for hydrocarbons, the reader is referred to the general published literature (see Chapter 5, Section II) and to texts dealing with this topic in more detail (e.g., Reference 5).

B. Gas-Solid Chromatography

Heats of sorption (H_a; affinity of solute for GSC sorbent packing) may be estimated (in the Henry's law region) when specific retention volumes of a solute have been determined at several column temperatures, using the expression

$$\ln V_g^o = -H_a/RT + C \qquad (17)$$

where H_a is in calories/mole, R is the gas constant (1.9872 cal/deg-mole), T is the column temperature (K), and C is the "y"-intercept in a linear regression plot of the specific retention volume vs. column temperature. GSC applications to the determination of thermodynamic data for hydrocarbons are given in Chapter 5, Section II.

REFERENCES

1. **Barker, P.,** *Computers in Analytical Chemistry,* Pergamon Press, New York, 1983.
2. **Zielinski, W. L., Jr.,** Ph.D. thesis, Georgetown University, 1972.
3. **Martire, D. E. and Pollara, L. Z.,** in *Advances in Chromatography,* Vol. 1, Giddings, J. C. and Keller, R. A., Eds., Marcel Dekker, New York, 1965, 335.
4. **Martire, D. E.,** in *Advances in Analytical Chemistry and Instrumentation,* Vol. 6, Purnell, J. H., Ed., Interscience, New York, 1968, 93.
5. **Laub, R. J. and Pecsok, R. L.,** *Physicochemical Applications of Gas Chromatography,* Interscience, New York, 1978.
6. **McGlashan, M. L. and Potter, D. J. B.,** *Proc. R. Soc. (London),* A267, 478, 1962; **Desty, D. H., Goldup, A., Luckhurst, G. R., and Swanton, W. T.,** in *Gas Chromatography,* van Swaay, M., Ed., Butterworths, London, 1962, 67.

Chapter 3

GAS CHROMATOGRAPHY OF HYDROCARBONS — GENERAL APPLICATIONS

Gas chromatography (GC) over the years has been the premier tool in the analysis of hydrocarbons. High-performance liquid chromotography (especially for polycyclic aromatic hydrocarbons), thin-layer chromatography, and other nonchromatographic disciplines are becoming increasingly applied to hydrocarbon analysis, but GC remains the main analytical technique. The reasons for this are quite understandable. The versatility of detectors and column substrate materials, the high resolving power of well-coated open-tubular columns, the ease for automation and data output, the stability and reliability of operating systems, the broad scope of applications potentials, and the thermal stability for volatilization of broad classes of hydrocarbons, all place GC into a class of its own for hydrocarbon analysis.

This is the first of six chapters covering literature reports on the use of GC of hydrocarbons. Chapter 4 covers publications specifically directed at the GC system (inlet systems, capillary columns, special column arrangements, detectors, gas chromatography-mass spectrometry (GC-MS), computers), as related to hydrocarbon studies. Chapter 5 cites special techniques (sampling, reaction/pyrolysis GC, hydrocarbon boiling point correlations with retention data, preparative-scale supercritical chromatography), and studies carried out to ascertain thermodynamic data on hydrocarbons. The wide breadth of hydrocarbon air and water pollution measurement activities is the subject of Chapter 6; while the final two chapters report on (1) separation materials used in and (2) tabulated reference retention data reported for GC studies on the various classes of hydrocarbons.

The present chapter initially shows applications of GC in the analysis of a variety of hydrocarbon mixtures (Figures 1 to 36*) and then treats general hydrocarbon analyses in categories such as petroleum, crudes, and gasoline, discusses general analyses by hydrocarbon class, and covers geological studies and the use of process GC stream analyzers. The material presented in this chapter predominately represents practical applications of the hydrocarbon industries as well as those studies that do not have the more specific focus of topic areas covered in Chapters 4 to 6.

I. EXAMPLES OF GC APPLICATIONS IN HYDROCARBON SEPARATIONS

The 36 figures shown in this chapter illustrate the broad scope of GC applications in the GC analysis of hydrocarbons. The chromatograms cover GC separations ranging from light to heavy hydrocarbons and provide examples of the current state-of-the-art for separations of specific hydrocarbon classes. Included are separations of standard mixtures of isomeric alkanes and alkenes, substituted aromatics, cyclics, and polycyclic aromatic hydrocarbons, as well as special applications for pyrolysis products, natural gas, commercial propane, kerosene jet fuel, crude oil, shale oil, gasoline, naphtha, and creosote. The chromatograms typify separations with various stationary phases using both packed and capillary GC columns.

II. GENERAL HYDROCARBON ANALYSIS

A. Petroleum
A number of overviews have been developed on the use of GC in the characterization of petroleum and petrochemicals. To illustrate, a recent text edited by Altgelt and Gouw,

---

* All Figures appear at the end of the text.

Chromatography in Petroleum Analysis,[1] described the application of GC and other chromatography techniques to the analysis of petroleum and its products. A review by Berezkin containing 81 references focused mainly on GC applications in petrochemistry and petroleum refining.[2] Ramond[3] also reviewed the use of GC in the petroleum industry. Camin and Raymond[4] discussed the development of chromatography in the petroleum industry for both bulk separations and analysis in a lengthy review covering 81 references. Forms of gas and liquid chromatography were described in detail in relation to the characterization and analysis of various refinery streams. The role of gas and liquid chromatography in the petrochemical industry was also the subject of a review by Kenlemans.[5] A review by Berezkin and Nametkin[6] on GC and petroleum chemistry discussed the combination of GC with physical and chemical analysis methods, reaction-GC, the use of computers for rapid data processing, stationary phases, column efficiencies, and faster retention times. In a recent review of the practice of GC in the People's Republic of China, Yue cited the extensive use of the technique in the petroleum industry.[7] Fraser[8] also reviewed the use of GC for petroleum analysis. Additional reviews to petroleum applications included a report[9] on the First Russian All-Union Conference of GC Applications in Petroleum Chemistry held in February 1973; a Russian text on the use of GC in studies of natural gases, petroleum, and condensates,[10] with 148 of the 220 references in Russian; and biannual reviews[8,11-15] on publications appearing in the petroleum literature that included GC applications. The determination of petroleum hydrocarbons containing 1 to 5 carbons by GC was reported by Kachlik-Olasinska,[16] as well as by Turowska and Pruszynska;[17] while Ruvinskii et al.[18] reported the analysis of aliphatic C_1 to C_8 hydrocarbons in fuel gases, petroleums, gasolines, and the gas space of tanks containing petroleum products. Analysis on petroleum and its products was the topic of the Second Conference on the Use of Gas Chromatography and Related Analytical Methods in the Petroleum and Petrochemical Industries.[19] Kuchhal[20] discussed the application of GC in petroleum hydrocarbon analysis. Bremer[21] reported the standardization of GC methods for studying lower hydrocarbon products of petroleum and gasoline. Applications to the determination of complex hydrocarbon mixtures derived from coal and petroleum,[22] and of hydrocarbon types in petroleum fractions[23] were also reported.

B. Crudes

Several reviews on crudes and heavy distillates were prepared covering the use of GC. A review of methods for the characterization of syncrudes from coal was published by Dooley and co-workers.[24] Trusell[25] discussed applications to heavy distillates, including the determinatin of alkanes, steranes, triterpanes, hydrocarbon types, and used lube oil. Adlard[26] reviewed analysis of heavy distillate residues and crudes. Rasmussen[27] mathematically treated chromatograms of crude oils obtained on a 100-ft Dexsil-300 SCOT column to achieve patterns characteristic of the oil; such patterns remained essentially unchanged after moderate weathering. Other workers determined many coker distillate components.[28] Abidova and co-workers[29] determined individual aromatic hydrocarbons in five distilled fractions of Uzbekistan oils boiling below 200°C and found that deeper strata oils were substantially richer in aromatics than shallower oils.

C. Gasolines

High-resolution GC for analysis of gasolines and napthas was described by Whittemore.[30] A high-resolution automatic system was reported for the analysis of gasoline-range hydrocarbon mixtures,[31] while a method for automated analysis of gasolines for hydrocarbon types was described by Ury.[32] Stavinoha[33] discussed hydrocarbon-type analysis of gasoline using stabilized olefin absorption and GC. Hydrocarbon-type analysis of gasoline by GC was also the subject of a report by Stavinoha and Newman.[34] Lulova et al.[35] studied the composition of individual C_5 to C_8 hydrocarbons in cracking gasolines. Ozeris[36] determined C_5 to C_{13}

hydrocarbons in regular Batman gasoline. Other workers reported the GC determination of dienes and aromatic hydrocarbons in pyrolysis gasoline.[37] Analysis of commercial gasolines and stocks for blending was studied as an approach to determining octane number and fractional composition.[38] A mathematical relationship was developed for determining optimum GC conditions for gasoline component separations,[39] while Laurgeau et al. reported a linear algorithm to evaluate octane numbers of unleaded and leaded gasolines based on chemical compositions determined by GC.[40] Block and co-workers[41] developed a method for analysis of individual alkane, olefin, cyclic, and aromatic constituents in methanol-derived gasolines in under 2 hr and demonstrated that such gasolines have much higher olefin and cyclic hydrocarbon contents.

D. Other General Hydrocarbon Analyses

While the use of high-performance liquid chromatography has grown rapidly in recent years for the analysis of aromatic hydrocarbons, GC remains the most used technique for the analysis of hydrocarbons per se. In addition to literature reviews covering the use of GC in hydrocarbon analysis that have appeared in alternate years by Bradley[8,11-13] and Mayer,[14,15] this topic has been addressed by Desty and Goldup[42] and by others. Petrov and Kolesnikova reported an earlier review on GC analysis of multicomponent hydrocarbon mixtures.[43] Quantitative aspects of hydrocarbon analyses have been considered by ASTM.[44] Kulikov and Sorokin[45] reported the GC analysis of hydrocarbons having boiling points below 200°C. Analyses of complex hydrocarbon mixtures were described by Mitra and co-workers[46] and by Weingaertner et al.,[47] while Buchta and Forster[48] classified hydrocarbon systems by GC analysis methods. Other workers reported the separation of lower saturated and unsaturated hydrocarbons[49] and of hydrocarbons, oxygenated hydrocarbons, and permanent gases.[50]

Other general applications of GC to hydrocarbon analysis focused on light and gaseous hydrocarbons. Such applications have been reviewed by Saha[51] and by Mindrup.[52] The latter updated the *Atlas of Gas Analysis by Gas Chromatography* covering 66 references classified and cross-indexed by sample type, with the inclusion of chromatograms and operating conditions. A one-step procedure was reported for the analysis of a mixture of light hydrocarbons and permanent gases.[53] Gas chromatographic methods for gaseous fuels were covered in a review[54] of the literature on solid and gaseous fuels over the period October 1974 to September 1976. Saha et al.[55] reported a generalized GC method for the analysis of gaseous hydrocarbons. Purcell and Gilson[56] summarized the official NGPA-ASTM method used for natural gas analysis and recommended three gas chromatographic methods and systems for the same application. Willis reported the analysis of off-gas from hydrocarbon pyrolysis.[57]

Other studies on light hydrocarbons focused specifically on hydrocarbons containing 1 to 5 carbon atoms. To illustrate: the degradation of insulating oils in power transformers was studied by determination of C-3 and lighter hydrocarbons;[58] the analysis of commercial C-4 fractions was reviewed (212 references);[59] Rah and Rhee[60] also reported the analysis of C-4 hydrocarbons; a combined gas-solid-liquid chromatography method was reported to separate all C-1 to C-4 hydrocarbons;[61] an automatic gas chromatograph was developed using Carle's H-transfer system to determine all hydrocarbons through *n*-pentane in under 40 min;[62] GC separation of C-4 and C-5 hydrocarbons at 50°C was described;[63] a method for the simultaneous GC separation of C-1 to C-4 hydrocarbons and volatile organic sulfur compounds was reported.[64]

The reliability and optimization of GC for hydrocarbon analysis also have been subjects of investigation. Sources of error in GC analysis in the petroleum industry have been discussed.[65] French,[66] at the International Conference on Benzole Producers, reported on the precision attainable in hydrocarbon analysis.[66] Tomi and co-workers[67] described optimum GC separations of some hydrocarbons. The use of precision GC was demonstrated for the identification of types of hydrocarbons.[68]

Miscellaneous hydrocarbon analyses have covered such studies as a GC investigation of heavy pyrolysis tar;[69] the determination of hydrocarbons in anhydrous hydrogen fluoride;[70] the analysis of hydrocarbons in a hydrogen atmosphere;[71] and a quantitative method for the analysis of aliphatic hydrocarbons and terpenes in tobacco.[72]

III. GENERAL ANALYSIS BY HYDROCARBON CLASS

This section covers general GC analyses of hydrocarbons sorted according to two principal chemical structure categories: (1) alkanes and unsaturated hydrocarbons; (2) aromatic, cyclic, and polycyclic aromatic hydrocarbons.

A. Alkanes and Unsaturated Hydrocarbons
1. Alkanes

Determinations of alkanes were reported by Eppert and co-workers,[73] Felscher,[74] and in a French patent by Matishev.[75] In other analyses, unusually abundant levels of C-24 to C-30 12-methylalkanes and C-26 to C-30 13-methylalkanes found in East Siberian crude oils were proposed to have been derived from 12,13-methylenetetracosanic acid in the original source material.[76] Sulatanov and Arustamova analyzed liquid *n*- and iso-alkanes.[77] Smith,[78] and Scacchi and Back,[79] reported methods for the analysis of *n*-alkanes. Shlyakhov and associates analyzed for isoprenoid alkanes in petroleum.[80] An apparatus was described for the analytical separation of methane, ethane, propane, and butanes in 7 sec.[81] Use of statistical techniques in the analysis of deuterium- and tritium-labeled propanes showed the possibility of determining isotopic ratios of the mixture as a whole.[82] Petrov and Krasavchenko[83] identified 24 C-9 to C-25 branched isoprenoid alkanes in gas oil fractions of crude oils. The *n*-alkane distribution through C-28 was determined in three Rechitsa crude oils.[84] The identities and levels of 37 monomethyl-substituted alkanes in three typical paraffin-base crude oils were determined and found to be similar to that found in the reaction mixture following liquid phase catalytic isomerization of 1-alkanes.[85] Column and carrier gas flow rates were varied simultaneously in a sequential simplex optimization of a separation of isomeric octane mixtures to demonstrate feasibility for such an approach.[86] Rapid analysis of trace hydrocarbon impurities in high-purity *n*-heptane and other pure *n*-alkanes has been described.[87] A method for the determination of normal and branched alkanes was reported by Toader and co-workers.[88]

2. Unsaturated Hydrocarbons

A method for the determination of trace quantities of C-3 to C-4 olefins in liquid petroleum gas was reported, based upon a simple comparison method of calibration.[89] 1,3-Butadiene was determined in isobutane-isobutene mixtures.[90] Carson and associates[91] discussed the problem of dimerization of diolefins with respect to GC analysis, particularly for 1,3-butadiene. In two Russian studies, Anosova and co-workers[92] described the determination of propylene and butadiene in a solution of aromatic hydrocarbons, while Seroshtan and Zakharova reported the measurement of alkadiene and alkyne impurities in C-1 to C-5 hydrocarbon fractions.[93] Specific methods were reported for analysis of acetylenes.[94] The solubility of acetylene hydrocarbons in kerosene was studied by a GC method,[95] while others[96] reported on the analysis of high-boiling acetylenic hydrocarbons.

B. Aromatic, Cyclic, and Polycyclic Aromatic Hydrocarbons
1. Aromatic and Cyclic Hydrocarbons

In general, GC studies on aromatic hydrocarbons have well-exceeded those on cyclic hydrocarbons. Engewald and co-workers[97] used gas-solid chromatography for the structural identification of cyclododeca-1,5,9-trienes. A review of GC applications to hydrocarbon analysis gave special attention to cyclic hydrocarbons.[98] A Russian patent was obtained for the determination of trace aromatic hydrocarbon impurities in gases.[99] Hanna et al.[100] described the quantitative identification of aromatic hydrocarbons in coker gasoline. Comparative GC test results were published[101] on benzene, toluene, ethylbenzene, and the three xylene isomers in gasoline. The effect of adsorption on errors in the chromatographic analysis of benzenes vapors was described by Lisenkov and associates.[102] Ottenstein and co-workers reported on the analysis of aromatic hydrocarbons;[103] 29 of 35 C-8 to C-13 aromatic hydrocarbons found in Uzbek petroleum were identified by Khodzhaev et al.[104] Egiazarov et al. discussed the GC determination of aromatic hydrocarbons in reforming gasoline fractions.[105] Other workers[106] reported on the sampling of stored samples for the analysis of trace aromatic hydrocarbon impurities in gases. The GC separation and identification of bicyclic aromatic hydrocarbons was studied in kerosene in the 200 to 280°C boiling point range.[107] A Russian study elaborated on the determination of C-6 to C-8 aromatic hydrocarbons.[108] Sojak and co-workers[109] described the determination of the principal aromatic hydrocarbons obtained from catalytic dehydrogenation of *n*-undecane. Peter and Torkos[110] used GC for measuring the purity of aromatic hydrocarbons. Roseira[111] reported on the chromatographic behavior of aromatic hydrocarbons and their derivatives. Duerbeck[112] discussed the separation of isomeric C-6 to C-10 aromatic hydrocarbons.

2. Polycyclic Aromatic Hydrocarbons

A method was reported for the isolation and determination of benzo[*a*]pyrene in shale oil with a detection limit of about 1 ppm.[113] Borwitzky and Schomberg[114] identified over 140 polycyclic aromatic hydrocarbons (PAH) in coal tar, which they suggested as an inexpensive and very complete mixture of these compounds. Bieri et al.[115] analyzed PAH and polycyclic aliphatic hydrocarbons in Atlantic outer continental shelf sediments at the parts-per-billion (ppb) level, suggesting that the PAH were the result of pyrogenic input, and citing the potential of [17]-hopanes as indicators of petroleum. Methods for enrichment, separation, isolation, identification, and determination of PAH in bitumens were reviewed.[116] Separations of PAH were reviewed by Zielinski and Janini,[117] citing the improved separation of PAH isomers when using high-temperature transition nematic liquid crystal phases as the stationary phase. Kasmatsu[118] applied GC to the analysis of nine crude oils for alkyl naphthalenes. PAH were also identified in carbon black.[119] Knowles[120] applied GC to the analysis of PAH in foods, while Grimmer and Boehnke[121] reported on PAH profile analysis in high-protein foods, oils, and fats. Selective determination of PAH and azulene in pyrolytic oil was described by Kusy;[122] other workers[123] studied PAH in crude and heavy fuel oils. Thomas and Lao[124] applied gas and high-performance liquid chromatography to the analysis of PAH in fossil fuel conversion processes. Lane and co-workers reported a method for the analysis of PAH, some heterocyclics and aliphatics using a single GC column.[125] The topic of PAH analysis by GC was considered by various investigators, including Beernaert[126] and Bhatia.[127] Grimmer and Boehnke described techniques for GC profile analysis of PAH in various systems, including: lubricating oil, cutting oil and fuel;[128] high-protein foods, fats, vegetable oils, plants, soils, and sewage sludge;[129] and lubricating oils.[130] A review on the GC determination of PAH was given by Pervukhina.[131]

IV. GEOLOGICAL STUDIES

GC has been found to be a useful technique for seeking locations of petroleum deposits and for intercomparing hydrocarbon compositions for different geologic sources of petroleum.

Schaefer and associates elucidated a unique method using hydrogen stripping followed by capillary GC to detect down to 0.01 ppb of C-2 to C-9 light hydrocarbons in 1-g rock samples and applied the method to the correlation of residual hydrocarbon content of the rocks with crude oil and natural gas deposits.[132,133] Overton and co-workers[134] identified key hydrocarbon components, using glass capillary columns coated with SE-52 for the analysis of organics extracted from sediments of three U.S. outer continental shelf areas; using spiking experiments, they were able to distinguish indigenous hydrocarbons from those added from petroleum. McKirdy and Horvath[135] used an integrated scheme for the analysis of oil, natural gas, and rocks to assist the search for hydrocarbon sources; they were able to infer composition and geological history from chromatograms of alkanes higher in carbon number than C-15, as well as to suggest whether an area was oil- or gas-prone from the C-2 to C-4 content of the natural gas. Levshunova and Telkova[136] analyzed light hydrocarbons in Lower Cretaceous formations of the eastern Caucasus following their isolation by thermal desorption and extraction, noting variations with depth and lithology. Two of three papers on GC in the oil industry (given at a joint meeting of the Chromatography and Electrophoresis Group, the Scottish Region, and the Chromatography Discussion Group of the Analytical Division of The Chemical Society, held November 1976 in Grangemorth, Scotland) discussed sources of error in GC analysis, and applications of GC in petroleum prospecting.[137] Safonova and Bulekova[138] studied C-13 to C-32 *n*-alkanes in geologically young Apsheron peninsula crude oils and in geologically older Prikumskii crudes following their isolation from saturate fractions by urea adduction. They found that: (1) the *n*-alkane distribution in younger crudes was somewhat irregular; (2) the *n*-alkane content in older crudes decreased with increasing carbon number; and (3) that overall *n*-alkane content increased with increasing geological age and with depth of the oil-bearing strata, with a concomitant decrease in alkane average molecular weight. Diamanti[139] reported on the analysis of *n*-alkanes in bituminous substances of carbonate formations by GC and their relation to gas and petroleum fields. Roucache and co-workers[140] examined several families of unsaturated hydrocarbons discovered in sedimentary rocks using GC and MS.

V. PROCESS ANALYZERS

Microprocessor and computer-controlled gas chromatographs have become the most widely used on-line process analyzers in the petroleum and petrochemical industries. A microprocessor-based GC system can generally handle up to six chromatographs and several process streams per chromatograph simultaneously, while a computer-controlled system can handle 4 to 16 or more chromatographs with each chromatograph handling several streams. The choice of microprocessor or larger computer has generally depended on the needs of the user. Due to the computing and memory capabilities of such systems, they can serve as the equivalent of three analyzers in one, providing compositional, gravity, and BTU outputs, in contrast to conventional analog chromatographic analyzers which predominantly provide only compositional information. In addition, data-system oriented chromatographic operations have the distinct advantages of greater reliability and less maintenance.

Jutila[141] surveyed a wide variety of analyzers, covering operating principles, sensitivities, detection methods, sampling and conditioning needs, and limitations. This survey covered gas and liquid chromatography, MS, UV, and IR spectrophotometry, atomic absorption, X-ray, neutron activation, and wet chemical analysis, with the finding that atomic absorption, neutron activation, and wet chemical methods were the most difficult to apply to on-line control. Bailey[142] surveyed microprocessor-based analyzers used by large chemical and petrochemical firms, covering a number of specific commercial analyzers that included GC systems with either flame-ionization or thermal conductivity detectors. Utterback[143] also provided a discussion of on-line process analyzers, including gas and liquid chromatography with various detectors, as well as the maintenance, calibration, and various factors promoting acceptance of computer-analyzer systems for on-line process control. Sandford[144] discussed analyzers as being faster and more reliable than previously, covering GC systems for monitoring catalytic crackers, ammonia production units, computer-controlled GC for high-speed multistream analysis, and a chromatographic analyzer having a pneumatic transmitter. Application of some of these analyzers to specific problems has been described for process and pollution control. Deans[145] discussed the laboratory gas chromatograph in the context of the total analysis requirements for process control of a petrochemical plant. A review of process control units developed by Villalobos[146] covered the basic elements of the process control technique, including sampling valves, programer-controller, column design, calibration, recorders, and interfacing several gas chromatographs to a minicomputer for controlling each chromatograph and processing all data. Applications included the determination of hydrocarbons in steam condensate. Lipavskii and Berezkin[147] reviewed the problems associated with the application of process GC in the research and control of petrochemical processes. Kosono[148] reviewed process control using gas chromatographs, process gas chromatographs, column systems, sampling devices, and applications in C-2, C-3, and xylene analyses. Kobayashi[149] reviewed the application of process gas chromatographs to ethylene plants and to trace hydrocarbons in liquid oxygen. Back-flush regrouping and helium ion detectors were also covered.

Other reports on the use of GC for specific process stream analysis/control applications have appeared. Stockinger et al.[150] developed an on-line computer-controlled GC system for monitoring compositional changes in a catalytic reformer pilot plant to evaluate catalysts. The chromatograph consisted of four high-resolution columns in two temperature zones and two flame-ionization detectors and had capability for temperature and flow programing for optimizing peak resolution. Up to 280 compounds were identified in some wide boiling range products; properties such as octane number, specific gravity, molecular weight, and vapor pressure were calculated. Stockinger[151] also reported a method for the quantitative analysis of gaseous products obtained in methanol conversion to gasoline using an on-line GC system. Grubaugh and Stobby[152] described an analyzer for determining ppm-level organics in strong HCl and HBr acid streams. A stripping column used a purge gas to scrub the hydrocarbons into the gas phase for chromatographic analysis. The advantage of this system was in its ability to separate sufficient quantities of the volatile components for analysis, while avoiding the hostile sample matrix that would otherwise prohibit direct analysis. Guillemin and Martinez[153] discussed a modified GC analyzer which reduced analysis time from 15 min to about 1 min. The columns contained 80 to 90 μm Spherosil of large specific area and a very thin stationary phase layer. The lowered mass transfer coefficients thus permitted the use of shorter columns and higher carrier gas flow rates. Several analyses could be made and averaged within the response time of the process, improving analytical reliability. Aliphatic hydrocarbons were determined in 30 sec. A computer was an essential component of this system since the chromatographic peaks occurred in about 1 sec. Annino and co-workers[154] used a totally pneumatic GC process analyzer for the analysis of hydrocarbon mixtures, which incorporated an orifice or capillary constriction in association with the detector. A commercial totally pneumatic GC analyzer was reported for use in hazardous areas and for closed-loop control, which automatically sampled, analyzed, and transmitted concentration signals by 3 to 15 psig signals.[155] Applications discussed included the control of distillation towers and the monitoring of chemical reactions. Pomerants and Simongauz[156] described the use of a gas chromatograph for automatic analysis of C-1 to C-6 hydrocarbons at 2-min intervals along a well bore. The system featured a flame-ionization detector, temperature programing, and high sensitivity. Sokolin and associates[157] discussed differences between laboratory and process GC functions and equipment, and process applications for the control of C-1 to C-4 separations and C-8 aromatic separations. Antar Petroles de L'Atlantique[158] patented a system utilizing either laboratory or process GC for the determination of total aromatics in light petroleum fractions. Analysis was based on the polarity of the stationary phase and the reversal of carrier gas flow following elution of the nonaromatic components. Baker and Johnson[159] described the role of supervisory computer-controlled gas chromatographs for the monitoring of 57 process streams in an olefin plant; 51 chromatographs contained peak markers and memory units, with the computer scanning each chromatograph every 3 sec to monitor sample injections. The total number of peaks were predefined for each chromatograph and stream, with data rejection by the computer when peak numbers did not exactly match the predetermined level. The report details automatic calibration, sample analysis, and preventive maintenance. Boyd[160] patented a GC output system for process control of a reforming unit in which integrated peak areas were calibrated for benzene, and a hexane and heptane isomer. The time interval of peak signals from a reactor effluent sample was handled by an optimizing module; the output signal varied the set-point of a temperature controller on the heater outlet stream. Lim and co-workers[161,162] reported the automatic analysis of C-1 to C-4 hydrocarbons using a process gas chromatograph. Terada[163] studied the use of process gas chromatographs in a facility designed for aromatic hydrocarbon production.

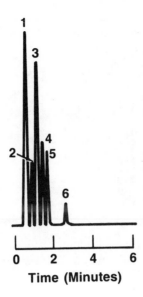

FIGURE 1. Analysis of light gases. Peaks: (1) methane, (2) ethane, (3) ethylene, (4) propane, (5) acetylene, (6) propylene. Column: 6-ft × 3.4-mm I.D. packed with 100 to 150 mesh Porasil B. Col. temp.: 55°C; detector, TC; carrier, He, 66 mℓ/min. (Reproduced by permission of Applied Science.)

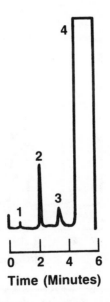

FIGURE 2. Impurities in ethane. Peaks: (1) 3 ppm methane, (2) 83 ppm acetylene, (3) 39 ppm ethylene, (4) ethane. Column: 6-ft × ⅛-in. s.s. packed with 80 to 100 mesh Carbosphere. Col. temp.: 225°C; FID; carrier, N₂, 45 mℓ/min. (Reproduced by permission of Alltech.)

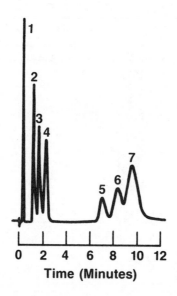

FIGURE 3.   Light hydrocarbons. Peaks: (1) methane, (2) acetylene, (3) ethylene, (4) ethane, (5) propyne, (6) propylene, (7) propane. Column: 5-ft × ¹/₈-in. s.s. packed with 60 to 80 mesh Carbosieve G. Col. temp.: 145 to 195°C at 6°/min and hold 5 min.; FID; carrier, N_2, 50 mℓ/min. Sample containing approximately 15 ppm of each component in a N_2 matrix. (Reproduced by permission of Supelco.)

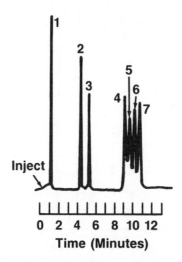

FIGURE 4.   Separation of C-1 to C-3 hydrocarbons. Peaks: (1) methane, (2) ethylene, (3) ethane, (4) cyclopropane, (5) propadiene, (6) propylene, (7) propane. Column: 3-ft × ¹/₈-in. s.s. packed with 80 to 100 mesh Spherocarb. Col. temp.: 70 to 350°C at 16°/min; FID; carrier, N_2, 35 mℓ/min. Butene-2 and butane appear as two broad peaks at about 280°C; the C-5 hydrocarbons are retained, even at 350°C. (Reproduced by permission of Analabs.)

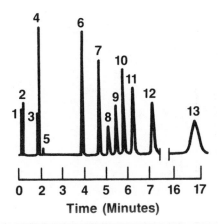

FIGURE 5.   Separation of C-1 to C-5 hydrocarbons. Peaks: (1) methane, (2) ethane, ethylene, and acetylene, (3) propane, (4) propylene, (5) propyne, (6) isobutane, (7) 1-butene, (8) *n*-butane, (9) isobutylene, (10) cis-2-butene, (11) trans-2-butene, (12) 1,3-butadiene, (13) *n*-pentane. Column: 6-ft × 2-mm glass packed with 80 to 100 mesh picric acid on Graphpac-GC. Col. temp.: 30°C; FID; carrier, He, 40 mℓ/min. (Reproduced by permission of Alltech.)

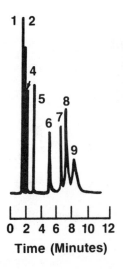

FIGURE 6.   Separation of C-1 to C-4 hydrocarbons. Peaks: (1) air, (2) methane, (3) ethane, (4) ethylene, (5) propane, (6) propylene, (7) isobutane, (8) *n*-butane, (9) acetylene. Column: 5-ft × ⅛-in. s.s. packed with 60 to 80 mesh alumina F-1. Col. temp.: 90°C; FID; carrier He, 22 mℓ/min. (Reproduced by permission of Alltech.)

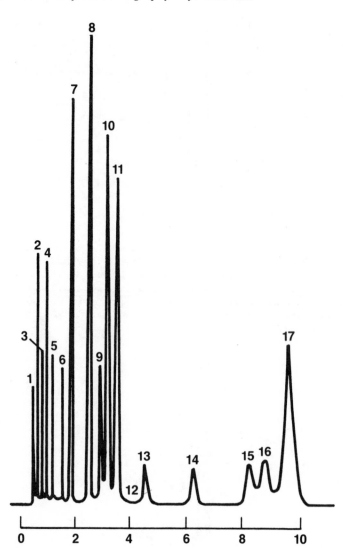

Time (Minutes)

FIGURE 7.   Separation of light isomeric hydrocarbons. Peaks: (1) methane, (2) ethane and ethylene, (3) acetylene (4) propane, (5) propylene, (6) isobutane, (7) *n*-butane, (8) butene-1, (9) isobutylene, (10) trans-butene-2, (11) cis-butene-2, (12) isopentane, (13) *n*-pentane, (14) pentene-1, (15) trans-pentene-2, (16) cis-pentene-2, (17) 2-methylpentane. Column: 1.5-m × 2.3-mm I.D. packed with 120 to 150 mesh Durapak *n*-octane/Porasil C. Col. temp.: 25°C; FID; carrier, N$_2$, 25 mℓ/min. (Reproduced by permission of Applied Science.)

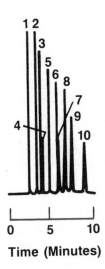

FIGURE 8.   Separation of isomeric butenes. Peaks: (1) air, (2) propane, (3) propene, (4) isobutane, (5) *n*-butane, (6) butene-1, (7) isobutylene, (8) trans-butene-2, (9) cis-butene-2, (10) 1,3-butadiene. Column: 24-ft × ¹/₈-in. s.s. packed with 20% BMEA on 60 to 80 mesh AW Chromosorb P. Col. temp.: 35°C; FID; carrier, He, 22 mℓ/min. (Reproduced by permission of Alltech.)

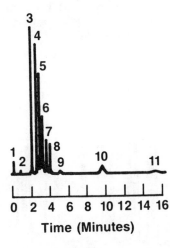

FIGURE 9.   Separation of C-3 to C-5 hydrocarbons. Peaks: (1) propane, (2) propylene, (3) isobutane, (4) butene-1, (5) *n*-butane, (6) isobutylene, (7) cis-butene-2, (8) trans-butene-2, (9) 1,3-butadiene, (10) isopentane, (11) *n*-pentane (relative mole-% concentrations: 1.0, 0.5, 36.5, 21.0, 13.0, 12.0, 6.0, 7.0, 1.0, 1.5, 0.5, respectively). Column: 2-m × ¹/₈-in. s.s. packed with 0.19% picric acid on 80 to 100 mesh Carbopack C. Col. temp.: 50°C; FID; carrier, N₂, 30 mℓ/min. (Sample: ASTM Section L Blend No. 6). (Reproduced by permission of Supelco.)

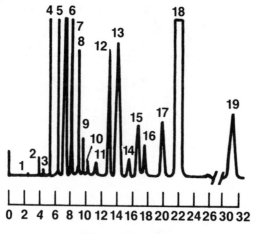

FIGURE 10.   Separation of C-1 to C-6 saturated and unsaturated hydro-carbons. Peaks: (1) ethane, (2) propane, (3) propylene, (4) isobutane, (5) *n*-butane, (6) butene-1 and isobutylene, (7) trans-butene-2, (8) cis-butene-2, (9) isopentane, (10) 1,3-butadiene, (11) *n*-pentane, (12) pentene-1, (13) 2-methyl-1-butene and trans-pentene-2, (14) cis-pentene-2, (15) 2-methyl-2-butene, (16) 2-methyl-1-pentene, (17) 3-methyl-1-pentene, (18) hexane, (19) 3-methylhexane. Column: 30-ft × $^1/_8$-in. s.s. packed with 23% SP-1700 on 80 to 100 mesh AW Chromosorb P. Col. temp.: 150°C; FID; carrier, He, 25 mℓ/min. (Sample: ASTM Section L Blend No. 6 plus C5s). (Reproduced by permission of Supelco.)

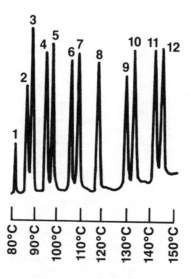

FIGURE 11.   Separation of C-10 to C-15 saturated and unsaturated hy-drocarbons. Peaks: (1) decane, (2) decene-1, (3) undecane, (4) undecene-1, (5) dodecane, (6) dodecene-1, (7) tridecane, (8) tridecene-1, (9) tetra-decane, (10) tetradecene-1, (11) pentadecane, (12) pentadecene-1. Column: 6-ft × 4-mm I.D. glass packed with 20% 1,2,3-tris(cyanoethoxy)propane on 80 to 100 mesh Gas-Chrom R. Col. temp.: 80 to 150°C at 5°/min; FID. (Reproduced by permission of Applied Science.)

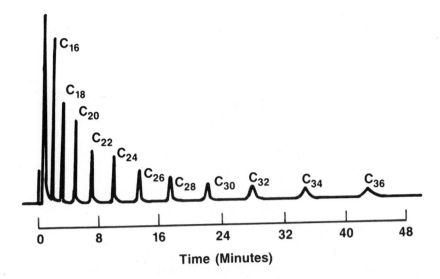

FIGURE 12. Separation of C-16 to C-36 normal hydrocarbons. Peaks: as indicated. Column: 6-ft × 4-mm I.D. glass packed with 3% Dexsil 300 on 80 to 100 mesh AW Chromosorb W. Col. temp.: 200 to 310°C at 10°/min; FID; carrier, N_2, 46 mℓ/min. (Reproduced by permission of Applied Science.)

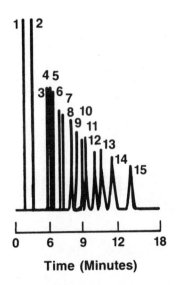

FIGURE 13. Separation of benzene and alkylbenzenes. Peaks: (1) benzene, (2) toluene, (3) ethylbenzene, (4) p-xylene, (5) m-xylene, (6) o-xylene, (7) isopropylbenzene, (8) styrene, (9) n-propylbenzene, (10) m-ethyl-toluene, (11) p-ethyl-toluene, (12) o-ethyl-toluene, (13) 1,3,5-tri-methylbenzene, (14) 1,2,4-trimethylbenzene, (15) 1,2,3-trimethylben-zene. Column: 6-ft × ¹/₈-in. s.s. packed with 5% AT-1200 and 1.75% Bentone-34 on 100 to 120 mesh HP Chromosorb W. Col. temp.: 75°C; FID; carrier, N_2, 20 mℓ/min. (Reproduced by permission of Alltech.)

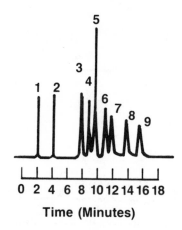

FIGURE 14.   Separation of benzene and alkylbenzene. Peaks: (1) benzene, (2) toluene, (3) ethylbenzene, (4) *p*-xylene, (5) *m*-xylene, (6) *m*-xylene, (7) isopropylbenzene, (8) styrene, (9) *n*-propylbenzene. Column: 6-ft × ¹/₈-in. s.s. packed with 5% SP-1200 and 1.75% Bentone-34 on 100 to 120 mesh Supelcoport. Col. temp.: 75°C; FID; carrier, N₂, 20 mℓ/min. Sample: 0.10 μℓ containing approximately equal volumes of each component. (Reproduced by permission of Supelco.)

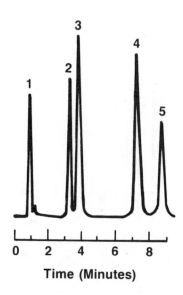

FIGURE 15.   Separation of methyl benzenes. Peaks: (1) benzene, (2) 1,3-dimethylbenzene, (3) 1,2-dimethylbenzene, (4) 1,2,4-trimethylbenzene, (5) 1,2,3-trimethylbenzene. Column: 6-ft × 3.4-mm I.D. packed with 2% Carbowax 20M on 80 to 100 mesh Porasil C. Col. temp.: 100°C; TC; carrier, He, 50 mℓ/min. (Reproduced by permission of Applied Science.)

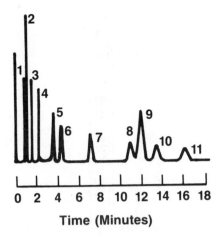

Time (Minutes)

FIGURE 16. Separation of aliphatic and aromatic hydrocarbons. Peaks: (1) *n*-hexane, (2) *n*-heptane, (3) *n*-octane, (4) *n*-nonane, (5) *n*-decane, (6) benzene, (7) toluene, (8) ethylbenzene, (9) *m*-xylene and *p*-xylene, (10) isopropylbenzene, (11) *o*-xylene. Column: 8-ft × ¹/₈-in. s.s. packed with 10% TCEP on 100 to 120 mesh AW Chromosorb P. Col. temp.: 80°C; FID; carrier, N_2, 20 mℓ/min. Sample: 0.2 μℓ containing approximately equal volumes of each component. (Reproduced by permission of Supelco.)

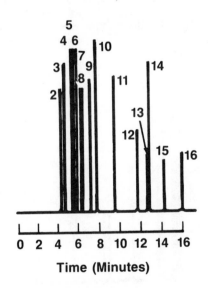

Time (Minutes)

FIGURE 17. Capillary separation of aliphatic, aromatic, and cyclic hydrocarbons. Peaks: (1) 2-methyl-butane, (2) *n*-pentane, (3) cyclopentane, (4) methyl cyclopentane, (5) benzene, (6) cyclohexane, (7) isopentane, (8) heptane, (9) methyl cyclohexane, (10) toluene, (11) octane, (12) ethylbenzene, (13) *p*-xylene, (14) *m*-xylene, (15) *o*-xylene, (16) nonane. Column: 50-m × 0.25-mm I.D. s.s. coated with squalane. Col. temp.: 91°C; FID, carrier, He, 20 psig. (Reproduced by permission of Foxboro.)

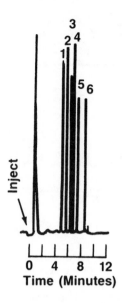

FIGURE 18.   Separation of pristane and phytane from normal-chain hydrocarbons. Peaks: (1) *n*-heptadecane, (2) pristane, (3) *n*-octadecane, (4)
phytane, (5) *n*-nonadecane, (6) *n*-eicosane. Column: 11-ft × ¹/₈-in. nickel-
200 packed with 30% eutectic salt mixture (54.5% KNO_3, 27.3% $LiNO_3$,
18.2% $NaNO_3$) on 60 to 80 mesh AW Chromosorb W. Col. temp.: 100
to 250°C at 8°/min; FID; carrier, N_2, 30 m𝓁/min. Sample: 0.5 μ𝓁 of mixture
containing a concentration of 1 mg each component/m𝓁 hexane. (Reproduced by permission of Analabs.)

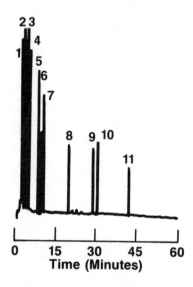

FIGURE 19.   Capillary separation of polycyclic aromatic hydrocarbons
and normal-chain hydrocarbons. Peaks: (1) fluorene, (2) *n*-octadecane, (3)
phenanthrene and anthracene, (4) *n*-eicosane, (5) fluoranthene, (6) *d*-do-
cosane, (7) pyrene, (8) *n*-tetracosane, (9) benz[*a*]anthracene, (10) chrysene, (11) *n*-hexacosane. Column: 15-m × 0.25-mm glass coated with
RSL-110LL. Col. temp.: 200°C; FID; carrier, H², 2 m𝓁/min. (Reproduced
by permission of Alltech.)

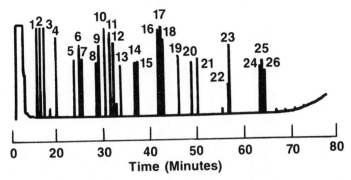

FIGURE 20. Capillary separation of alkyl benzenes and polycyclic aromatic hydrocarbons. Peaks: (1) *o*-xylene, (2) isopropylbenzene, (3) *n*-propylbenzene, (4) indane, (5) trimethylbenzene, (6) naphthalene, (7) benzothiophene, (8) 2-methylnaphthalene, (9) 1-methylnaphthalene, (10) triisopropylbenzene, (11) biphenyl, (12) dimethylnaphthalene, (13) hexamethylbenzene, (14) trimethylnaphthalene, (15) fluorene, (16) dibenzothiophene, (17) phenanthrene, (18) anthracene, (19) 1-methylphenanthrene, (20) fluoranthene, (21) pyrene, (22) benz[*a*]anthracene, (23) chrysene, (24)benzo[*e*]pyrene, (25) benzo[*a*]pyrene. Column: 30-m × 0.25-mm coated with SE-54. Col. temp.: 40°C for 3 min, programmed to 320°C at 4°/min; FID; carrier, H$_2$, 1.5 mℓ/min. (Reproduced by permission of Alltech.)

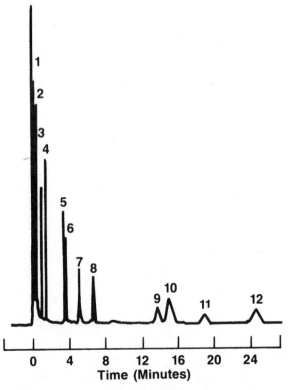

FIGURE 21. Separation of polycyclic aromatic hydrocarbon isomers on a high temperature transition liquid crystal. Peaks: (1) phenanthrene, (2) anthracene, (3) fluoranthene, (4) pyrene, (5) benzo[*mno*]fluoranthene, (6) triphenylene, (7) benz[*a*]anthracene, (8) chrysene, (9) benzo[*k*]fluoranthene, (10) benzo[*e*]pyrene, (11) perylene, (12) benzo[*a*]pyrene. Column: 6-ft × 2-mm I.D. glass packed with 2.5% *N,N'*-bis[*p*-phenylbenzylidene]-1,1'-bi-*p*-toluidine (BPhBT) on 100 to 120 mesh HP Chromosorb W. Col. temp.: 270°C; FID; carrier, He, 20 mℓ/min. (Reproduced by permission of Analabs.)

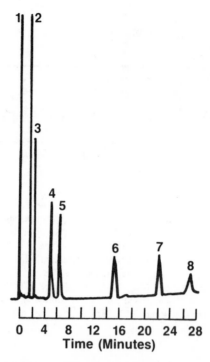

FIGURE 22.   Separation of 3 to 4 ring polycyclic aromatic hydrocarbon isomers on a high temperature transition liquid crystal. Peaks: (1) solvent, (2) phenanthrene, (3) anthracene, (4) fluoranthene, (5) pyrene, (6) triphenylene, (7) benz[a]anthracene, (8) chrysene. Column: 4-ft × $^1/_8$-in. s.s. packed with 2.5% N,N'-bis[p-methoxyphenyl]-1,1'-bi-p-toluidine (BMBT) on 100 to 120 mesh HP Chromosorb W. Col. temp.: 200°C for 2 min, programmed to 250°C at 2°/min; FID; carrier, N_2, 37.5 mℓ/min. Sample concentration: 7 μg of each component/ mℓ CS_2. (Reproduced by permission of Analabs.)

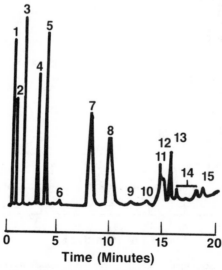

FIGURE 23.   Separation of natural gas components. Peaks: (1) methane, (2) ethane, (3) propane, (4) isobutane, (5) n-butane, (6) neo-pentane, (7) isopentane, (8) n-pentane, (9) 2,2-dimethylbutane, (10) 2,3-dimethylbutane, (11) 2-methylpentane, (12) 3-methylpentane, (13) n-hexane, (14) C7s, (15) n-heptane. Column: 6-ft × $^1/_8$-in. s.s. packed with 80 to 100 mesh Chemipack C-18. Col. temp.: 30°C for 10 min, programmed to 200°C at 10°/min; carrier, He, 30 mℓ/ min. (Reproduced by permission of Alltech.)

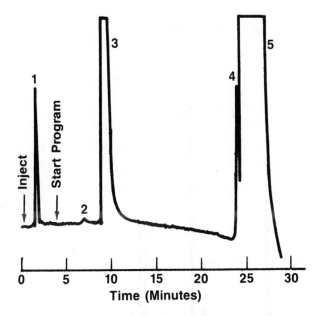

FIGURE 24.   Analysis of commercial grade propane. Peaks: (1) methane (0.5%), (2) ethylene (0.01%), (3) ethane (8%), (4) propylene (0.5%), (5) propane (91%). Column: 3-ft × $^1/_8$-in copper packed with 80 to 100 mesh Spherocarb. Col. temp.: 150 to 250°C at 5°/min; FID; carrier, He. Sample: 0.5 mℓ injected using gas sampling valve. (Reproduced by permission of Analabs.)

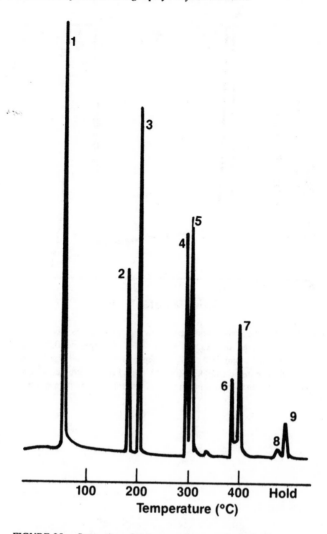

FIGURE 25.   Separation of light gases from pyrolysis. Peaks: (1) methane, (2) ethene, (3) ethane, (4) propene, (5) propane, (6) butene-1, (7) *n*-butane, (8) pentene-1, (9) *n*-pentane. Column: 4-ft × ¹/₈-in. s.s. packed with 80 to 100 mesh Spherocarb. Col. temp.: 50°C to methane elution, then increased to 400°C at 20°/min and held until elution of pentane; carrier, He, 55 mℓ/min. (Reproduced by permission of Analabs.)

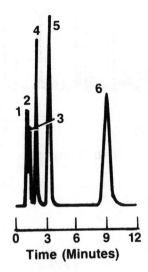

FIGURE 26. Separation of paraffin-olefin-aromatic mixture. Peaks: (1) neohexane, (2) *n*-hexane, (3) hexene-1, (4) cyclohexane, (5) cyclohexene, (6) benzene. Column: 6-ft × ¹/₈-in. s.s. packed with 10% Carbowax 400 on 80 to 100 mesh HP Chromosorb W. Col. temp.: 30°C; carrier, N_2, 12 mℓ/min. (Reproduced by permission of Alltech)

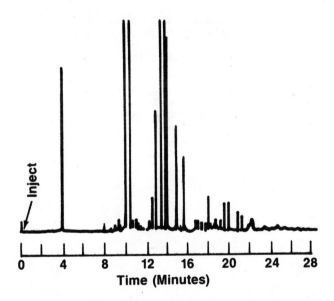

FIGURE 27. Capillary separation of a methyl naphthalene distillate fraction. Column: 50-m × 0.25-mm I.D. s.s. coated with OV-101. Col. temp.: 160°C; FID; carrier, He, 22 psig. Sample: 0.4 µℓ; split ratio, approximately 1000:1. (Reproduced by permission of Foxboro.)

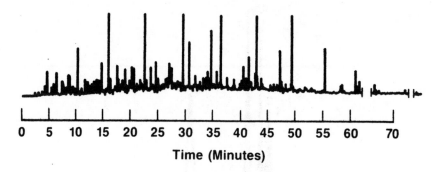

Time (Minutes)

FIGURE 28.   Capillary separation of kerosene jet A. Column: 30-m × 0.25-mm I.D. coated with SP-2100. Col. temp.: 50 to 2000°C at 2°/min; FID; carrier, He, 20 cm/sec. Sample: 0.1 μℓ; split ratio, 100:1. (Reproduced by permission of Supelco.)

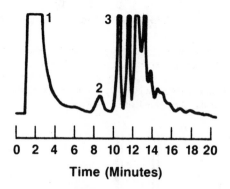

Time (Minutes)

FIGURE 29.   Analysis of benzene in gasoline. Peaks: (1) C-1 to C-12 aliphatic hydrocarbons, (2) benzene, (3) toluene. Column: 10-ft × ¹/₈-in. s.s. packed with 35% BC-150 on 100 to 120 mesh AW-DMCS Chromosorb P. Col. temp.: 110°C for 8 min, then programmed to 220°C at 32°/min and hold; FID; carrier, N_2, 20 mℓ/min. Sample: 0.5 μℓ gasoline. (Reproduced by permission of Supelco.)

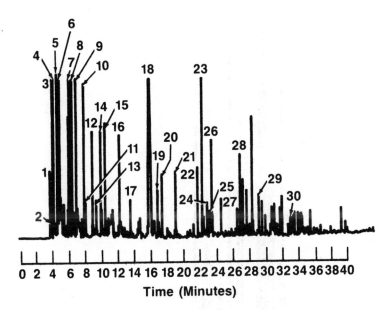

FIGURE 30.   Wide-bore capillary analysis of gasoline. Peaks: (1) air, (2) propane, (3) isobutane, (4) *n*-butane, (5) isopentane, (6) *n*-pentane, (7) 2-methylpentane, (8) 3-methylpentane, (9) *n*-hexane, (10) methylcyclopentane, (11) 2,4-dimethylpentane, (12) benzene, (13) cyclohexane, (14) 2-methylhexane, (15) 3-methylhexane, (16) *n*-heptane, (17) 2,4-dimethylheptane, (18) toluene, (19) 2-methylheptane and 4-methylheptane, (20) 3-methylheptane, (21) *n*-octane, (22) ethyl-benzene, (23) *m*-xylene and *p*-xylene, (24) 2-methyloctane, (25) 3-methyloctane, (26) *o*-xylene, (27) *n*-nonane, (28) propylbenzene, (29) *n*-decane, (30) *n*-undecane. Column: 60-m × 0.75-mm I.D. coated with bonded SE-30 (SPB-1) at a film thickness of 1.00 μm. Col. temp.: 35°C for 12 min, then programmed to 200°C at 4°/min and hold; TC; carrier, He, 5 mℓ/min (He make-up: 30 mℓ/min). Sample: 0.4 μℓ gasoline, splitless injection. (Reproduced by permission of Supelco.)

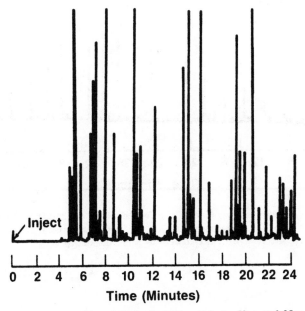

FIGURE 31.   Capillary analysis of gasoline. Column: 50-m × 0.25-mm I.D. s.s. coated with OV-101. Col. temp.: 40°C for 4 min, programmed to 160°C at 5°/min; FID; carrier, He, 22 psig. Sample: 1.0 μℓ; split ratio, approximately 1000:1. (Reproduced by permission of Foxboro.)

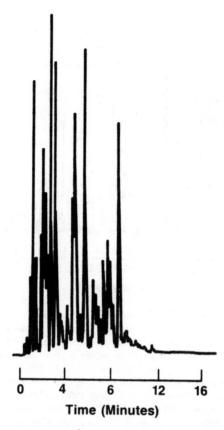

Time (Minutes)

FIGURE 32.   Analysis of naphtha. Column: Series 6000 column containing 3% OV-101. Col. temp.: 28 to 100°C at 3°/min; FID; carrier, N_2, 12 mℓ/min. Sample: 1.0 $\mu\ell$ naphtha. (Reproduced by permission of Applied Science.)

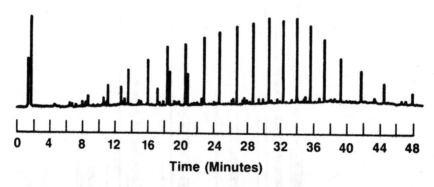

Time (Minutes)

FIGURE 33.   Capillary analysis of shale oil. Column: 30-m × 0.25-I.D. coated with SE-54, Col. temp.: 100°C for 2 min, programmed to 270°C at 5°/min; FID; carrier, H_2, 42 cm/sec. Sample: 0.1 $\mu\ell$ shale oil; split ratio, 67:1. (Reproduced by permission of Supelco.)

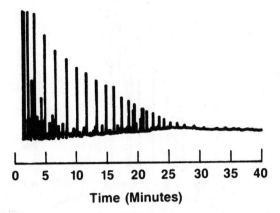

Time (Minutes)

FIGURE 34.   Capillary analysis of Kuwait crude oil. Column: 25-m ×
0.25-mm fused silica coated with SE-30. Col. temp.: 60°C for 2 min,
programmed to 280°C at 10°/min; FID; carrier, He, 30 psig. (Reproduced
by permission of Foxboro.)

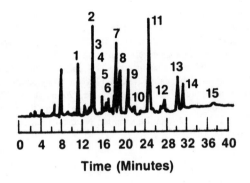

Time (Minutes)

FIGURE 35.   Analysis of creosote. Peaks: (1) naphthalene, (2) 2-meth-
ylnaphthalene, (3) 1-methylnaphthalene, (4) biphenyl, (5) 2,6-dimethyl-
naphthalene, (6) 2,3-dimethylnaphthalene, (7) acenaphthalene, (8)
dibenzofuran, (9) fluorene, (10) methylfluorenes, (11) phenanthrene and
anthracene, (12) carbazole, (13) fluoranthene, (14) pyrene, (15) chrysene.
Column: 10-ft × $^1/_8$-in. s.s. packed with 10% SP-2100 on 100 to 120
mesh Supelcoport. Col. temp.: 100 to 300°C at 6°/min; FID; carrier, N$_2$,
20 mℓ/min. (Reproduced by permission of Supelco.)

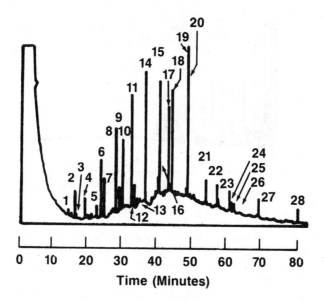

FIGURE 36.   Analysis of polycyclic aromatic hydrocarbons in an ambient air sample. Peaks: (1) C17, (2) phenanthrene, (3) anthracene, (4) C18, (5) "phthalate", (6) C19, (7) "phthalate", (8) C20, (9) fluoranthene, (10) pyrene, (11) C21, (12) 1,2-benzofluorene, (13) 2,3-benzofluorene, (14) C22, (15) C23, (16) benz[a]anthracene, (17) chrysene, (18) C24, (19) C25, (20) diethylhexylphthalate, (21) C26, (22) benzo[k]fluoranthene, (23) C27, (24) benzo[e]pyrene, (25) benzo[a]pyrene, (26) perylene, (27) C28, (28) C29. Column: 35-m × 0.25-mm glass coated with SE-30 HL. Col. temp.: 170 to 235°C at 2°/min; FID; carrier, N_2, 2 mℓ/min. (Reproduced by permission of Alltech.)

REFERENCES

1. **Altgolt, K. H. and Gouw, T. H., Eds.,** *Chromatography in Petroleum Analysis,* Marcel Dekker, New York, 1979.
2. **Berezkin, V. G.,** *J. Chromatogr.,* 91, 559, 1974.
3. **Ramond, D.,** *Ind. Petrol.,* 40, 29, 1972.
4. **Camin, D. L. and Raymond, A. J.,** *J. Chromatogr. Sci.,* 11, 625, 1974.
5. **Kenlemans, A. I. M.,** *Erdoel Kohle, Erdgas, Petrochem. Brennst.-Chem.,* 27, (2), 96, 1974.
6. **Berezkin, V. G. and Nametkin, N. S.,** *J. Chromatogr.,* 65, 85, 1972.
7. **Cram, S. P., Risby, T. H., Field, L. R., and Yue, W. L.,** *Anal. Chem.,* 52, 324R, 1980.
8. **Fraser, J. M.,** *Anal. Chem.,* 49, 231R, 1977.
9. **Anon.,** *Khim. Tekhnol. Topl. Masel,* 3, 61, 1974.
10. **Kolesnikora,** *Gas Chromatography in Investigations of Natural Gases, Petroleum, and Condensates,* Nedra, Moscow, 1972.
11. **Terrell, R. E.,** *Anal. Chem.,* 53, 88R, 1981.
12. **Fraser, J. M.,** *Anal. Chem.,* 51, 211R, 1979.
13. **Fraser, J. M.,** *Anal. Chem.,* 47, 169R, 1975.
14. **King, R. W.,** *Anal. Chem.,* 45, 169R, 1973.
15. **King, R. W.,** *Anal. Chem.,* 43, 162R, 1971.
16. **Kachlik-Olasinska, B.,** *Nafta (Katowice),* 28 (4), 168, 1972.
17. **Turowska, A. and Pruszynska, E., Gax,** *Woda Tech. Sanit.,* 47, 2, 1973.
18. **Ruvinskii, L. Y., Cherezova, V. V., and Akchurin, F. G.,** *Uspekhi Gaz Khromatogr.,* 5, 198, 1978.
19. **Tesarik, K.,** *Chromatographia,* 3, 196, 1970.
20. **Kuchhal, R. K.,** *Chromatogr. Polym. Pet. Petrochem., (Pap. Symp. Workshop),* 33, 1979.

21. **Bremer, H.,** *Arch. Med. Sadowej Kryminol.,* 28 (1), 17, 1978.
22. **Saha, N. C. and Mitra, G. D.,** *Proc. Symp. Chem. Oil Coal,* p. 236, 1972.
23. **Deur-Siftar, D. and Svob, V.,** *Nafta (Zagreb),* 23 (4), 163, 1972.
24. **Dooley, J. E., Thompson, C. J., and Scheppele, S. E.,** in *Analytical Methods of Coal and Coal Products,* Vol. 1, Carr, C., Jr., Ed., Academic Press, New York, 1978, 467.
25. **Trusell, F. C.,** *Chromatogr. Sci.,* 11, 91, 1979.
26. **Adlard, E. R.,** *Chromatogr. Sci.,* 11, 137, 1979.
27. **Rasmussen, D. V.,** *Anal. Chem.,* 48, 1562, 1976.
28. **Roubicek, V. and Splichal, B.,** *Hunt. Listz,* 32, 351, 1977.
29. **Abidova, Z., Kh., Belopol'skaya, S. I., Sokol'nikova, M. D., Sagidova, F. Z., and Akhmedova, M. K.,** *Dokl. Resp. Nauchno-Tekh. Konf. Neftekhim,* 1, 220, 1974.
30. **Whittemore, I. M.,** *Chromatogr. Sci.,* 11, 41, 1979.
31. **Adlard, E. R., Bowen, A. W., and Salmon, D. G.,** *J. Chromatogr.,* 186, 207, 1979.
32. **Ury, G. B.,** *Anal. Chem.,* 53, 481, 1981.
33. **Stavinoha, L. L.,** *J. Chromatogr. Sci.,* 13, 72, 1975.
34. **Stavinoha, L. L. and Newman, F. M.,** AD Rep. No. 748446, *U.S. National Technical Information Service, Springfield, Va.,* 1972.
35. **Lulova, N. I., Leont'eva, S. A., and Fedosova, A. K.,** *Usp. Gazov. Khromatogr.,* 2, 141, 1970.
36. **Ozeris, S.,** *Chim. Acta Turc.,* 1, 33, 1973.
37. **Milina, R. and Pankova, M.,** *Chem. Tech. (Leipzig),* 31, 212, 1979.
38. **Chachulski, J., Czarnik, L., and Dettloff, W.,** *Nafta (Katowice, Pol.),* 35, 131, 1979.
39. **Dracheva, S. I. and Bryanskaya, E. K.,** *Khim. Tekhnol. (Kiev),* 101, 35, 1978.
40. **Laurgeau, C., Espian, B., and Barras, F.,** *Rev. Inst. Fr. Pet.,* 34, 669, 1979.
41. **Block, M. G., Callen, R. B., and Stockinger, J. H.,** *J. Chromatogr. Sci.,* 15, 504, 1977.
42. **Desty, D. H. and Goldup, A.,** in *Chromatography,* 3rd ed., Heftmann, E., Ed., Van Nostrand-Reinhold, New York, 1975, 915, 1978.
43. **Petrov, A. A. and Kolesnikova, L. P.,** *Zh. Anal. Khim.,* 27, 1050, 1972.
44. ASTM Manual on Hydrocarbon Analysis, *3rd ed., American Society for Testing and Materials, Philadelphia,* 1978.
45. **Kulikov, V. I. and Sorokin, M. E.,** *Zh. Anal. Khim.,* 30, 1594, 1975.
46. **Mitra, G. D., Mohan, G., and Sinha, A.,** *J. Chromatogr.,* 91, 633, 1974.
47. **Weingaertner, E., Guer, T., and Bayuness, O.,** *Fresenius' Z. Anal. Chem.,* 254, 28, 1971.
48. **Buchta, H. and Forster, L.,** *Erdoel. Erdgas Z.,* 86, 139, 1970.
49. **Churacek, J., Komarkova, H., Komarek, K., and Dufka, O.,** *Chromatographia,* 3, 465, 1970.
50. **Vovelle, C., Foulatier, R., Guerin, M., and Delbourgo, R.,** *Method Phys. Anal.,* 6, 277, 1970.
51. **Saha, N. C.,** *Chromatogr. Polym. Pet. Petrochem. (1978 Pap. Symp. Workshop),* 126, 1979.
52. **Mindrup, R.,** *J. Chromatogr. Sci.,* 16, 380, 1978.
53. **Nand, S. and Sarkar, M. K.,** *J. Chromatogr.,* 89, 73, 1974.
54. **Hattman, E. A., Schultz, H., and McKinstry, W. E.,** *Anal. Chem.,* 49, 176R, 1977.
55. **Saha, N. C., Jain, S. K., and Dua, R. K.,** *J. Chromatogr. Sci.,* 16, 323, 1978.
56. **Purcell, J. E. and Gilson, C. P.,** *Chromatogr. Newsl.,* 1, 45, 1972.
57. **Willis, E. D.,** *Anal. Chem.,* 44, 387, 1972.
58. **Rao, K. A. and Iyer, R. M.,** *Indian J. Technol.,* 16, 43, 1978.
59. **Schoellner, R., Platzdasch, K., and Ulber, G.,** *Z. Chem.,* 17, 321, 1977.
60. **Rah, S. C. and Rhee, H. K.,** *Hwahak Konghak,* 17, 73, 1979.
61. **Al-Thamir, W. K., Lanb, R. J., and Purnell J. H.,** *J. Chromatogr.,* 142, 3, 1977.
62. **Carle Instruments Ltd.,** *Process Eng. (London),* 17, May 1977.
63. **Di Corcia, A. and Samperi, R.,** *J. Chromatogr.,* 117, 199, 1976.
64. **Raulin, F. and Toupance, G.,** *J. Chromatogr.,* 90, 218, 1974.
65. *Proc. Anal. Div. Chem. Soc.,* 14, 108, 1977.
66. **French, K. H. V.,** *J. Chromatogr.,* 67, 237, 1972.
67. **Tomi, P., Liteanu, C., and Andreeson, G.,** *Analusis,* 9, 135, 1981.
68. **Cramers, C. A., Rijks, J. A., Pacakova, V., and Ribeiro-de-Andrade, I.,** *J. Chromatogr.,* 51, 13, 1970.
69. **Kugucheva, E. E., Berents, A. D., Gulovskaya, L. D., and Mukhina, T. N.,** *Pr-vo Nizsh. Olefinov. M.,* 54, 1978.
70. **Cottom, W. P. and Stelz, D. E.,** *Anal. Chem.,* 52, 2073, 1980.
71. **Wagner, J. H., Lillie, C. H., Dupuis, M. D., and Hill, H. H.,** *Anal. Chem.,* 52, 1614, 1980.
72. **Severson, R. F., Ellington, J. J., Arrendale, R. F., and Snook, M. E.,** *J. Chromatogr.,* 160, 155, 1978.
73. **Eppert, G., Ludwig, H., and Schinke, I.,** *J. Prakt. Chem.,* 321, 570, 1979.

74. **Felscher, D.,** *Chem. Tech. (Leipzig),* 30, 147, 1978.
75. **Matishev, V. A.,** French Patent 2,362,804, 1978.
76. **Makushina, V. M. and Aref'ev, O. A.,** *Neftekhimiya,* 18, 847, 1978.
77. **Sulatanov, N. T. and Arustamova, L. G.,** *Azerb. Neft. Khoz.,* 5, 34, 1974.
78. **Smith, E.,** *Anal. Chem.,* 47, 1874, 1975.
79. **Scacchi, G. and Back, M. H.,** *J. Chromatogr.,* 114, 255, 1975.
80. **Shlyakhov, A. F., Koreshkova, R. I., and Telkova, M. S.,** *J. Chromatogr.,* 104, 337, 1975.
81. British Gas Corporation, British Patent 1,325,733, 1970.
82. **Genty, C.,** *Anal. Chem.,* 45, 505, 1973.
83. **Petrov, A. A. and Krasavchenko, M. I.,** *Neftekhimiya,* 13, 779, 1973.
84. **Kozlov, N. S., Egiazarov, Yu. G., Agabekova, L. A., Cherches, B. Kh., and Savon'kina, M. G.,** *Dokl. Akad. Nauk Beloruss, S.S.R.,* 15, 46, 1971.
85. **Krasavchenko, M. I., Zemskova, E. K., Mikhnovskaya, A. A., Pustil'nikova, S. D., and Petrov, A. A.,** *Neftekhimiya,* 11, 803, 1971.
86. **Morgan, S. L. and Deming, S. N.,** *J. Chromatogr.,* 112, 267, 1975.
87. **Nigam, R. N. and Moolchandra, R.,** *Anal. Chem.,* 43, 1683, 1971.
88. **Toader, M., Voicu, F., and Sandulescu, D.,** *Rev. Chim (Bucharest),* 21, 568, 1970.
89. **Cuevas, P. A. and Guzman, G.,** *Rev. Inst. Mex. Pet.,* 11, 90, 1979.
90. **Lukashenko, I. M. and Musaev, I. A.,** *Neftekhimiya,* 13, 164, 1973.
91. **Carson, J. W., and Lege, G. J.,** *J. Chromatogr. Sci,* 12, 49, 1974.
92. **Anosova, V. I., Korablina, N. V., and Shchegoleva, L. N.,** *Khim. Prom-sti.,* 3, 12, 1979.
93. **Seroshtan, V. A. and Zakharova, N. V.,** *Zh. Anal. Khim.,* 34, 1166, 1979.
94. **Pies, R. and Markova, M.,** *Z. Anal. Chem.,* 275, 27, 1975.
95. **Bliznyuk, N. L., Voronkov, A. P., Mislavskaya, V. S., and Mushii, R. Ya.,** *Khim. Tekhnol. (Kiev),* 1, 47, 1975.
96. **Il'in, D. T., Zorya, L. A., Shugyakovskii, G. M., and Tolstobrova, S. A.,** *Khim. Prom. (Moscow),* 47, 867, 1971.
97. **Engewald, W., Graefe, J., Kiselev, A. V., Shcherbakova, K. D., and Welsch, T.,** *Chromatographia,* 7, 229, 1974.
98. **Petrov, A. A. and Kolesnikova, L. P.,** *Zh. Anal. Khim.,* 27, 1050, 1972.
99. **Vitenberg, A. G. and Tsibul'skava, I. A.,** U.S.S.R. Patent 697,921, 1979.
100. **Hanna, A. H., Mahmoud, B. H., and Guichard, N.,** *Egypt. J. Chem.,* 19, 693, 1976.
101. **Anon.,** *Sekiyu Gakkai Shi,* 20, 1143, 1977.
102. **Lisenkov, V. F. and Kogan, L. A.,** *Zh. Fiz. Khim.,* 49, 1350, 1975.
103. **Ottenstein, D. M., Bartley, D. A., and Supina, W. R.,** *Anal. Chem.,* 46, 2225, 1974.
104. **Khodzhaev, G. Kh., Belopol'skaya, S. I., Abidova, Z. Kh., and Akhmetov, M. K.,** *Uzb. Khim. Zh.,* 16, 48, 1972.
105. **Egiazarov, Yu. G., Potapova, L. L., Smol'skii, A. M., and Savchits, M. F.,** *Khim. Tekhnol. Topl. Masel,* (9), 43, 1980.
106. **Tsibul'skii, V. V., Tsibul'skaya, I. A., and Yaglitskaya, N. N.,** *Zh. Anal. Khim.,* 34, 1364, 1979.
107. **Pankova, M., Milina, R., Belcheva, R., and Iranov, A.,** *J. Chromatogr.,* 137, 198, 1977.
108. **Mel'kanovitskaya, S. G.,** *Gidrokhim. Mater.,* 53, 153, 1972.
109. **Sojak, L., Majer, P., Krupcik, J., and Janak, J.,** *J. Chromatogr.,* 65, 143, 1972.
110. **Peter, I. and Torkos, L.,** *Magy. Kem. Lapja,* 26, 595, 1971.
111. **Roseira, A. N.,** *Rev. Brasil. Technol.,* 2, 1, 1971.
112. **Duerbeck, H. W.,** *Fresenius' Z. Anal. Chem.,* 251, 108, 1970.
113. **Hurtubise, R. J., Skar, G. T., and Poulson, R. E.,** *Anal. Chim Acta,* 97, 13, 1978.
114. **Borwitzky, H. and Schomburg, G.,** *J. Chromatogr.,* 170, 99, 1979.
115. **Bieri, R. H., Cueman, M. K., Smith, C. L., and Su, C.-W.,** *Inst. J. Environ. Anal. Chem.,* 5, 293, 1978.
116. **Neumann, H. J. and Kaschani, D. T.,** *Wasser, Luft. Betr.,* 21, 648, 1977.
117. **Zielinski, W. L., Jr. and Janini, G. M.,** 2nd Proc. ORNL Workshop Exposure Polynucl. Aromat. Hydrocarbons Coal Convers. Processes, 1977.
118. **Kasmatsu, K.,** *Bull. Jpn. Pet. Inst.,* 17, 28, 1975.
119. **Qazi, A. H. and Nau, C. A.,** *Am. Ind. Hyg. Assoc. J.,* 36, 187, 1975.
120. **Knowles, M. E.,** *Proc. Anal. Div. Chem. Soc.,* 15, 153, 1978.
121. **Grimmer, G. and Boehnke, H.,** *J. Assoc. Off. Anal. Chem.,* 58, 725, 1975.
122. **Kusy, V.,** *Chem. Prum.,* 30, 413, 1980.
123. **Utashiro, S. and Takada, T.,** *Kenkyu Hokoku - Kaiyo Hoan Daigakko, Dai-2-bu,* 25 (2), 1, 1980.
124. **Thomas, R. S. and Lao, R. C.,** *2nd Proc. ORNL Workshop Exposure Polynucl. Aromat. Hydrocarbons Coal Convers. Processes,* 77, 1977.

125. **Lane, D. A., Moe, H. K., and Katz, M.,** *Anal. Chem.,* 45, 1776, 1973.
126. **Beernaert, H.,** *J. Chromatogr.,* 173, 109, 1979.
127. **Bhatia, K.,** *Anal. Chem.,* 43, 609, 1971.
128. **Grimmer, G. and Boehnke, H.,** *IARC Publ. 29 (Environ. Carcinog: Sel. Methods Anal.),* 3, 155, 1979.
129. **Grimmer, G. and Boehnke, H.,** *IARC Publ. 29 (Environ. Carcinog: Sel. Methods Anal.),* 3, 163, 1979.
130. **Grimmer, G. and Boehnke, H.,** *Chromatographia,* 9, 30, 1976.
131. **Pervukhina, I. V.,** *Soyuzev,* 97, 55, 1975.
132. **Schaefer, R. G., Leythaeuser, D., and Weiner, B.,** *J. Chromatogr.,* 167, 355, 1978.
133. **Schaefer, R. G., Weiner, B., and Leythaeuser, D.,** *Anal. Chem.,* 50, 1848, 1978.
134. **Overton, E. B., Bracken, J., and Laseter, J. L.,** *J. Chromatogr. Sci.,* 15, 169, 1977.
135. **McKirdy, D. M. and Horvath, Z.,** *Sci. Aust. Technol.,* 14(2), 12, 1976.
136. **Levshunova, S. P. and Telkova, M. S.,** *Tr., Vses. Nauchno-Issled Geologorazved. Neft. Inst.,* 138, 115, 1973.
137. Proc. Anal. Div. Chem. Soc., 14, 108, 1977.
138. **Safonova, G. I. and Bulekova, L. M.,** *Khim. Tekhnol. Topl. Masel,* 14(10), 21, 1969.
139. **Diamanti, F.,** *Bul. Shkencave Nat.,* 32(4), 75, 1978.
140. **Roucache, J., Boulet, R., DaSilva, M., and Fabre, M.,** *Rev. Inst. Fr. Pet.,* 32, 981, 1977.
141. **Jutila, J.,** *Instrum. Technol.,* 26(7), 38, 1979.
142. **Bailey, S. J.,** *Contr. Eng.,* 26(3), 61, 1979.
143. **Utterback, V. C.,** *Chem. Eng.,* 83, 141, 1978.
144. **Sandford, J.,** *Instrum. Control Syst.,* 50,(3), 25, 1977.
145. **Deans, D. R.,** *Gas Chromatogr. Proc. Int. Symp. (Europe),* (9), 199, 1972.
146. **Villalobos, R.,** *Anal. Chem.,* 47, 983A, 1975.
147. **Lipavskii, V. N. and Berezkin, V. G.,** *J. Chromatogr.,* 91, 583, 1974.
148. **Kosono, Y.,** *Keiso,* 14, 20, 1971.
149. **Kobayashi, S.,** *Keiso,* 14, 40, 1971.
150. **Stockinger, J. H., Callen, R. B., and Kaufman, W. E.,** *J. Chromatogr. Sci.,* 16, 418, 1978.
151. **Stockinger, J. H.,** *J. Chromatogr. Sci.,* 15, 198, 1977.
152. **Grubaugh, K. W. and Stobby, G. E.,** *Anal. Chem.,* 50, 377, 1978.
153. **Guillemin, C. L. and Martinez, F.,** *J. Chromatogr.,* 139, 259, 1977.
154. **Annino, R., Curren, J. Kalinowiski, R., Karas, E., Lindquist, R., and Prescott, R.,** *J. Chromatogr.,* 126, 301, 1976.
155. **Anon.,** *Chem. Week,* 118(19), 56, 1976.
156. **Pomerants, L. I. and Simonguaz, S. E.,** *Prikl. Geofiz.,* (72), 223, 1973.
157. **Sokolin, G. F., Lulova, N. I., and Lipavskii, V. N.,** *Neftepererab. Neftekhim (Moscow),* (2), 37, 1972.
158. Antar Petroles de L'Atlantique, French patent 2,177,142, 1973.
159. **Baker, W. J. and Johnson, M. L.,** *Chem. Eng. Progr.,* 68(10), 46, 1972.
160. **Boyd, D. M., Jr.,** U.S. Patent 3,501,700, 1970.
161. **Lim, C. R., Choe, G. H., Li, U. M., and Sin, D. R.,** *Punsok Hwahak,* 1, 16, 1981.
162. **Lim, C. R., Li, U. M., and Choe, G. H.,** *Punsok Hwahak,* 4, 7, 1980.
163. **Terada, O.,** *Aromatikkusu,* 23, (1), 17, 1971.

Chapter 4

GAS CHROMATOGRAPHY OF HYDROCARBONS — SPECIFIC STUDIES

Literature discussed in this chapter covers hydrocarbon studies which have focused principally upon chromatographic column sample inlet systems, parallel and serial arrangements of chromatographic columns, capillary columns, conventional and novel chromatographic detectors, gas chromatography-mass spectrometry (GC-MS), and the use of microprocessors and computers in the gas chromatographic analysis and monitoring of hydrocarbons.

I. INLET SYSTEMS

Proper sample introduction into a gas chromatograph is a critical aspect of obtaining high quality chromatographic results. Excessive and undesirable zone broadening of the sample components at the chromatographic inlet due to inefficient inlet design will result directly in deterioration of peak sharpness in the resultant chromatogram, regardless of the efficiency of the column itself. Hence, extensive efforts have been made in the design and evaluation of sample inlet systems in order to maximize the quality of the chromatography by mininizing sample zone broadening in the chromatograph sample inlet. This is particularly an important area of concern when analysis of highly complex mixtures is needed, which is often the case in hydrocarbon analysis. Highly complex mixtures generally lead to the selection of capillary columns, for which special attention to inlet system optimization must be given. Studies that have been reported concerning the design and evaluation of inlet systems for various applications in hydrocarbon analyses are given below.

Much effort has been expended in the evaluation of inlet systems for capillary GC. Brander[1] evaluated a glass capillary system for C-11 to C-36 hydrocarbon separations using glass inserts in the capillary injector which allowed for direct, split, and splitless modes of operation without removal of the capillary column. Branching of the column was facilitated by using ferrules, and good retention data precision was attributed to a thermostated pneumatic system and precise column oven temperature control. Evrard and co-workers[2] described a sampling system for high-temperature isothermal capillary GC based upon the use of heated sample vials and a precolumn (1 to 12% OV-1 on 80/100 mesh Gas-Chrom Q). The vial, which contained the sample diluted in 0.1 to 0.2 mℓ of volatile solvent, was placed in a loading chamber and heated under a nitrogen purge. The sample vapors condensed on the precolumn operated in a back-flush mode. The sample components (C-14 to C-36 *n*-alkanes) were then transferred to a 25-m OV-101 capillary analytical column by means of a wall-coated capillary trap. The device was controlled by a programable electromechanical processor. Blanks were free of ghost peaks when thermally prestripped vials were used. Galli et al.[3] reported an efficiently cooled on-column injector, evaluated with C-18 to C-38 alkanes, that eliminated discrimination of compounds in a given sample by isolating the end of the column from the oven temperature. The cooling was obtained by an air stream entering the area of the capillary fitting and flowing along the outside of the capillary into the oven. No observable band broadening occurred due to the cold spot at the column inlet. Problems caused by insufficient cooling were discussed. Grob and Neukom[4] studied the effect of the syringe needle on the precision and accuracy of vaporizing sample injections of a C-9 to C-44 alkane mixture and stream-splitting on a capillary column. Discrimination of high-boiling components in the injector was attributed to selective evaporation of the sample from the needle. The level of sample component discrimination was measured and correlated with material remaining in the needle for various injection procedures. It was concluded that the sample should be pulled back into the barrel of the syringe and the needle allowed to heat in the injector before sample injection. Solvent flush may be preferred for samples containing components sensitive to the hot metal surface of the needle. Sonchik and Walker[5] reported an inlet for trace analysis using capillary GC. The inlet consisted of a 11.4-cm length glass-lined stainless-steel tube for direct sample injection. A 1-ppm cyclohexane peak was resolved from injection of 99% pure *n*-hexane on a squalane SCOT column using this inlet; comparative analyses of a C-6 to C-13 *n*-alkane mixture on an OV-17 Scot column between the direct injection inlet and a commercially available splitter inlet showed little difference in column efficiency. The reported inlet was cited as useful for wide boiling range hydrocarbon mixtures. Thermally labile compounds are appropriate since a high inlet temperature is not needed; a C-7 to C-16 *n*- alkane mixture was analyzed using an inlet temperature of 150 to 250°C compared to 300°C needed with a splitter inlet. Tejedor[6] also reported on direct injection into capillary columns for trace analysis using MS for component identification. No loss in resolution or excessive loss of peak symmetry was noted; the method employed can be routinely used without efficiency loss or deterioration of the column. Evaluations were performed using a mixture containing *n*-hexane, methylcyclohexane, and cyclohexane and several polar components.

Ullrich and Dulson[7] developed a simple all-glass system comprised of two injection ports and a distributing device for transferring sample components onto two analytical columns according to their polarity. Application of the system in analyses of commercial gasoline and auto exhaust was demonstrated. Vitenberg and associates[8] reported a device for injection of a gas in equilibrium with a liquid. Gaspar and co-workers[9] evaluated a fluid logic gate for injecting narrow sample plugs. Stolyarov[10] reported a modification of the internal normalization method for use in sample splitting. Jennings[11] reported on the linearity of a glass inlet splitter, evaluated with a C-6 to C-19 hydrocarbon mixture. Borba de Oliveira and Alfonso[12] described the design of a device for chromatographic injection of pressurized liquid samples, particularly applicable to liquified petroleum gas and natural gasoline. The sampler consisted of a double-walled chamber with coolant circulated through the annular space around the inner chamber to maintain the sample in the liquid state. Stockwell and Sawyer[13] developed an automated chromatograph having a fully automatic liquid injector. The injector was a modification of one designed for preparative use. Samples were handled and mixed with an internal standard; the mixture was then forced through a capillary onto the column using an over-pressure of the carrier gas. One of the most innovative sampling systems developed was the fluidic monostable system of Gasper et al.,[14] which could provide variable injection band widths as fast as a few milliseconds and should be particularly appropriate to automated high-resolution and high-speed separations since there were no moving parts. Sampling of gaseous hydrocarbons in soils was automated by Smith and Harris[15] by combining an automated syringe and valve inlet assembly.

Frankiewicz and Williams[16] maintained chromatographic column efficiency with large-volume low-pressure gas sampling down to 0.25 torr for compounds having short retention times by using a modified version of Umstead's injection valve system.[17] The secondary helium supply was modified by adding a ballast tank close to a 100-mℓ (or 500-mℓ) sample loop; a single-stage regulator was substituted for a two-stage regulator. Little column efficiency loss was observed over 0.23 to 760 torr during injection of helium samples containing 85 ppm each of CH_4, C_2H_6, C_2H_4, C_3H_8, and C_3H_6, using such large loops. A solids injector was used by di Lorenzo[18] in the analysis of polycyclic aromatic hydrocarbons in carbon black, in which the carbon black samples were injected directly into the chromatograph. Evrard and Guiochon[19] described a technique using a short 4-mm I.D.-packed precolumn for the automated analysis of traces of high-boiling alkanes in a volatile matrix. Following sample injection into the precolumn and evaporation of the volatile material, the high-boiling constituents were transferred to the cooled analytical capillary column by progressive heating. Durbin and Fruge,[20] working with systems using low dead-volume catharometers, 0.023 to

0.051 in. diameter high-speed columns of finely divided Porapak S or Chromosorb 102, and sample loops, found that column plate number had a semiquantitative linear relationship with optimum sample loop length. Sample loop use in separating C_2H_4 from C_2H_6 in a volatile mixture containing CO_2 and CH_4 increased sensitivity by a factor of 50 over that obtained without the loop. Pease[21] reported a solid injector system which allowed on-column injection from a dry ice cooled sample holder without column gas flow interruption, and which also allowed for sample precooling. The system was used for wax, pitch, coke (extracted from cracking catalysts), marine oil spill samples, and hydrocarbon-contaminated soil and sand. Umstead[17] described a technique for sampling subambient pressure systems, such as a tubular flow reactor operated at 2 torr. A 100-mℓ gas sampling loop was used, with the sample compressed with a high volume secondary He stream to force it into the chromatograph as a narrow band. Good resolution was obtained for a five-component C-1 to C-3 hydrocarbon mixture. Dunlop and Pollard[22] described a GC injection system for routine analysis of light liquid hydrocarbons (C-2 to C-8 hydrocarbon streams; C-5 and heavier hydrocarbons), based upon encapsulation in indium tubing. Reproducibility with this technique was found to be superior to syringe sampling. This injection system was designed to improve the durability of commercial sample introduction systems using indium tubing, and to be intrinsically safer than high-pressure liquid sampling valves.

II. CAPILLARY COLUMNS

Narrow-bore open-tubular ("capillary") columns, wall- or support-coated, offer resolution capacities for up to several hundred components per single injection of a complex sample. Such capacities are unparalleled by any other separation technique. Capillary columns have great use, therefore, in the analysis of complex hydrocarbon mixtures, or for the separation of hydrocarbon constituents that are otherwise difficult or intractable by other techniques, including packed GC columns.

In a 1975 review on hydrocarbons, Bradley[23] noted that the most popular stationary phase used for capillary GC of hydrocarbons was squalane. More recently, additional phases have been employed that have extended the selection of useful phases. Much effort and attention has been given to the preparation of capillary columns during the past decade, and useful commercial columns have become more prevalent. The need to further improve resolution while ensuring thermal stability, useful column lifetimes, and ease of preparation will continue to occupy the attention of researchers employing capillary columns. To illustrate the development of high-resolution columns with phases other than squalane, silanized glass capillary columns coated with polysiloxanes such as SE-30 or SE-52 could be programmed up to 350°C without any appreciable stationary phase loss.[24] The recent development of flexible fused silica capillaries appear to offer an enhanced inertness to borosilicate glass capillaries.[25] The observation that capillary column GC has attained distinct and significant advantages over packed column use has been well-documented and illustrated in various reviews.[26,27] A recent text by Jennings[28] provided detailed treatment of the theory and practice of capillary GC using glass capillaries. A comprehensive discussion of the durability and lifetime of a large number of glass capillary columns has been given by Grob.[29] Investigators are referred to these and the many other excellent treatments of capillary GC that have appeared by experts in this technique in recent years.

The hydrocarbon literature dealing with capillary GC given below is divided into two principal categories: general hydrocarbon analyses; and analyses by hydrocarbon class.

A. General Hydrocarbon Analyses

Lochmueller et al.[30] intercompared two thermal conductivity detectors and a flame-ionization detector for use with open-tubular columns and presented analysis results for a premium grade gasoline. Rooney[31] discussed the advantages of glass capillary columns in petroleum products analysis, using a 60-m SP-2100 coated column with flame ionization detection. He intercompared H_2, He, and N_2 carrier gases, attributing the lower viscosity of H_2 to the shorter retention times observed when using it. Toth[32] compared the efficiencies of packed steel and glass capillary columns having rough or smooth inner surfaces, using different stationary phases at 50 and 100°C in separations of *n*-alkane and aromatic hydrocarbon mixtures. He concluded important parameters to be peak resolution per unit column length, effective number of plates per unit column length, rate of generation of effective plate number, column type, stationary phase, and partition ratio.

Schieke and Pretorius[33] described the use of "whisker-walled" open-tubular glass columns for hydrocarbon separations. Block and co-workers[34] reported methodology for the analysis of hydrocarbons resulting from methanol conversion. McGill and associates[35] obtained enhanced separations by connecting two straight capillary end pieces of silanized soda glass to the capillary column using medical-grade heat-shrinkable Teflon®; improved resolution and the significance of the capillary exit position were demonstrated for a North Sea alkyl fraction on an OV-101 column. A rapid response, sensitive thermal conductivity detector (TCD) for use with glass capillary columns was reported by Schirrmeister[36]. Its response was similar to that of a flame ionization detector, and its sensitivity similar to a conventional TCD using a packed column. The detector inlet and outlet were pressure limited by a constricted capillary between the column and the detector, and by a needle valve between the outlet and a vacuum pump. Gas entering the TCD was expanded 50-fold. Improved peak shapes were obtained using this system for a hydrocarbon mixture.

High-molecular-weight hydrocarbons were analyzed by Talalaev and associates[37] using a stainless-steel poly(methylphenyl)siloxane capillary column, while Rijks and co-workers[38] separated complex hydrocarbon mixtures on serially coupled polar and nonpolar high-resolution capillaries. Capillary columns for petroleum hydrocarbon analyses were prepared by a dynamic coating method and their performance was intercompared with packed columns.[39] Separations of hydrocarbon fractions were stressed in a detailed review[40] of capillary GLC separations.

Squalane-coated capillaries were used by a number of workers for the analysis of complex hydrocarbon mixtures. Shafter[41] reported on over 190 paraffinic and naphthenic hydrocarbons in straight-run naphtha from five U.S.S.R. crudes. A 90-m squalane capillary was used by Nakamura[42] to assist the identification of 200 components in catalytically cracked gasoline from 200 peaks observed and also employed a 200-m by 0.2-mm I.D. squalane column to separate all dodecenes and the principal alkyl aromatics generated by catalytic dehydrogenation of dodecane.

Dupuy et al.[44] used a 50-m by 0.25-mm I.D. stainless-steel MS-550-coated capillary for the analysis of C-6 to C-7 cycloalkyl chlorides, cycloolefins, methylcycloalkanes, and bicyclo-[n.1.0.] alkanes. A report of the use of capillary GLC for the recognition of crude oils was prepared by Douglas.[45]

A versatile OV-101 capillary column (10 m by 0.01 in. I.D.) was used to analyze petroleum fractions having end points above 1000°F and high-boiling waxes up to n-C58, as well as for the separation of iso-butane from *n*-butane at lower analytical temperatures. The column had an exceptional ability to resolve isomers of high molecular weight, as well, and could also be used for simulated distillation of wide-boiling range mixtures.[46] Petrovic and Kapor[47] reported on the use of a selective liquid phase mixture (*n*-hexadecane, *n*-hexadecene, and KEL) for capillary GLC analysis of straight-run gasolines and reformates to determine the composition of C-5 to C-8 hydrocarbons in straight-run gasolines (C-5 to C-7 in the non-aromatic fraction) and C-6 to C-9 hydrocarbons in reformates (aromatic fraction). The method had an estimated uncertainty of 2% for components present above 1%, 5% for components present at 0.1 to 1.0%, and 10% for components present at 0.01 to 0.1%. Cramers and co-workers[48] evaluated micropacked columns for applications to petroleum analyses, using 0.6 to 0.8-mm I.D. columns up to 15 m long with homogeneous packing densities and up to 50,000 theoretical plates. The columns were prepared by packing the coiled column under pressure with vibration, using carefully sized supports precoated with a candidate stationary phase. Applications to the separation of hydrocarbon mixtures containing low-boiling alkanes, alkenes, and alkyldienes up to C-8, benzene, neo-pentane, and methylcyclohexanes at 70 and 150°C were demonstrated. Novotny and associates[49] used Dexsil and carborane polymers as phases on silylated glass capillary columns (0.3 mm I.D. by 28 to 35 m) for the efficient separation of petroleum fractions containing hydrocarbons up to C-60 and other components of molecular weight up to 800, using column temperatures up to 350°C. Robinson et al.[50] reported on a method for the rapid analysis of gasoline hydrocarbons using an uncoated open-tubular column at 250°C connected in series to a coated column (N,N'-bis[2-cyanoethyl]formamide) at 50°C. Application was made to the determination of saturated, olefinic, and aromatic gasoline hydrocarbons. Following elution of the saturated hydrocarbons from the coated column, the olefinic species were sorbed on Chromosorb P (coated with 40% $Hg(ClO_4)_2$ and 15% $HClO_4$ by weight), and the aromatics were backflushed through the coated column for rapid determination. A mean uncertainty of 1.6% was obtained for the determination of the hydrocarbons in a synthetic mixture. Walker and Wolf[51] also used a dual column approach, coupling a 5-ft capillary column packed with Porasil B to the front end of a 50-ft capillary column coated with OV-101, for the analysis of hydrocarbons from phytane pyrolysis. The resultant complex mixtures containing C-1 to C-20 hydrocarbons could be resolved without subambient temperature programing or backflushing.

Schomburg[52] provided a general description of the practical problems associated with capillary GC applications. Examples of the use of capillary columns for applications in the field of petroleum chemistry were described by Mathews and co-workers[53] in the separation of 39 components in 28 to 114 °C range of petroleum distillates in an analysis time of about 20 min; Bloch and co-workers[54] in the analysis of 200 components in the reaction products from the conversion of methanol conversion to gasoline; and by Pacakov and Kozlik in the separation of 54 components derived from the catalytic decomposition of C-6 hydrocarbons.[55] Dielman et al.[56] used an OV-1 glass capillary suitable for the analysis of components with low volatility and applied this to the separation of high-molecular-weight hydrocarbons.

In terms of the reliability of capillary columns, Kaiser[57] asserted that such columns can provide retention precisions of less than 0.1 retention index units, while Grob[58] described the durability and lifetimes of a wide variety of glass capillary columns.

The use of binary liquid phases in capillary columns was reported by Mitooka[59] for 69 saturated C-5 to C-8 hydrocarbons from petroleum naphtha. Gallegos[60] and Sidorov and co-workers[61] reported on the analysis of aromatic hydrocarbons, while Gouw and associates[62] found that short (e.g., 10-m) capillary columns were useful and versatile for the analysis of hydrocarbons in petroleum fractions having end points above 530°C. Polyphenylether-coated capillaries were used for the separation of aromatics and aliphatics in the 350 to 400°C boiling fraction of tars.[63] Robinson et al.[64] described the rapid analysis of hydrocarbons in gasoline.

In two related studies, Sojak and co-workers[65] reported on the use of nematic and smectic liquid crystal phases as selective coatings for the separation of hydrocarbon isomers. DiSanzo and associates[66] coupled preparative liquid chromatography with glass capillary GC for the separation and analysis of shale oil hydrocarbons. Kumar et al.[67] described the isothermal analysis of naphtha hydrocarbons, while Rijks and Cramers[68] presented an overview of the use of high precision capillary GC for hydrocarbon separations.

B. Analyses by Hydrocarbon Class
1. Alkanes and Unsaturated Hydrocarbons
a. Alkanes

Lukac[69] studied the separation of isotopic methanes using an etched glass capillary column partly coated with a thin layer of active silica to improve the resolution per unit time (expressed as R^2/t) for separating isotopic pairs. Johansen[70] used short glass capillaries for the separation of a variety of *n*-alkanes, including pristane and phytane, while Nygren[71] coupled short class capillaries with flow programming to obtain rapid analyses of hydrocarbons (e.g., the separation of C-10 to C-18 alkanes in 3 min). Oshima and co-workers[72] established conditions for the analyses of monomethyl alkanes in *n*-alkane mixtures. Using a simple, high-precision time differentiating printer constructed from standard equipment, Nillson[73] found that capillary columns afforded the best precision for the measurement of Kovats retention indices for *n*-alkanes. Column performance in terms of coupling resolution and retention time was intercompared for packed columns and coated capillaries in analyses of C-7 to C-16 *n*-alkanes by Ettre.[74] An application of packed capillary columns (2-m by 1-mm I.D.) was reported by Talalaev and associates[75] for the analysis of C-23 to C-48 *n*-alkanes derived from a crude oil distillate by urea inclusion. Petrovic and Vitrovic[76] employed a 50-m by 0.1-mm I.D. stainless steel Apiezon L capillary column for direct determination of C-9 to C-14 *n*-alkanes in kerosene fractions, with quantification achieved by comparing chromatogram peak heights for the samples analyzed with those on a reference standard chromatogram; it was concluded that this method afforded a coefficient of variation of 1 to 2%, a relative error in the order of 2 to 3%, and was more rapid than a molecular sieve paraffin adsorption method. Pollack[77] reported that capillary columns coated with silylated or unsilylated Dexsil 400-GC resulted in small baseline drifts when they were operated at column temperatures greater than 325 or 225°C, respectively, but could be used for short periods of time at 350 to 400°C or 225 to 300°C, respectively; applications to the separation of C-10 to C-20 alkanes were shown. Raverdino and Sassetti[78] used an SE-30 capillary column and retention index data to identify postional or geometric alkane isomers derived from the "Olex" process. Kozlov and Afanasiev[79] and Talalayev and co-workers[80] described the use of capillary column GC for the analysis of high-molecular-weight alkanes in petroleum.

b. Unsaturated Hydrocarbons

An improved GC analysis of α-olefins was reported by Johansen[81] using an all-glass system with columns coated with either OS-138 or nitroterephthalic acid-modified Carbowax 20M. Orav and co-workers[82] evaluated GC separations of cis-trans isomers of C-10 to C-12 alkenes at 60 to 80°C using 0.25-mm by 100-m stainless-steel capillaries coated with 1,2,3-tris(2-cyanoethoxy)propane; retention index values and their temperature and structural increments were correlated with isomer structure and compared with data obtained from capillary columns of lower polarity. Rang and associates[83] described the separation of C-6 to C-14 *n*-alkanes on polyphenyl ether and also evaluated the separation of *n*-alkynes.[84] Sojak et al.[85] used a 200-m by 0.2-mm squalane capillary to separate all 85 theoretically possible linear alkenes up to C-14 and to study dehydrogenation products of undecane.[86] In an evaluation of silver nitrate mixtures as column coatings for the separation of olefin isomers, Zlatkis and de Andrade[87] obtained their best results using a coating mixture ratio for 1,2-bis(2-cyanoethoxy)ethane and $AgNO_3$ of 1 mℓ to 2 g; application of this coating afforded separations of alkenes and several cis-trans alkene isomer pairs using a column temperature of 60°C or temperature programing. Welsch and associates[88] evaluated capillary columns for separations of octynes, octenes, octadiynes, and octenynes and found a reliable correlation of retention index increments with structures of the unsaturated C-8 isomers. Orav and Eisen reported on separations of alkenes, alkynes, and cyclenes,[89] while Sojak and co-workers also discussed capillary column separations of alkenes[90] and olefins from C-6 to C-14.[91]

Metitzow and Fell[92] described the capillary column separation of all positional and geometric isomers of C-8 to C-11 n-olefins. In order to determine the dead volume of a capillary GC system, a relationship was developed for the accurate calculation of the air peak based on the precise measurement of the retention times of the 2,4-hexadienes.[93]

2. Aromatic Cyclic, and Polycyclic Aromatic Hydrocarbons
a. Aromatic and Cyclic Hydrocarbons

Aromatic hydrocarbons have been more extensively studied by capillary GC than have cyclic hydrocarbons. To illustrate, Chmela and co-workers[94] used poly-m-phenylether as a stationary phase in both capillary and packed columns for the analysis of aromatic hydrocarbons. Anderson and associates[95] described a novel dual glass capillary column system which was evaluated with alkylaromatic mixtures; the two columns, located in separate ovens, were connected by a heated transfer line, allowing the use of a flow circuit for changing direction of carrier gas flow, use of the transfer line as an intermediate trap, and the use of different column polarities for optimizing separations. The design of the system and the selection of operating parameters are discussed. Retention indices of a series of alkylbenzenes were determined on several stationary phases by Engewald and Wennrich.[96] Sojak et al. reported several studies on the separation of alkylbenzenes using capillary columns (e.g., Reference 97). Bartle[98] evaluated techniques for obtaining a desired film thickness in capillary column coatings and studied the effect of film thickness on the capacity ratio for benzene. Individual aromatic hydrocarbons were determined in 120 to 200°C boiling range fractions of gasoline by Diskina and co-workers[99] using a 50-m tricresylphosphate capillary. A retention index scale based on alkylbenzenes was developed by Louis[100] for the identification of aromatic hydrocarbons separated on a 50-m by 0.25-mm stainless-steel capillary coated with 1,2,3-tris(2-cyanoethoxy)propane. Krupcik and associates[101] used correlations between solute structure and retention data on two different capillary columns (polyethylene glycol and squalane) to identify C-11 alkylbenzene isomers from their retention data. The method was based on the assumption of a linear relationship between solute free energies and the differences in retention indices between the two stationary phases for C-6 to C-10 alkylbenzenes, with an experimental verification of this relationship. An analytical application was made of the use of two 0.01-in. I.D. capillary columns connected in series (a 300-ft column of 1,2,3-tris(2-cyanoethoxy)propane, followed by a 150-ft DC-550 column) by Stuckey[102] for identifying 23 aromatic hydrocarbons in the 375 to 435°F fraction of crude oil. The saturated hydrocarbons present were separated from the aromatics on the first column, with the second column separating 38 aromatic hydrocarbons eluting after tridecane. Shlyakhov and co-workers[103] used capillary GC for the analysis of bicyclic aromatic hydrocarbons in petroleum.

Rang and associates, as a follow-up to capillary GC studies on n-alkenes and n-alkynes, reported a study of the separation behavior of monosubstituted cycloalkanes.[104] A 1000-ft by 0.01-in. I.D. squalane capillary column was used by Lindeman[105] for the analysis of naphthas for C-9 and C-10-alkylcyclohexenes. All of the nongeminally substituted C-9 isomers and all but six of the nongeminally substituted C-10 isomers were prepared by reduction of the corresponding alkylbenzenes, with MS used to confirm the identity of the products. Retention indices and relative retention data were calculated for all the isomers; these data and data available for the geminally substituted cyclohexanes were used to identify 20 components from the isomerization of n-propylcyclohexane over aluminum chloride. Vanek and co-workers[106] examined the retention of saturated tricyclic isomers on capillary columns of Carbowax 20M and SE-30, successfully correlating retention indices with homomorphic factors, retention index differences between the two stationary phases, and structure of the tricyclics. These correlations were useful for differentiating between two or more of the stereoisomers in mixtures, for predicting retention data, and for determining the stereo-configuration of unknown stereoisomers. Gallegos[107] used a 150-ft by 0.02-in. I.D. OV-17 capillary column in a GC-MS system for the identification of 36 components in the saturated fraction of Green River shale, including two tetracyclic and 11 tricyclic terpanes. Kozlov and Afanasiev,[108] and Soulages and Brieva[109] reported studies of naphthenes using glass capillary columns.

b. Polycyclic Aromatic Hydrocarbons

Blomberg and associates[110] reported on the preparation of glass capillary columns containing bonded phenylsilicone for the separation of polycyclic aromatic hydrocarbons (PAH). Kodama[111] applied the use of capillary columns to the analysis of alkylnaphthalene mixtures used as encapsulating solvents for pressure-sensitive copying paper, while Tesarik and co-workers[112] evaluated capillary column conditions for the separation of naphthalene and biphenyl homologs. Grimmer and associates issued a series of reports[113-115] on the sampling and analysis of PAH in automotive exhausts based on preliminary separation and enrichment of the PAH on Sephadex® LM-20, followed by analysis on a 50-m by 0.5-mm stainless-steel OV-101 or OV-17 capillary column. A total of 59 PAH were identified by retention data. Application to the analysis of PAH in industrial flue gases using this procedure is also suggested. Grimmer reported on the use of a cyclohexane-methanol solvent partition technique for the concentration of PAH from air particulate matter, automobile exhaust, and cigarette smoke, followed by analysis of the PAH on an OV-101 capillary column.[116] Mostecky et al.[117] used three different capillary columns for the determination of C-12 alkylnaphthalenes and methylbiphenyls in aromatic fractions from various sources. A total of 18 dimethylnaphthalenes, ethylnaphthalenes, and methylbiphenyls were identified in 250 to 280°C pyrolysis oil derived from naphtha cracking and in a 130 to 160°C coal tar fraction on the basis of retention data on Apiezon L, *m*-bis(*m*-phenyloxyphenoxy)benzene, and polyethyleneglycol adipate capillary columns at 180, 200, and 190°C, respectively. The 2,6- and 2,7-dimethylnaphthalenes, unseparable by capillary GC, were determined by UV spectrophotometry.

Bjorseth[118] and Wauters et al.[119] reported on the development of routine methods for the analysis of PAH. Lee and co-workers[120] proposed a new retention index scale for PAH based on four PAH internal standards and illustrated its application to 10 PAH using an SE-52 glass capillary column. Lee and associates also reported earlier studies of PAH separations using capillary columns.[121,122] Various other workers have evaluated the use of capillary columns for the separation of PAH components, including Onuska et al.,[123] Doran and McTaggart[124] and Hutzinger and co-workers.[125]

Bruner and associates[126] evaluated and compared various GC columns relative to open-tubular columns and reported the use of the latter for PAH separations. Sojak and Barnoky described the use of open-tubular columns for the separation and analysis of naphthalenes and biphenyls.[127] Borwitzky and Schomburg[128] reported on the retention behavior of PAH on glass capillary, while Stenberg and co-workers[129] elaborated on the GC separation of high-molecular PAH in samples from different sources using temperature-stable glass capillary columns. The GC separation and determination of PAH from industrial processes was reported by Lao and Thomas.[130] Bjorseth and Eklund[131] described a unique analytical system for the analysis of PAH on glass capillary columns, based on simultaneous detection with flame-ionization and electron capture detectors.

Other studies on PAH are included in the section of Chapter 6 concerning PAH as air pollutants.

III. SPECIAL GC SYSTEMS OR COLUMN ARRANGEMENTS

A wide variety of system arrangements have been used and, in fact, been necessitated, for the resolution of multicomponent hydrocarbon mixtures. Due to the complexity of such mixtures or the need to resolve overlapping peaks, extensive use has been made of multiple column systems.

In the simplest case, this has involved the determination of retention data on two or more stationary phases having different selectivities. If two hydrocarbons have significantly different vapor pressures at the typical column operating temperatures, they can generally be separated from each other on a nonpolar silicone or hydrocarbon stationary phase. Two hydrocarbons having similar vapor pressures may be difficult-to-impossible to separate on such phases. However, thermodynamically, if two hydrocarbons have similar vapor pressures but different chemical functionalities (e.g., saturated vs. unsaturated paraffins vs. aromatics vs. polycyclics), various stationary phases can be used which take advantage of significantly different Gibbs free energies of mixing of the two hydrocarbons in a "selective" stationary phase to effect their separation. Such phases can take advantage of π-bonding interactions (for aromatic or unsaturated hydrocarbons), geometric molecular length-to-breadth ratio differences between the hydrocarbons (liquid crystal stationary phases to separate isomers of polycyclic aromatic hydrocarbons), differential hydrocarbon-sorbent interactions (gas-solid chromatography), or the like.

Another specialized system technique that has grown in popularity in recent years is the use of "peak-cutting", in which peak components in a section of the chromatogram are systematically transferred from one column to a second column for independent resolution and measurement of one or more peaks in the chromatogram window cut taken from the first column.

Other system designs used to improve separations and measurements have been the use of precolumns, subtraction chromatography, or the use of automated or cyclic chromatographs.

A sampling of the literature illustrating some of the more prominent of these activities over the 1970 to 1980 time-frame is provided in this section.

A. Use of Multiple Column Systems

Grizzle and Coleman[132] described a GC system using two columns (OV-1 and 1,2,3-tris(2-cyanoethoxy)propane (TCEP)) in series for the analysis of benzene and toluene in crude oil and other fossil fuels, with backflushing used to remove higher molecular weight hydrocarbons; a precision and accuracy of 2 and 4% relative, respectively, were claimed. Lychagin and Sidorova[133] described a three-column chromatograph for the analysis of petroleum and hydrocarbon gas mixtures. Saha and co-workers[134] reported on a parallel column system (oxydipropionitrile and phenyl isocyanate, respectively, both on Porasil C) positioned in a dual-column chromatograph for a generalized analytical method for 16 C-1 to C-5 light alkanes and alkenes. A dual concentric column system was developed by Rendl and associates[135] in which the inner column and the annular space around it were packed with two different materials. This novel configuration was used for the separation of hydrocarbon and permanent gas mixtures. Willis[136] reported that a porous polymer and a molecular sieve column could be used in parallel, series, or series-by-pass modes for the separation of permanent gases and light hydrocarbons. Bertsch[137] elaborated on two-dimensional chromatographic approaches for dealing with samples of high complexity, as well as for the analysis of trace and ultra-trace components in mixtures.

Lukac[138] described a parallel column system using a capillary column coated with Apiezon L and a packed column of phenylisocyanate on Porasil C, followed by a column of squalane on Gas-Chrom P, for a single-injection analysis of C-1 to C-10 hydrocarbons produced from gas-phase radiolysis of *n*-pentane. Ullrich and Dulson[139] used a chromatograph containing two injection ports for two-channel analysis of gasoline and air samples, in which the sample is introduced into two columns of different polarity and simultaneous detection is made with a dual flame-ionization detector. They used this system approach to determine retention data and peak shifts corresponding to retention index increments, specific for hydrocarbon substructural groups, for identification purposes. Chizhkov and Yushina[140] employed an unheated eight-position switching valve for three different operational modes: using a single column with normal and reversed carrier gas flow for the separation of benzene, cyclohexane, and m-xylene from decane, undecane, and dodecane; using two columns operated in parallel or in series; or using two to four columns operated with a circulating carrier gas flow. The latter was used for a ten-cycle separation of benzene-deuterobenzene mixtures using two columns. A parallel column system was used by Shykles[141] for the separation of both permanent gases and light hydrocarbons. Luke and McTaggart[142] described a system employing a $10\times$ molecular sieve precolumn followed by a $13\times$ molecular sieve analytical column for the separation of cyclic and aliphatic hydrocarbons in the presence of aromatic hydrocarbons. The aromatics were retained by the precolumn and recovered by backflushing, while the hydrocarbons of interest were separated on the analytical column.

Carson and associates[143] described the analysis of 28 C-1 to C-5 hydrocarbons in butadiene process streams using two columns (20% dibutylmaleate and 10% bis(2-methoxyethoxy)ethyl ether on Chromosorb P) connected in series. Stavinoha and Newman[144] described a method for the isolation and measurement of gasoline aromatic hydrocarbons using a multioven, multicolumn system, which was subsequently improved by Stavinoha[145] by including an internal standard in the procedure. Petkova and co-workers[146] used a two-column system for the direct determination of cyclopentadiene and dicyclopentadiene in pyrolysis naphtha to obtain improved results over a previous method that relied on the complete conversion of cyclopentadiene to the dimer. A two-column system (12-ft by 0.125-in. column of 15% 7-ring polyphenylether on 60 to 70 mesh Anakrom ABS, and a 5-ft by 0.125-in. column of 1.5% SE-30 on F-20 alumina) was employed by Willis[147] for the separation of C-1 to C-7 hydrocarbons in the off-gas from hydrocarbon pyrolysis. Peterson and co-workers[148] used a two-column system (90-mm by 4.8-mm column of molecular sieve 5Å, and a 1-m by 3-mm I.D. column of molecular sieve $13\times$) for the selective separation of naphthenic hydrocarbons and iso and n-alkanes in naphthas. Deans and Scott[149] studied the characteristics of using two different columns in series for hydrocarbon separations, reporting a number of methods for adjusting such characteristics. These workers described a simple apparatus that permitted the ratio of the gas velocities to be changed by adjusting the pressure at the junction between the two columns to alter the retention of the eluted components. They applied this system to the analysis of aromatic hydrocarbons in naphtha using a two-column system, and to the separation of an 11-component C-4 hydrocarbon mixture.

Three columns (column A: 20% bis(2-methyloxyethyl) adipate on Chromsorb P; column B: a 1-m column of Porapak Q, operated at 20°C; column C: a 5-m column of Porapak Q, operated at −78°C) were used by Churacek and co-workers[150] in various combinations for the separation of 21 components of a mixture containing rare and permanent gases and low-molecular-weight saturated and unsaturated hydrocarbons; the combination of columns A and C separated the permanent gases and C-1 to C-5 hydrocarbons, while the combination of columns B and C separated the rare gases and the permanent gases. Deans and associates[151] reported the use of a three-column system (a primary column of Porapak T, connected in series to two columns in series (Porapak S and $13\times$ molecular sieve)) for the analysis of light gases using isothermal temperatures between 35 to 100°C, describing the design of the apparatus and its operation (including column switching and backflushing). Malan and Brink[152] also described a three-column system for the analysis of inorganic gases and light hydrocarbons using columns of DC-200, dibutyl maleate, and Porapak Q. This system was well-suited for computer interfacing since no negative peaks or base-line disturbances were noted in their work. Marchio[153] employed a dual-channel system for the analysis of mixtures of permanent gases and C-1 to C-2 hydrocarbons. This system was comprised of two complete GC systems at 50°C, with separate gas-sampling valves and thermal conductivity detectors:

one channel, having a single column, measured hydrogen with nitrogen as the carrier gas; the second channel, having two columns linked in series, measured the C-2 hydrocarbons and higher components, using helium as the carrier gas. This dual-channel approach could be applied to the analysis of sample mixtures in 16 to 18 min. A rapid two-stage GC method was reported by Vigalok and associates[154] for the determination of aromatics in gasolines: the first stage was a column of 1,2,3,4-tetrakis(cyanoethoxy) neopentane for the retention of aliphatic and aromatic hydrocarbons at 80 to 140°C; the second stage was a column of 10% *p,p'*-azoxyphenetole at 140°C. Mator[155] patented a GC system for the automatic analysis of hydrocarbons in which two columns were arranged to allow for backflushing and component isolation within the system. Using this system, aromatic hydrocarbons were isolated by the first column, while *n*-alkanes were separated from naphthenic hydrocarbons on the second column. Total paraffinic content was determined by multiplying the measured *n*-alkane content by an exponential factor, while the naphthenic hydrocarbon content was determined by difference.

A number of workers have pioneered the development and application of two-dimensional GC analysis, including Schomburg et al.,[156] Kaiser,[157] and Rijks and co-workers.[158] Such studies have illustrated the feasibility and power of two-dimensional GC separations. Rijks and group[158] illustrated that column-to-column transfers can be achieved without the need for interim trapping, while Schomburg and associates[156] reported that the trapping technique has the advantage of decreasing peak widths on the second column, enriching trace components, allowing for improved precision in retention index measurements, and permitting the determination of retention data on two different stationary phases for component identification.

Shykles[159] described the precise determination of hydrogen and C-1 to C-5 hydrocarbons by the simultaneous use of columns in parallel. Robinson[160] reported a method for rapid hydrocarbon-type analyses for gasoline samples using a dual-column system, and Bertsch[161] reviewed two-dimensional techniques for the separation of complex samples from an experimental standpoint.

B. Other Special GC System Arrangements

Clark employed an automated gas chromatograph for the determination of aliphatic, naphthenic, and aromatic hydrocarbons in gasoline fractions.[162] Berezkin[163] reviewed the technology of substractive gas chromatography. Chizhkov and co-workers[164] described a procedure on the simultaneous analysis and preparative separation of binary hydrocarbon mixtures using a circulation gas chromatograph having two columns arranged in series, reporting superior separation efficiency to that obtained using thermal distillation columns of the same dimensions employed in the chromatograph. Reid[165] described a chromatograph which could be used in a recycling or circulating mode, in which the column "length" could be selectively altered by recycling to obtain results comparable with peak resolutions that could be obtained using longer columns, extending the effective column length up to three times its actual length. Application was reported to the separation of *m*- and *p*-xylene. This work also reported that peak heart-cutting was feasible on a quantitative basis. Houghton[166] described a sample trapping and re-injection system and procedure using a thin-walled stainless-steel trap (packed with glass wool or a suitable chromatographic column packing) attached to the GC column outlet. Following sample component trapping, the trap was cooled with a liquid nitrogen jacket, coupled to a syringe needle and connected to carrier gas. The syringe needle of the trap assembly was then inserted into the septum of the GC inlet port and the trap was heated through its jacket to 220°C to obtain sharp, near-symmetrical peaks for the re-injected, trapped components. Application was shown for C-11 to C-15 hydrocarbons.

Deur-Siftar and associates[167] employed a short precolumn of 5Å molecular sieve for the retention of *n*-alkanes and *n*-olefins to simplify the analysis of hydrocarbons in petroleum fractions. An intercolumn detector technique was reported by Mieure[168] for the continuous monitoring of the effluent from a precolumn using specially controlled carrier flows. Mitooka[169] described the use of a multichannel column containing three parallel tubes, each having a flow control valve and different mixed tube packings, connected to a single injection port and a single detector. This system was used for the determination of retention data for a number of organic classes at 120°C, including C-6 to C-9 aromatic hydrocarbons, naphthalene derivatives, and styrenes. Evrard and Guiochon[170] reported on a sampling device for the determination of high-boiling trace components in low-boiling solvents. The trace components were trapped on a large-bore precolumn and thermally transferred to the analytical column by temperature programing the precolumn after elution of the solvent. Excellent results were reported using 100-μℓ samples containing 50 to 100 ng of *n*-C-18 to C-28 hydrocarbons.

Dorokhov and associates[171] described a two-stage GC system consisting of two chromatographs coupled in series through a six-way valve for the analysis of unstable catalytic reformates: the first GC was used for the analysis of aromatic hydrocarbons in the reformates; the second GC was used for the analysis of the C-2 to C-6 paraffinic reformate fraction. Use of this system shortened the total analysis time to 1 hr. A totally automated GC system was reported by Boer and Van Arkel[172,173] for the analysis of straight-run and process hydrocarbon streams in the boiling range of 0 to 200°C, with the system providing results in percent paraffins, naphthenes, and aromatics per carbon number during 2 hr of unattended operation. The separation of paraffinic and naphthenic elution regions of equal carbon number was quantitative through C-9, semiquantitative through C-10, and separations which were still useful through·C-11. Using the system for the analysis of naphtha samples, the system could perform a sequence of separations on a polar, a nonpolar, and a 13 × molecular sieve column using flow-switching and hold-up of sample fractions. For accurate analyses, all separations had to be carried out on a single sample in a closed system with a single detector. A separate module also allowed for the differentiation between five- and six-ring naphthenes and *n*- and iso-paraffins. Bagirov and co-workers[174] described the use of continuous GC with solid adsorbent packings for the separation and analysis of hydrocarbon gases. Burgett and Green[175] reported the development of a totally automated GC system for the analysis of carbon monoxide, methane, and total hydrocarbons.

IV. DETECTORS

The majority of studies on GC detectors for the characterization and analysis of hydrocarbons has been on the flame-ionization detector (FID) and the growing application of gas chromatographs coupled to mass spectrometers (see the following section) with few investigations on the thermal-conductivity detector (TCD). In addition, several new detector systems have been developed and characterized, and increased activities have occurred in the coupling of GC studies with other instrumental systems (e.g., spectrometric and liquid chromatographic systems). The two subsections following cover activities on detection systems for general hydrocarbon analyses and detection systems used for the analysis of specific hydrocarbon classes.

A. General Hydrocarbon Analyses
1. Conventional and Other GC Detectors
a. Thermal Conductivity (TCD)
Schirrmeister[176] described a rapid-response, sensitive TCD for use with capillary columns, in which the detection response was similar to that of an FID and the sensitivity similar to that of a conventional TCD used with packed columns. By pressure-limiting means at the TCD inlet and outlet (using a constricted capillary between the GC column and the TCD,

and a needle valve between the TCD outlet and a vacuum pump), carrier gas entering the TCD was expanded by a factor of 50. Improved peak shapes for components of a hydrocarbon mixture were obtained using this TCD with a glass capillary column. Rosie and Barry[177] published a comprehensive review (82 references) of the basic principles, assembly, performance, and applications of TCDs. Pecsar and co-workers[178] described the performance of a reduced-volume TCD having an order of magnitude higher sensitivity than a conventional volume TCD that was compatible with capillary columns and 0.125-in I.D. packed columns. Applications were shown to the analysis of samples of 0.1% ethane and propane in helium on a Porapak Q column at 200°C, and to the analysis of complex samples of petroleum-derived hydrocarbons on a capillary column. Use of the detector with SE-30 packed columns revealed a minimum detectable level of 1.4, 2.7, 3.5, and 4.1 ppm for *n*-octane, *n*-decane, *n*-dodecane, and *n*-tetradecane, respectively. Careful temperature and carrier gas flow control were required. Chowdhury and Karasek[179] described the use of a single thermistor bead TCD which had a lower detection limit of 60 ppb for hydrocarbons.

b. Flame-Ionization (FID)

Olacsi and associates[180] standardized the analysis of natural gas components using TCD and FID for the determination of hydrocarbons up to C-13. Proksch et al.[181] evaluated the FID response of hydrocarbons and perfluorocarbons and found similar FID responses for high carbon number perfluorocarbons and the corresponding hydrocarbons, with response ratios close to unity, independent of operating conditions, provided that higher than usual hydrogen flow rates to the FID were used. Driscoll and co-workers[182] reported on the combined use of a photoionization detector and an FID for the characterization of hydrocarbon structures in complex mixtures. Williams and Eaton[183] described the purification of air for its use as an organic-free carrier gas with an FID in the NRL Total Hydrocarbon Analyzer, in which methane and halogen-containing hydrocarbons were removed by catalytic oxidation in a stainless-steel air purifier containing palladium-on-alumina catalyst at 325°C. The characteristics of the FID were discussed by Blades,[184] who proposed mechanisms for the constancy of response per carbon atom, showing application to FID response data for C-1 to C-4 hydrocarbons.

Wojdala and Guichard[185] calculated FID response factors from solute physical properties (e.g., refractive index, density, molecular weight) and reported that C/H ratios could be determined with a relative error of 5% for hydrocarbons. Reviews on the FID have included an assessment of the characteristics and operational variables,[186] the mechanism of flame ionization, the types of flames commonly used, and factors influencing the production of ions in the flame.[187] McWilliam[188] concluded that the response linearity of the FID at high solute concentrations was dependent on hydrogen flow rate to the flame jet and that notable nonlinearity may occur at high hydrogen/nitrogen carrier gas flow ratios. On the other hand, Rossiter[189] observed that the FID response in the overloaded nonlinear solute concentration region was linearly related to the logarithm of sample size. Folmer and Haase[190] and Grant and Clarke[191] studied the quantitative effects of instrument control on the analytical precision of the FID, finding that estimates can be made on the precision of instrument control needed to attain set precision levels in analysis.

c. Other GC Detectors

Schwarz and co-workers[192] reported an oscillating slit mechanism for the determination of hydrogen isotope ratios using a microwave-induced plasma detector. Undeuterated and deuterated hydrocarbons eluted from a GC column in a helium carrier gas stream were exposed to a low-pressure, microwave discharge, resulting in the fragmentation of the hydrocarbons and the generation of intense atomic emission signals sent to a monochromator having an oscillating exit slit, such that the hydrogen and deuterium signals were alternately passed to the photomultiplier (the output of which was resolved by two lock-in amplifiers). It was observed that the ratio of the two signals and the hydrogen/deuterium ratio in the hydrocarbon mixtures were rectilinear over more than an order of magnitude. Application was shown to the technique of isotope dilution analysis in which predeuterated naphthalene was added to a water sample containing traces of naphthalene and benzene, and the hydrocarbons were extracted with an organic solvent for GC analysis with this detector. Thiede and Ehrlich[193] described a commercial digital GC that could accommodate FID, flamephotometric, and electron-capture detectors. The electrometer offered a dynamic range of six orders of magnitude; calibration curves (with a square-root function) were linear over three orders of magnitude. Application was shown to the determination of 17 hydrocarbons in a petroleum distillate with a relative standard deviation not exceeding 2.6%. Carson and associates[194] evaluated hydrocarbon molar response factors for GC detection and reported that the observed factors were in good agreement with published data. Lasa et al.[195] described a ^{63}Ni argon ionization detector that was sensitive to 1 pg/sec for benzene and pentene. Janghorbani and Freund[196] evaluated a quartz crystal coated with squalane (or other stationary phases) as a partition detector and showed its application as a GC detector for alkane, benzene, and cyclohexane mixtures. The detector signals could be coupled to an on-line data acquisition system. When this detector was coated with Carbowax stationary phases, it could be used for the detection of sulfur dioxide in nitrogen and sulfur compounds in pulp mill stack gases.

Rezl and co-workers[197] coupled a conventional GC containing a katharometer detector directly to a previously developed carbon-hydrogen analyzer for elemental analysis of carbon and hydrogen contents of hydrocarbons on a submicro scale, with an absolute precision of less than 0.3% Umstead and associates[198] evaluated the response characteristics of a catalytic ionization detector for a variety of organic molecular structures, including hydrocarbons, and found that the degree of ionization on a platinum catalyst in an oxygen atmosphere at 400 to 900°C was a strong function of the solute molecular structure.

2. Coupling With Other Instrumental Systems

The combination of GC with other instrumental techniques (e.g., IR and Raman spectrophotometry, mass spectrometry (see section GC-MS, following), liquid chromatography) has become a common occurrence. The current literature base has grown in this area and has become formidable, indeed. To illustrate the magnitude of activities in this area, almost 600 papers were published on applications of combined instrumentation during the 2-year period of 1970 to 1971. A representation of studies using instrumental techniques coupled to the GC analysis of hydrocarbons during the 11-year period of 1970 to 1980 is described below.

Yokoyama and co-workers[199] intercompared analytical results from ^{13}C-NMR, proton-NMR, and GC for fractions of hydrocarbon oil derived from coal hydrogenation and found that the ^{13}C results generally confirmed the results from proton-NMR and GC analysis. Griffiths[200] reported techniques for enhancing the sensitivity of the interface between a GC and a rapid-scanning Fourier-transform infrared (IR) spectrophotometer using a single-beam system with a triglycine sulfate detector for measuring the interferogram. Submicrogram amounts of solutes eluting from the GC could be identified from the FT-IR spectra. A double-beam system with a cooled mercury-cadmium telluride detector was also designed to further enhance detection sensitivity without limiting the sensitivity by digitization noise. Gallaher and Grasselli[201] reported on procedures and equipment whereby IR and NMR spectra could be obtained for the same solute eluting from the GC. De Haseth and Isenhour described a procedure for direct reconstruction of gas chromatograms from single-scan GC-IR interferograms.[202] Louw and Richards[203] described a simple directly combined GC-IR system for the identification of low-molecular weight hydrocarbons.

Crowley and co-workers[204] and DiSanzo[205] characterized hydrocarbons in shale oil using GC and liquid chromatography (LC), covering the separation and analysis of alkanes, alkenes, and aromatics. GC and silica gel LC were used by Sapozhnikova and associates[206] for the examination of cracked gasolines and naphthas. Arich and co-workers[207] used LC combined with GC to resolve a gas oil fraction for the identification of over 400 components at concentration levels above 0.02%. Quantitative determinations were made of about 35 components having concentrations exceeding 1% in the gas oil fraction.

Torradas et al.[208] described a method based on distillation and LC to obtain fractions from a crude oil, which they subsequently analyzed by GC, MS, IR, and NMR to characterize the original crude oil. Greinke and Lewis[209] employed a combination of GC, UV, and MS to identify volatile hydrocarbons derived from petroleum pitches in the environment. Fujita[210] reviewed instrumental techniques for the analysis of petroleum and petroleum products using GC, IR, NMR, and MS. Machida[211] reviewed the laboratory use of GC and UV and visible spectrophotometry in petrochemical plants.

B. Use of Detectors by Hydrocarbon Class

Studies on the use of detectors (excluding MS, see Section V) for specific classes of hydrocarbons are given below.

1. Alkanes and Unsaturated Hydrocarbons
a. Alkanes

Agrawal and Joshi[212] examined distillate waxes from Ankleshwar crude oil tank bottoms by physical properties correlation data, isothermal GC, proton NMR, and ^{13}C FT-NMR, reporting that the results of all their analyses were consistent with the finding that the composition of the bottoms was predominantly n-alkanes. Carson and associates[213] determined detector molar response values on a thermal conductivity detector for C-5 to C-8 alkanes relative to benzene in two hydrocarbon blends, each containing 68 components. Their results indicated a fixed relative response of 16 units between alkanes in a homologous series within the aliphatic and cyclic hydrocarbons. It was anticipated that a similar response difference would be found for other homologous series. Folmer described a simple method for separating n- and iso-alkanes by using a 1:1 splitter at the end of an analysis column, with half the effluent flow directed to one side of a dual-FID through a 5Å molecular sieve column and the other half directed to the other side of the FID through a dummy absorber (Celite) column; the differential FID response signal corresponded to the detection of the n-alkanes.[214] Stuckey[215] applied Folmer's method to analyses of n-alkanes in kerosines and crude oil heavy distillates.

Carruth and Kobayashi[216] calibrated an FID detector by the accurate metering of gaseous or liquid alkanes into the helium carrier gas stream and into the detector using a micropump maintained at constant temperature. This study found that the FID signal vs. the helium carrier gas flow rate was rectilinear with the alkane hydrocarbon metering rate over a range of 10^{-11} to 10^{-8} mol/sec. Gawlowski and associates[217] evaluated the argon ionization detector for monitoring trace levels of methane and ethane. Uden and co-workers[218] described a dual gas density balance detector for the on-the-fly determination of molecular weights of long-chain alkanes, using an algorithm for the calculation of molecular weights having an imprecision of less than 0.4 amu. In a study of the effect of carrier gases on detection response using an FID, it was found that the use of ammonia as the carrier gas resulted in a twofold improvement in separation speed for n-alkanes, enhancement of the FID sensitivity, and a reduction in the tailing of amine peaks.[219]

b. Unsaturated Hydrocarbons

Poznyak and associates[220] described a flowthrough UV detector for the determination of olefins by GC. The method was based on the ozonolysis of the eluting olefins in a reaction cell located prior to the UV detector, with measurement of the unreacted ozone at 254 nm. The reliability of this detection system depended on maximum conversion of the olefins to ozonides. Conditions affording good results involved the use of a sample size of 3 $\mu\ell$, a reaction cell length of 2.5 cm, and flow rate of an ozone-oxygen mixture 25 μM in ozone of 1.5 mℓ/min, and a GC carrier gas flow rate of 1.5 mℓ/min. Guglya and Korobeinik[221] reported on the GC determination of low concentrations of unsaturated hydrocarbons in natural gas using a pyroelectric detector. Unsaturated hydrocarbons in the range of 10^{-3} to 10^{-4} vol% could be determined in natural gas from several U.S.S.R. gas fields containing 1 to 7.5% C-2 to C-5 alkanes. Optimum operating conditions and minimum detection limits were obtained for ethylene, acetylene, propylene, butylene, hexene, and cyclohexene. Nakajima and Sakai[222] developed a catalytic reaction detector for the GC analysis of unsaturated hydrocarbons, based on the continuous monitoring of the temperature in and above a catalytic bed while the GC column effluent is passing through it. When an eluted unsaturated hydrocarbon reacts with the catalyst, an exothermic or endothermic response is detected. Detection sensitivities for unsaturated hydrocarbons were comparable to those obtained with a thermal conductivity detector. The linearity of detector response with alkene concentration was shown for 1-pentene.

2. Aromatic and Polycyclic Aromatic Hydrocarbons
a. Aromatic Hydrocarbons

Bjorseth and Eklund[223] used simultaneous FID-electron capture detection in the capillary GC analysis of aromatic hydrocarbons for identifying the hydrocarbons on the basis of FID/ECD ratios. Selucky and co-workers[224] used a combination of GC and high-performance liquid chromatography (HPLC) data on aromatic mixtures to determine that mono- and di-aromatics (and often, di- and tri-aromatics) could be differentiated from alkyl- and cycloalkylaromatics. The HPLC analysis (silica gel column; heptane mobile phase; UV and differential refractometer detectors) afforded the determination and relative distribution of hydrocarbon classes in crudes, and the boiling point distribution of the saturates in 10 to 12 min. The combination of GC and HPLC was used to fingerprint samples of automotive and aviation gasolines, lube oils, and other refinery stream products. D'Orazio[225] employed a Raman multichannel spectrometer as a GC detector for the analysis of aromatic hydrocarbons (0.25-mm I.D. by 15-m OV-101 column). The eluted hydrocarbons were passed through the focus of an argon-ion laser beam in the spectrometer, which was directed to the entrance slit of a single-grating monochromator, and, via an objective, to a 500-channel array detector. Conditions were reported for the determination of benzene and for the detection limit of toluene in acetone. The resolution of fractions of cyclohexane and benzene was measured, and *ortho-* and *meta*-xylene were readily identified in a mixture of toluene, benzene, and acetone. Houpt and Baalhuis[226] employed a second-derivative UV spectrophotometer as a GC detector for the determination of aromatic hydrocarbons. The radiating wavelength was adjusted to that of a UV absorption peak of the aromatic to be measured, for obtaining a second derivative signal which was selective for UV absorption at the adjusted wavelength. It was demonstrated that mixtures of benzene (252.9 nm) and toluene (266.9 nm) could be selectively resolved in this manner. Albert and Kaplan[227] connected a ^{14}C radioactivity detector in parallel with an FID to analyze liquid products derived from catalytic hydrocarbon cracking reactions using feedstocks containing ^{14}C-labeled 2-methylnaphthalene. Most of the radioactivity was contained in the resultant aromatics, and cracking severity at 35.2% conversion was sufficient to prevent demethylation. The cracking reactions resulted in both alkylation and/or trans-methylation, and the production of greater quantities of ^{14}C-methyl-labeled than ^{14}C-ring-labeled products. Lowering the conversion to 11.5% resulted in less gasoline components and traces of hydro-cracked products; however, significant alkylation was noted. These results were made possible through the use of this dual radioactivity-FID detection system.

b. Polycyclic Aromatic Hydrocarbons

Brown and Searl[228] reported a review on the use of GC and UV spectroscopy for the analysis of polycyclic aromatic hydrocarbons (PAH). Cooney and Winefordner[229] evaluated instrumental effects on the limits of detection with various gas-phase fluorescence detectors, using various excitation sources with a quartz flow-through detector cell. These workers reported limits of detection for anthracene, pyrene, chrysene, and phenanthrene. They concluded that optimum systems either should have an intense UV source or a conventional source with an efficient UV excitation monochromator. Greinke and Lewis[230] employed GC followed by UV spectrometry for the identification of four methyl-substituted and five nonsubstituted PAH. Robinson and Goodbread[231] described a selective UV-fluorescence detector for the GC analysis of PAH compounds. Beroza and Bowman[232] patented an apparatus for coupling a GC to a spectrophotofluorometer and other devices by means of a flowing liquid surface, in which the components eluting from the GC column were absorbed by a flowing solvent stream which was subsequently monitored by the spectrophotofluorometer. Chromatograms were reported for the detection of five PAH, some of the PAH having a detectable limit of less than 1 ng at a noise level of 0.01 relative intensity. Burchfield and associates[223] reported the development of a gas-phase fluorescence GC detector for the analysis of PAH pairs (e.g., perylene and benzo[a]pyrene) which could not be separated by conventional GC stationary phases. Gupta and Kumar[234] reviewed the literature on the identification and measurement of PAH in petroleum and related products. Their review covered various techniques used for determining carcinogenic petroleum PAH using chromatographic and solvent extraction methods for sample isolation and concentration, chromatographic methods for separating the PAH components, and GC, UV, and spectrophotofluorometry for the identification and quantification of individual PAH. Searl et al.[235] reported an analytical method for determination of PAH components in coke oven effluents based on GC and UV. This method was suitable for routine plant laboratory use for the quantification of fluoranthene, pyrene, benz[a]anthracene, chrysene, benzo[a]pyrene, and benzo[e]pyrene. The method was validated by mass spectrometric and fluorescence analysis.

Spectrophotofluorometry has been cited as being a far superior selective and sensitive GC detector than a UV absorption detector for the analysis of PAH components due to their fluorescent properties. The fluorescence GC detector also has been used Katlafsky and Dietrich,[236] and by Mulik et al.[237] for the analysis of PAH components. Burchfield and associates[238] employed molecular emission (especially UV fluorescence) for the sensitive and selective GC analysis of PAH. Grinberg and co-workers[239] employed a UV detector for the GC analysis of individual naphthalene hydrocarbons in characterizing the composition of jet fuels. Panalaks[240] identified and determined PAH components in smoked and charcoal-broiled food products using a combination of GC and HPLC.

V. GC-MS

Without a doubt, the most significant development in the area of GC-MS since 1970 has been the development and incorporation of data systems for the analysis of hydrocarbons, as well as for the analysis on complex organic mixtures, in general. The coupling of the computer to GC-MS systems has afforded techniques in data interpretation and control that are intractable without the use of the computer. While improvements have been made in GC-MS interfacing and GC and MS electronics that have improved sensitivity, selectivity, and resolution, the addition of data systems have made possible powerful techniques such as reconstructed gas chromatograms, multiple selective ion monitoring, isotope dilution, chromatogram and mass spectrogram substractive methods, mass spectral library searches, and other data manipulation techniques that offered a simplicity and quantifiability not previously attainable by GC-MS analysis.

This section discusses hydrocarbon analyses using GC-MS systems, with and without coupled data systems, studies involving GC-MS interfaces, and the application of GC-MS to the analysis of hydrocarbons, in general, and for the analysis of specific classes of hydrocarbons.

A. General Hydrocarbon Analyses

A computerized fast-scanning GC-MS system was developed by Hedfjaell and Ryhage[241] which used a pre-processor to group an appropriate number of sample signals from an analog-to-digital convertor to obtain a maximal digital filtering effect from different MS scan speeds. Employing a capillary GC column with this system with a 0.7-sec scan speed for the m/e range of 5 to 500 and a repetitive frequency of 1.4 sec afforded good quality spectra for GC peaks which eluted as rapidly as a few seconds. Heller[242] described a conversational MS system and its application for structural determinations, covering its search options, software programs for generating and searching mass spectral files, and a mass spectral file containing 1800 reference mass spectra from API Research Project 44. Justice and Isenhour[243] employed a complex-valued nonlinear discriminant function with low-resolution MS for the analysis of hydrocarbons. The function was constructed on a single pass through a training set of 387 hydrocarbons from API Research Project 44 and required the storage of only one spectrum at a time. Application of the function for distinguishing between C_nH_{2n} and non-C_nH_{2n} hydrocarbons was illustrated. An early review on the use of computers in GC-MS appeared by Karasek,[244] who also published reviews of GC-MS-computer systems featuring interactive consoles[245] and dual-disc based systems.[246]

A number of studies on interfacing GC and MS systems appeared during 1970 to 1980. Examples of these included a device described by Thome and Young[247] for the direct coupling of glass capillary columns to a MS inlet. This interface design permitted high efficiency transfer of eluted sample components to the MS and provided a leak-free connection with low dead volumes for obtaining an MS detection sensitivity comparable to an FID. The design also allowed for control of carrier flow rate to the MS ion source and a mechanism for protecting the ion source in the event of column breakage. Henderson and Steel[248] coupled a dual-column GC directly to a double-focusing high-resolution MS for the separation and analysis of low-molecular-weight gas mixtures of CH_4, C_2H_2, C_2H_4, C_2H_6, C_3H_8, and C_4H_{10}. Henneberg et al.[249,250] reported that the use of an open split GC-MS interface allowed for the dilution of high peak concentrations, enrichment by trapping, and on-line hydrogenation. This interface was well-suited for glass capillary columns and offered the advantages of high yield, the use of flexible column parameters, and suppression of large solvent peaks, and an inherent simplicity in that no vacuum-tight seals of the column to the MS or any special geometric considerations at the end of the capillary column were necessary. The basic concepts and methods of sampling and interfacing GC-MS systems and the relative merits of interfacing were reviewed by Flath.[251]

Following are reports of the use of GC-MS for general hydrocarbon analyses.

Larskaya and Khramova[252] used GC and MS, separately and in combination, to analyze a 200 to 500°C fraction of Sokolova crude oil, with the subsequent identification of almost 100 individual hydrocarbons. Bruner and associates[253] described the conversion of a conventional MS to a GC-MS system; the combined system was evaluated for GC column efficiency and MS sensitivity and resolution using a C_6 to C_9 hydrocarbon mixture. Aczel[254] addressed the role of high-resolution MS in the petroleum industry, particularly for the analysis of aromatic and polar hydrocarbon fractions in petroleum and coal products. Aleksandrov and co-workers[255] employed GC-MS for the analysis of paraffinic and naphthenic hydrocarbons having the same GC retention times. This method was based on an evaluation of the summed intensities of the MS peaks of the characteristic mass ions and was applied to the qualitative and quantitative analysis of C_6 to C_9 hydrocarbons in a 62 to 105°C fraction from petroleum. The fundamental bases and modes of operation of GC-MS was the subject of reviews by Schuy and Hunneman[256] and by Staszewski.[257] Swansiger and Dickson[258] described the direct coupling of a capillary column of a GC to an MS for the successful analysis of C_{11} to C_{14} reformed petroleum products. Kagler[259] provided a comprehensive review of the development, instrumentation, and application of GC-MS for the analysis of petroleum hydrocarbons.

Protic et al.[260] reported on mass spectrometric and GC data in hydrocarbon analysis. Termonia and co-workers[261] described the analysis of C_6 to C_{20} hydrocarbons using high-resolution GC-MS. Khadzhieva and associates[262] employed a capillary column GC-MS system using both electron-impact and chemical-ionization for the analysis of the hydrocarbon fraction of Bulgarian rose oil. Ciccioli and co-workers[263] described the use of graphitized carbon in gas-liquid-solid chromatography and GC-MS for the characterization of high-boiling hydrocarbon mixtures. Hatch and Burnaby[264] evaluated the relative rate constants for reactions of CH_5^+ and $C_2H_5^+$ with hydrocarbons using GC-MS with chemical ionization. Kubelka et al.[265] assessed the advantages of GC-MS and mass fragmentography in the analysis of complex hydrocarbon mixtures. Kuras[266] also examined the utility of GC-MS for the analysis of complex mixtures of hydrocarbons. Gibert and Oro employed GC-MS for the determination of potential contaminant hydrocarbons in moon rock samples.[267]

B. Analyses by Hydrocarbon Class

1. Saturated and Unsaturated Hydrocarbons

Mitera[268] employed GC-MS analysis for the identification of alkyl derivatives of diamantane. Kovachev[269] described the isolation and GC-MS identification of paraffinic hydrocarbons from Bulgarian oil shales, while Albaiges et al.[270] used GC-MS for the identification of geochemically significant isoalkanes. Spivakovskii et al.[271] reported on the calculation of retention indices of saturated hydrocarbons from their structural formulas for their combined identification by GC-MS.

Raverdino and Sassetti[272] employed capillary GC and GC-MS for the identification of positional and geometrical isomers of C_8 to C_{13} n-alkene mixtures derived from the "Olex" process. Analytical methods were developed for use with or without prior separation of the mixtures into cis-, trans-, and α-isomer fractions on a silver nitrate-impregnated alumina column using liquid chromatography. GC analyses employed 100-m by 0.25-mm I.D. stainless-steel capillary columns coated with DC-550, using a flame-ionization detector, with portions of the isomer fractions converted to chloro-derivatives prior to GC analysis for further identification. The authors recommended direct GC analysis without derivatization for determining β-isomers and direct IR analysis for α- and geometric-isomers. About 80% of the alkene components in the mixtures could be identified by GC as their chloro-derivatives and about 95% could be identified by GC analysis of the oxidation products of the alkenes following their reaction with osmium tetraoxide. Doering and co-workers[273] reported on the identification of over 80% of the higher boiling hydrocarbons present in the circulating dimethylformamide stream of a butadiene extraction plant using capillary GC-MS, in which dimethylformamide was used to extract butadiene from the C_4 fraction from the pyrolysis of naphtha or gas oil. The dimethylformamide stream was analyzed directly, rather than first separating the hydrocarbon fraction. Capillary GC-MS was also employed by Gallegos and co-workers[274] for the analysis of pyrolysis naphtha, with the identification of 1,3-butadiene, isobutene, 1-butene, 1-butyne, 1,4-pentadiene, cyclopentadiene, cis-1,3-pentadiene, cyclopentane, and 2-methyl-1,3-pentadiene, based on the use of m/e lines of 65, 67,

69, 71, and 77. This method was also used for the identification of dicyclopentadiene, cross-dimers of cyclopentadiene, methylcyclopentadiene, and several C_9 aromatic species.

2. Aromatic, Cyclic, and Polycyclic Hydrocarbons

DiCorcia et al.[275] used GC-MS with a packed 5-m column of graphitized carbon black modified with tetranitrofluorenone for the analysis of premium-grade gasoline in which hydroxide ions generated in the MS by reaction of N_2O with hexane resulted in the abstraction of hydrogens of alkylbenzenes (but not of saturated hydrocarbons). GC-MS analysis was used with a 100-m by 0.3-mm I.D. glass capillary column coated with Apiezon L for the analysis of industrial dodecylbenzenes by Otvos and associates.[276] Their analysis led to the identification of a number of linear alkylbenzenes and the detection of low levels of alkyl-indanes and -tetralins, but branched isomers could only be analyzed qualitatively. Vykhres-tyuk and co-workers[277] described the GC-MS analysis of an aromatic fraction from aviation kerosine using a 100-m by 0.3-mm capillary column coated with dinonyl phthalate in which over 100 components were separated by GC and identified by MS analysis. Henneberg and associates[278] discussed the need for computers for GC-MS to assist the determination of mass spectra that contain particularly useful information and outlined the use of a GC-MS computer system for the analysis and identification of impurities in diisopropylbenzene. Iida and Okada[279] employed methane-chemical ionization GC-MS for the identification of 17 trace impurities in toluene using a packed GC column of either polyethylene glycol 400 or squalene. Kubelka[280] used GC-MS and mass fragmentography for the analysis of aromatic hydrocarbons in Romashkino petroleum fractions and products.

A special report[281] described a study of 60 paleozoic and mesozoic oils, with specific attention given to steranes and pentacyclic triterpanes. Oil samples were initially passed through a Florisil column to remove the asphaltenes, with the steranes and triterpanes further purified and concentrated by silica gel liquid chromatography and adduction prior to analysis by GC-MS. The isolation and analytical methodology also could be applied to samples of sediment extracts. Lindeman[282] reported retention data and mass spectra from GC-MS anal-ysis of cis- and trans-isomer mixtures prepared from the corresponding alkyl cyclohexanes. Gallegos[283] described a GC-MS method for the identification of 23 mono- and one di-aromatic phenylcycloalkane isolated from Green River Shale. Gallegos[284] earlier had reported the identification of 36 individual cyclic components in the saturated fraction of shale from the same source using GC-MS with a 150-ft by 0.02-in. ID capillary column coated with OV-17. The components identified included two tetracyclic and 11 tricyclic terpanes, two 5-β-steranes, 5-α-pregnane, and several branched paraffins. Arpino and co-workers[285] de-scribed the analysis of bicyclic and tricyclic hydrocarbons which had been isolated from a petroleum fraction, using capillary GC-MS.

Vorob'eva et al.[286] employed GC-MS for the characterization of polycyclic C_{14} to C_{26} naphthenes in a 200 to 420°C fraction from Siva crude oil and proposed a scheme for the formation of the naphthenes based on the conversion of specific aliphatic hydrocarbons. Nishishita and co-workers described the functioning of an Hitachi RMU-5B MS system coupled to a GC and a data system having a programmed data processor for the analysis of aromatic hydrocarbons.[287] This method employed the MS base peak ion for bezene as a standard for the determination of benzene, naphthalene, tetrahydronaphthalene, acena-phthene, fluorene, phenanthrene, and anthracene. Liao and Browner[288] reported a method for the analysis of polycyclic aromatic hydrocarbons in poly(vinylchloride) smoke particulates based on preliminary isolation of the hydrocarbons using HPLC, followed by their separation and determination by GC-MS. Schultz and associates[289] used a GC-MS with glass capillary columns to characterize polycyclic aromatic hydrocarbons and aliphatic hydrocarbons in fractions derived from solvent-refined coal. Schultz[290] also described this work in greater detail in his 1979 Ph.D. dissertation. Lee and co-workers[291] reported on the use of combined glass capillary column GC-mixed charge exchange chemical ionization MS for the analysis of polycylic aromatic hydrocarbon isomers. Janini and associates[292] developed several new high temperature-transition nematic liquid crystals as novel GC stationary phases and showed their application for vastly improved separations of three- to six- ring isomers of polycyclic aromatic hydrocarbons using GC and GC-MS.

Additional applications of GC-MS for the analysis of polycyclic aromatic hydrocarbons and other hydrocarbon classes are included in Chapter 6, Environmental Studies on Hydrocarbons: Air Pollution; Water Pollution.

VI. USE OF COMPUTERS

Along with capillary column technology and GC-MS, there has been no greater positive impact felt in the GC of hydrocarbons, and in gas chromatographic analysis in general in recent years than from the development of computer systems. This development has seen a marked refinement in hardware components and software applications, without which current capabilities in data handling and data manipulations would be wholly intractable. Aside from the recognized unique power afforded by the coupling of computers to GC-MS systems, GC capabilities with interactive computer coupling have led to analytical activities in rapid real-time collection of retention data, deconvolution of overlapped GC peaks, assistance in liquid phase selection, the prediction of solute structures from retention data, automated GC compositional analyses, the use of multi-GC systems, and the routine production of analytical reports, to mention several of the more active areas of application.

Following are examples of activities in the use of computers in the GC analysis of hydrocarbons.

Glajch and co-workers[293] described a computerized data collection system for facilitating the comparison of data in the packed column analysis of pyrolyzed coal tars and petroleum pitches. Gaspar et al.[294] reported a computer-controlled automatic GC system equipped with a fluid-logic sampling gate as a sampling device and showed its application to the analysis of light gasoline using a squalane GC column, Inoue et al.[295] described a GC-computer system for the determination of paraffins, olefins, naphthenes, and aromatics (PONA analysis) in heavy naphtha with an analysis time equivalent to 5% that required using the conventional sulfuric acid-aniline method. Norman and Fells[296] developed a system in which the amplified FID signal was fed to a four-channel FM tape recorder, with the tape played back through a PDP-8 computer A-D converter. The digitized signal was then processed by a computer program for the analysis of mixtures of methane, ethane, ethylene, and acetylene. Bordet and Gourlia[297] developed a patented method for the rapid analysis of multicomponent hydrocarbon mixtures based on the use of short GC columns and computer-assisted peak deconvolution. Bartoli and associates[298] delineated the hardware and software components of an automated GC-computer system for hydrocarbon analysis in which the software program allowed for variable background substraction and resolution of overlapped GC peaks.

Lekova and Gerasimov[299,300] described a computer method for the selection of an optimum GC stationary phase for hydrocarbon separations. The method is based on the calculation of the degree of separation (R) between two adjacently eluting peaks using an ALGOL computer language program, with the resultant computer selection of the three most suitable phases for a given hydrocarbon separation. The method was demonstrated in the GC analysis of hydrocarbon mixtures on 37 hydrocarbon and silicone stationary phases. The resolution criteria were based on the use of relative retention data. Macnaughtan and co-workers[301] developed a principal component method for deconvoluting overlapped peaks in a study of the effect of overlapped GC peaks on quantitative GC analysis using a series of *n*-alkanes and benzene and perdeuterobenzene. Badinska and Gerasimov[302] developed a computer technique of selecting the most suitable GC stationary phase for hydrocarbon analysis, while

Follain[303] employed computer capabilities for real-time processing and interpretation of complex chromatograms. Chastrette and associates[304] developed a computer routine for predicting structure of GC analytes from their retention data. Yakimenko and Chebonenko[305] developed computer software to allow for accurate and precise computations of the fractional composition of petroleum fractions. Guichard-Loudet also reported a computer method for the identification of hydrocarbons from GC retention data.[306] A critical performance evaluation was carried out by Malan and Brink[307] on the use of the Hewlett-Packard 3360A GC Data Processing System in a petroleum refinery laboratory environment, with the finding that this system can process outputs from seven GC instruments analyzing 55 to 65 samples per day. Schomberg and co-workers[308] evaluated a time-shared computer system which could provide automated evaluation of qualitative and quantitative GC analytical data (together with other on- and off-line computation tasks), and applied this system to the GC analysis of hydrocarbon mixtures.

McGarry[309] described a computerized monitoring system using an IBM System/7. The monitoring system allowed one person to continuously monitor the outputs from eight GC instruments, as well as the periodic outputs from an additional five GC instruments. The overall operation provided accurate compositional analyses of hydrocarbon mixtures (resulting in a savings of 10 min per analysis), and carbon-balance calculations (saving about 2.5 hr/day). The use of this monitoring system produced a labor savings of $23,000/year and afforded more accurate and rapid GC analysis data and a consolidation of laboratory functions. A special report[310] described the use of two GC data systems (employing the Perkin-Elmer PEP system) at the Gulf Research and Development Company's process research laboratory for the analysis of hydrocarbon products from catalytic reactions of liquid or gaseous feedstock in 15 to 30 min. This rapid analysis feature permitted the evaluation of up to 16 new catalyst formulations per day, compared to a previous procedure requiring at least 2 days. The GC data system could accommodate up to eight GC instruments and offered complete automation from sample injection to report generation. It was estimated that the break-even point on system cost was less than ten analyses per day. Haynes[311] described the on-line coupling of a computer system to GC, high-and low-resolution mass spectrometer, and emission spectrometer systems to provide combined analyses of various types into a single report (e.g., analyses of gasoline by packed- and capillary-column GC and by MS. Guichard[312] described a system used by Institut Francais du Petrole (IFP) for automating its analytical laboratories via an IRIS 50 central computer. The automation involved a variety of systems (X-ray fluorescence and diffraction, slow- and fast-scanning MS, GC, LC, NMR, direct-reading emission spectrometry, atomic absorption, IR, and 200 conventional sensors for monitoring pressure, flow, temperature, and other factors in pilot plants). The hardware and software were designed by IFP, and the GC-computer interface was controlled from a command module in the laboratory. Two options of the system were feasible: one for research and method adjustments; one for routine analyses. Guichard described the GC automation in detail and provided results on analysis time, reproducibility, and precision.

Uetani and associates[313] described the hardware and software of a laboratory GC-computer system based on an TOSMAC-7000/20 computer for use in a petrochemical plant. This system provided scanning, peak detection, peak area calculation, and hydrocarbon component calculation functions, and generated reports of these analyses.

Castello and Parodi[314] developed a Fortran IV computer program for the automated computation of GC retention indices using either normal paraffins or other homologous series as reference materials for the treatment of retention data from isothermal or programed-temperature GC operations. The computer program was designed for use by GC analysts having no knowledge of computer programing.

A number of reviews were published on the use of computerized systems for the acquisition and processing of digital data derived from gas chromatographs, with special emphasis on dedicated digital integrators produced by U.S. manufacturers. These include reports issued by Derge,[315] Hettinger and Hubbard,[316] Kahn and Gill,[317] Leung and Gill,[318] and Sullivan and Silverman.[319] Caesar[320] compared retention indices vs. adjusted relative retention data for the identification of GC peaks using a satellite IBM 1800 processor coupled to an IBM 360-65 computer, with the finding that the use of retention indices provided more reliable results.

Kulikov[321] described graphic methods for use with GC data of hydrocarbons. Lekova and Gerasimovf[322] developed a computer-based method for the selection of suitable GC stationary phases and a computer program and showed the application of the program to hydrocarbon mixtures. Badinska and Gerasimov[323] also described a computer method for the selection of a GC liquid phase during the GC separation of hydrocarbon mixtures.

Some researchers evaluated the use of computers for use in GC analyses of particular classes of hydrocarbons. Examples of these are given below.

Lauer[324] described a computer-assisted GC system for off-line analyses of natural gas, town gas, and process gases from liquefaction and cracking plants, based on the use of a parallel-column GC containing a thermal conductivity detector and a FID. The injected sample was split between the two columns, with peak areas obtained by an integrator and component concentrations calculated with a Combitron S computer equipped with two punched-tape readers. Application was shown to the analysis of mixture containing C_1 to C_5 normal and branched-chain alkanes, nitrogen, carbon monoxide, and carbon dioxide. Terauchi and co-workers[325] reported on a computerized process GC system in a petroleum complex producing normal alkanes. The computer output provided GC peak elution order numbers, relative retention times, peak areas, and calculated components concentrations. Introduction of this system into the petroleum complex resulted in an improvement in reproducibility of analysis from 2 to 8% to about 0.1%, and a reduction in the number of operators needed to carry out such analyses.

Alabuzhev[326] discussed integration errors, signal resolution, analog-to-digital conversion, instrument stability, and programing operations in an automated GC system and described a system of three GC instruments and a mass spectrometer for the analysis and control of Al-Cr catalyzed dehydrogenation of butane. Holderith and associates[327] developed an experimental approach based on simplex algorithms for determining GC analysis conditions which afforded acceptable resolution in a minimum amount of analysis time. Their method was applied to the separation of methylated benzene mixtures containing *m*- and *p*-xylene, and 1,2,4- and 1,3,5-trimethylbenzene. Schomberg and co-workers[328] elaborated on the differences between GC data-processing in industrial vs. research laboratory environments, the evaluation of GC chromatograms on the basis of peak location, height, and area, and described the computerized GC systems at the Max-Planck-Institut fuer Kohlenforschung, covering on-line analog-to-digital conversion and computer programs. An application of the capability of this facility was illustrated in the complete identification of a sample of alkyl- and alkenyl-benzenes from retention index data and the use of nonane and decane as double internal standards. Landowne and Morosani[329] reported the development of an automatically programed, on-line system using a time-shared computer having a capacity of up to 16 chromatographs. The data options of the computer program ranged from routine quality control to complex method development. The computer program was validated using mixtures of toluene, isomeric xylenes, and aliphatic alcohols.

Computerized GC systems have now become commonplace and afford a power in real-time data handling and manipulation that is unparalleled. Future developments in the application of computers to GC analysis will undoubtedly involve improvements in minimizing analytical uncertainties, providing structural information on separated components, and the coupling of data outputs from ancillary analytical techniques.

Other reported applications and research involving the use of computers in hydrocarbon analyses may be found in the sections on GC-MS (Section V), retention data (Chapter 8), hydrocarbon boiling point correlations with GC retention data (Chapter 5, Section I.D) and process stream analyses (Chapter 3, Section IV).

REFERENCES

1. **Brander, B.,** *Analusis,* 7, 505, 1979.
2. **Evrard, E., Mercier, M., and Bal, M.,** *J. High Resolut. Chromatogr. Chromatogr. Commun.,* 2, 216, 1979.
3. **Galli, M., Trestianu, S., and Grob, K.,** *J. High Resolut. Chromatogr. Chromatogr. Commun.,* 2, 366, 1979.
4. **Grob, K. and Neukom, H. P.,** *J. Resolut. Chromatogr. Chromatogr. Commun.,* 2, 15, 1979.
5. **Sonchik, S. M. and Walker, J. Q.,** *J. Chromatogr. Sci.,* 17, 277, 1979.
6. **Tejedor, J. N.,** *Afinidad,* 34, 739, 1977.
7. **Ullrich, D. and Dulson, W.,** *Chromatographia,* 10, 537, 1977.
8. **Vitenberg, A. G., Bukaeva, I. L., and Dimitrova, Z. S.,** *Chromatographia,* 8, 693, 1975.
9. **Gaspar, G., Arpino, P., and Guiochon, G.,** *J. Chromatogr. Sci.,* 15, 256, 1977.
10. **Stolyarov, B. V. and Vitenberg, A. G.,** *Chromatographia,* 9, 3, 1976.
11. **Jennings, W. G.,** *J. Chromatogr. Sci.,* 13, 185, 1975.
12. **Borda de Oliveira, D. and Alfonso, A.,** *J. Chromatogr. Sci.,* 12, 109, 1974.
13. **Stockwell, P. B. and Sayer, R.,** *Anal. Chem.,* 42, 1136, 1970.
14. **Gaspar, G., Arpino, P., and Guiochon, G.,** *J. Chromatogr. Sci.,* 15, 256, 1977.
15. **Smith, K. A. and Harris, W.,** *J. Chromatogr.,* 53, 358, 1970.
16. **Frankiewicz, T. C. and Williams, F. W.,** *J. Chromatogr. Sci.,* 14, 63, 1976.
17. **Umstead, M. E.,** *J. Chromatogr. Sci.,* 12, 106, 1974.
18. **di Lorenzo, A.,** *Chim. Ind. (Milan),* 55, 573, 1973.
19. **Evrard, E. and Guiochon, G.,** *Chromatographia,* 5, 587, 1972.
20. **Durbin, D. E. and Fruge, J.,** *Anal. Chem.,* 44, 1502, 1972.
21. **Pease, E. C.,** *Anal. Chem.,* 45, 1584, 1973.
22. **Dunlop, A. S. and Pollard, S. A.,** *Anal. Chem.,* 43, 1344, 1971.
23. **Bradley, M. P. T.,** *Anal. Chem.,* 47, 189R, 1975.
24. **Grob, K., Grob, G., and Grob, K., Jr.,** *J. High Resolut. Chromatogr. Chromatogr. Commun.,* 2, 31, 1978.
25. **Dandeneau, R. D. and Zerenner, E. H.,** *J. High Resolut. Chromatogr. Chromatogr. Commun.,* 2, 351, 1979.
26. **Cram, S. P. and Yang, F. J.,** *Ind. Res. Dev.,* 20 (4), 89, 1978.
27. **Grob, K. and Grob, G.,** *J. High Resolut. Chromatogr. Chromatogr. Commun.,* 2, 109, 1979.
28. **Jennings, W. G.,** *Gas Chromatography with Glass Capillary Columns,* Academic Press, New York, 1978.
29. **Grob, K.,** *Chromatographia,* 8, 423, 1975.
30. **Lochmueller, C. H., Gordon, B. M., Lawson, A. E., and Mathieu, R. J.,** *J. Chromatogr. Sci.,* 16, 523, 1978.
31. **Rooney, T. A.,** *Ind. Res. Dev.,* 20, 143, 1978.
32. **Toth, J.,** *Magy. Kem. Lapja,* 32, 194, 1977; *Int. Chem. Eng.,* 19, 259, 1979.
33. **Schieke, J. D. and Pretorius, V.,** *J. Chromatogr.,* 132 (2), 231, 1977.
34. **Block, M. G., Callen, R. B., and Stockinger, J. H.,** *J. Chromatogr. Sci.,* 15 (11), 504, 1977.
35. **McGill, A. S., Parsons, E., and Smith, A.,** *Chem. Ind. (London),* 11, 456, 1977.
36. **Schirrmeister, H.,** *J. Chromatogr.,* 137, 13, 1977.
37. **Talalaev, E. I., Sergienko, S. R., and Ovezova, A. A.,** *Izv. Akad, Nauk. Turkm. S.S.R., Ser. Fiz.-Tekh., Khim. Geol. Nauk.,* 4, 55, 1973.
38. **Rijks, J. A., Van den Berg, J. H. M., and Diependaal, J. P.,** *J. Chromatogr.,* 91, 603, 1974.
39. **Gustavo Ober, A. and Didyk, B. M.,** *Scientia (Valparaiso),* 39, 84, 1973.
40. **Sidorov, R. I., Denisenko, A. N., Ivanova, M. P., Reznikov, S. A., Khvostikova, A. A., and Borovskaya, I. S.,** *Zh. Anal. Khim.,* 27, 1013, 1972.
41. **Shefter, V. E. and Kristol, L. D.,** *Neftepererab. Neftakhim (Moscow),* 5 (11), 1972.
42. **Nakamura, M.,** *Sekiyu Gakkai Shi,* 16, 51, 1973.

43. **Sojak, L. and Hrivnak, J.,** *J. Chromatogr. Sci.,* 10, 701, 1972.
44. **Dupuy, W. E., Hudson, H. R., and Karam, P. A.,** *J. Chromatogr.,* 71, 347, 1972.
45. **Douglas, A. G.,** *J. Chromatogr. Sci.,* 9, 742, 1971.
46. **Gouw, T. H., Wittemore, I. M., and Jentoft, R. E., Jr.,** *Anal. Chem.,* 42, 1394, 1970.
47. **Petrovic, K. and Kapor, M.,** *Tehnika (Belgrade),* 26, 353, 1971; *Chem. Abstr.,* 75, 8172b, 1971.
48. **Cramers, C. A., Rijks, J., and Bocek, P.,** *J. Chromatogr.,* 65, 29, 1972.
49. **Novotny, M., Segura, R., and Zlatkis, A.,** *Anal. Chem.,* 44, 9, 1972.
50. **Robinson, R. E., Coe, R. H., and O'Neal, M. J.,** *Anal. Chem.,* 43, 591, 1971.
51. **Walker, J. Q. and Wolf, C. J.,** *Anal. Chem.,* 42, 1652, 1970.
52. **Schomburg, G.,** *J. High Resolut. Chromatogr. Chromatogr. Commun.,* 2, 461, 1979.
53. **Mathews, R. G., Torres, J., and Schwartz, R. D.,** *J. High Resolut. Chromatogr. Chromatogr. Commun.,* 1, 139, 1978.
54. **Bloch, M. G., Callen, R. B., and Stockinger, J. H.,** *J. Chromatogr. Sci.,* 15, 504, 1977.
55. **Pacakov, V. and Kozlik, V.,** *Chromatographia,* 11, 266, 1978.
56. **Dielman, G., Meier, S., and Rapp, U.,** *J. High Resolut. Chromatogr. Chromatogr. Commun.,* 2, 343, 1979.
57. **Kaiser, R. E.,** *Chromatographia,* 7, 92, 1974.
58. **Grob, K.,** *Chromatographia,* 8, 423, 1975.
59. **Mitooka, M.,** *Bunseki Kagaku,* 21. 1447. 1972.
60. **Gallegos, E. J.,** *Anal. Chem.,* 45, 1399, 1973.
61. **Sidorov, R. I., Denisenko, A. N., Ivanova, M. P., Reznikov, S. A., Khvostikova, A. A., and Borovskaya, I. S.,** *Zh. Anal. Khim.,* 27, 1013, 1972.
62. **Gouw, T. H., Whittemore, I. M., and Jentoft, R. E.,** *Anal. Chem.,* 42, 1394, 1970.
63. **Pichoer, H., Ripperger, W., and Schwarz, G.,** *Erdoel Kohle,* 23, 91, 1970.
64. **Robinson, R E., Coe, R. H., and O'Neal, M. J.,** *Anal. Chem.,* 43, 591, 1971.
65. **Sojak, L., Kraus, G., Ostrovsky, I., Kralovicova, E., and Krupcik, J.,** *J. Chromatogr.,* 206 (3), 463, 1981; 206 (3), 475, 1981.
66. **DiSanzo, F. P., Uden, P. C., and Siggia, S.,** *Anal. Chem.,* 52, 906, 1980.
67. **Kumar, P., Sarowha, S. L. S., and Gupta, P. L.,** *Analyst (London),* 104 (1241), 788, 1979.
68. **Rijks, J. A. and Cramers, C. A.,** *Chromatographia,* 7, 99, 1974.
69. **Lukac, S.,** *J. Chromatogr.,* 166, 287, 1978.
70. **Johansen, N.,** *Chromatogr. Newsl.,* 5, 141, 1977, *A.A.,* 6J32.
71. **Nygren, S.,** *J. High Resolut. Chromatogr. Chromatogr. Commun.,* 2, 319, 1979.
72. **Oshima, Y., Ohnuma, H., and Akai, Y.,** *Sekiyu Gakkai Shi,* 18 (6), 497, 1975.
73. **Nilsson, O.,** *Chromatographia,* 10, 5, 1977.
74. **Ettre, L. S.,** *Chromatographia,* 8, 355, 1975.
75. **Talalaev, E. I., Sergienko, S. R., Aidogdyev, A., and Parienova, N. M.,** *Izv. Akad. Nauk. Turkm. S.S.R., Ser. Fiz. Tekh., Khim. Geol. Nauk.,* 1973, 55.
76. **Petrovic, K. and Vitrovic, D.,** *J. Chromatogr.,* 65, 155, 1972.
77. **Pollock, G. E.,** *Anal. Chem.,* 44, 634, 1972.
78. **Raverdino, V. and Sassetti, P.,** *J. Chromatogr.,* 169, 223, 1979.
79. **Kozlov, S. P. and Afanasiev, M. I.,** *Usp. Gaz. Khromatogr.,* 3, 118, 1973.
80. **Talalayev, Y. I., Serghiyenko, S. R., and Ovezova, A. A.,** *Izv. Akad. Nauk. Turkm. S.S.R., Ser. Fiz. Technol. Khim. Geol. Nauk.,* 1973 (4), 55.
81. **Johansen, N. G.,** *Chromatogr. Newsl.,* 6 (2), 17, 1978.
82. **Orav, A., Kuningas, K., Rang, S., and Eisen, O.,** *Eesti NSV Tead. Akad. Toim., Keem.,* 29, 18, 1980.
83. **Rang, S., Kuningas, K., Orav, A., and Eisen, O.,** *Chromatographia,* 10, 55, 1977.
84. **Rang, S., Kuningas, K., Orav, A., and Eisen, O.,** *J. Chromatogr.,* 128, 59, 1976.
85. **Sojak, L., Hrivnak, J., Krupcik, J., and Janak, J.,** *Anal. Chem.,* 44, 1701, 1972.
86. **Sojak, L., Druscova, A., and Janak, J.,** *Ropa Uhlie,* 14, 238, 1972; *A.P.I.A.,* 19, 9198, 1972.
87. **Zlatkis, A. and de Andrade, I. M. R.,** *Chromatographia,* 2 (7), 292, 1969.
88. **Welsch. Th., Engewald, W., and Berger, P.,** *Chromatographia,* 11, 5, 1978.
89. **Orav, A. and Eisen, O.,** *Izv. Akad. Nauk. Est. S.S.R., Ser. Geol.,* 21, 39, 1972.
90. **Sojak, L., Hrivnak, J., Ostrovsky, I., and Janak, J.,** *J. Chromatogr.,* 91, 613, 1974.
91. **Sojak, L., Hrivnak, J., Majer, P., and Janak, J.,** *Anal. Chem.,* 45, 293, 1973.
92. **Metitzow, W. and Fell, B.,** *Erdol Kohle,* 25, 311, 1972.
93. **Ladon, A. W. and Sandler, S.,** in *Column Chromatography,* Kovats, E. Sz., Ed., Swiss Chemists' Association, 1970, 160.
94. **Chmela, Z., Cap, L., and Adamek, J.,** *Acta Univ. Palacki. Olomuc., Fac. Rerum Nat.,* 53 (Chem. 16), 163, 1977; *C.A.,* 90, 33597m, 1979.
95. **Anderson, E. L., Thomason, M. M., Mayfield, H. T., and Bertsch, W.,** *J. High Resolut. Chromatogr. Chromatogr. Commun.,* 2, 335, 1979.

96. **Engewald, W. and Wennrich, L.,** *Chromatographia,* 9, 540, 1976.
97. **Sojak, L., Janak, J., and Rijks, J. A.,** *J. Chromatogr.,* 142, 177, 1977.
98. **Bartle, K. D.,** *Anal. Chem.,* 45, 1831, 1973.
99. **Diskina, D. E., Lazareva, I. S., and Zarubina, M. I.,** *Khim. Tekhnol. Topl. Masel,* 17 (6), 61, 1972.
100. **Louis, R.,** *Erdoel Kohle, Erdgas, Petrochem. Brennst.-Chem.,* 25, 582, 1972.
101. **Krupcik, J., Liska, O., and Sojak, L.,** *J. Chromatogr.,* 51, 119, 1970.
102. **Stuckey, C. L.,** *J. Chromatogr. Sci.,* 9, 575, 1971.
103. **Shlyakhov, A. F., Novikova, N. V., and Koreshkova, R. I.,** *Zavod. Lab.,* 45, 103, 1979.
104. **Rang, S., Orav, A., Kuningas, K., and Eisen, O.,** *Chromatographia,* 10, 115, 1977.
105. **Lindeman, L. P.,** *Amer. Chem. Soc. Div. Pet. Chem. Prepr.,* 16, A85, 1971.
106. **Vanek, J., Podrouzkova, B., and Landa,S.,** *J. Chromatogr.,* 52, 77, 1970.
107. **Gallegos, E. J.,** *Anal. Chem.,* 43, 1151, 1971.
108. **Kozlov, S. P. and Afanasiev, M. I.,** *Usp. Gaz Khromatogr.,* 3, 118, 1973.
109. **Soulages, N. L. and Brieva, A. M.,** *J. Chromatogr.,* 101, 365, 1974.
110. **Blomberg, L., Buijten, J., Gawdzik, J., and Wannman, T.,** *Chromatographia,* 11, 521, 1978.
111. **Kodama, T.,** *Aichi-ken Kogai Chosa Senta,* 4, 114, 1976.
112. **Tesarik, K., Fryeka, J., and Ghyezy, S.,** *J. Chromatogr.,* 148, 223, 1978.
113. **Grimmer, G. and Boehnke, H.,** *Fresenius' Z. Anal. Chem.,* 261, 310, 1972.
114. **Grimmer, G., Hildebrandt, A., and Boehnke, H.,** *Erdoel Kohle, Erdgas, Petrochem. Brennst. Chem.,* 25, 442, 1972.
115. **Grimmer, G., Hildebrandt, A., and Boehnke, H.,** *Erdoel Kohle, Erdgas, Petrochem. Brennst. Chem.,* 25, 531, 1972.
116. **Grimmer, G., Hildebrandt, A., and Boehnke, H.,** *Erdoel Kohle, Erdgas, Petrochem. Brennst. Chem.,* 25, 339, 1972.
117. **Mostecky, J., Popl, M., and Kriz, J.,** *Anal. Chem.,* 42, 1132, 1970.
118. **Bjorseth, A.,** *Anal. Chim. Acta,* 94, 21, 1977.
119. **Wauters, E., Sandra, P., and Verzele, M.,** *J. Chromatogr.,* 170, 125, 1979.
120. **Lee, M. L., Vassilaros, D. L., White, C. M., and Novotny, M.,** *Anal. Chem.,* 51, 768, 1979.
121. **Lee, M. L., Bartle, K. D., and Novotny, M.,** *Anal. Chem.,* 47, 540, 1975.
122. **Novotny, M., Lee, M. L., and Bartle, K. D.,** *J. Chromatogr. Sci.,* 12, 606, 1974.
123. **Onuska, F. I., Wolkoff, A. W., Combra, M. E., and Larose, R. H.,** *Anal. Lett.,* 9, 451, 1976.
124. **Doran, T. and McTaggart, N. G.,** *J. Chromatogr. Sci.,* 12, 715, 1974.
125. **Hutzinger, O., Safe, S., and Zander, M.,** *Analabs Res. Notes,* 13 (3), 1, 1973.
126. **Bruner, F., Ciccoli, P., Bertoni, G., and Liberti, A.,** *J. Chromatogr. Sci.,* 12, 758, 1974.
127. **Sojak, L. and Barnoky, L.,** *Ropa Uhlie,* 16, 654, 1974.
128. **Borwitzky, H. and Schomburg, G.,** *Erdoel Kohle, Erdgas, Petrochem.,* 33, 94, 1980.
129. **Steinberg, U., Alsberg, T., Blomberg, L., and Waennman, T.,** in Polynucl. Aromat. Hydrocarbons, 3rd Int. Symp. Chem. Biol. Carcinog. Mutagen., Ann Arbor, Mich., 1979, 313.
130. **Lao, R. C. and Thomas, R. S.,** in Polynucl. Aromat. Hydrocarbons, 3rd Int. Symp. Chem. Biol. Carcinog. Mutagen, Ann Arbor, Mich., 1979, 429.
131. **Bjorseth, A. and Eklund, G.,** *J. High Resolut. Chromatogr. Chromatogr. Commun.,* 2, 22, 1979.
132. **Grizzle, P. L. and Coleman, H. J.,** *Anal. Chem.,* 51, 602, 1979.
133. **Amerkhanov, I. M., Lychagin, A. V., and Sidorova, V. A.,** *Nefteprom Delo.,* 1978 (11); Chem. Abstr., 90, 89736y, 1979.
134. **Saha, N. C., Jaim, S. K., and Dua, R. K.,** *J. Chromatogr. Sci.,* 16, 323, 1978.
135. **Rendl, T. W., Anderson, J. M., and Dolan, R. A.,** *Am. Lab.,* 12, 60, 1980.
136. **Willis, D. E.,** *Anal. Chem.,* 50, 827, 1978.
137. **Bertsch, W.,** *J. High Resolut. Chromatogr. Chromatogr. Commun.,* 1, 187, 1978.
138. **Lukac, S.,** *Chromatographia,* 12, 17, 1979.
139. **Ullrich, D. and Dulson, W.,** *Chromatographia,* 10, 537, 1977.
140. **Chizhkov, V. P. and Yushina, G. A.,** *Zh. Anal. Khim.,* 31, 16, 1976.
141. **Shykles, M.,** *Anal. Chem.,* 47, 949, 1975.
142. **Luke, L. A. and McTaggert, N. G.,** British Patent 1,355,335, June 5, 1974.
143. **Carson, J. W., Ewald, F., et al.,** *J. Chromatogr. Sci.,* 10, 737, 1972.
144. **Stavinoha, L. L. and Newman, F. M.,,** *J. Chromatogr. Sci.,* 10, 583, 1972.
145. **Stavinoha, L. L.,** *J. Chromatogr. Sci.,* 11, 515, 1973.
146. **Petkova, T., Dimov, N., and Boeva, N.,** *God. Nauchnoizled. Inst. Koksokhim. Neftospreab,* p. 279, 1971; Chem. Abstr., 79, 68366k, 1973.
147. **Willis, D. E.,** *Anal. Chem.,* 44, 387, 1972.
148. **Peterson, R. M. and Rodgers, J.,** *Chromatographia,* 5, 13, 1972.
149. **Deans, D. R. and Scott, I.,** *Anal. Chem.,* 45, 1137, 1973.

150. **Churacek, J., Komarkova, H., Komarek, K., and Dufka, O.**, *Chromatographia*, 3, 465, 1970.
151. **Deans, D. R., Hickle, M. T., and Peterson, R. M.**, *Chromatographia*, 4, 279, 1971.
152. **Malan, E. and Brink, B.**, *Chromatographia*, 4, 178, 1971.
153. **Marchio, J. L.**, *J. Chromatogr. Sci.*, 9, 432, 1971.
154. **Vigalok, R. V., Tsybulevskii, A. M., and Vigdergauz, M. S.**, *Neftepeerab. Neftekhim. (Moscow), 1972* (3), 17; *Chem. Abstr.*, 76, 156383g, 1972.
155. **Mator, R. T., Petrocelli, J. A., and Puzniak, T. J.**, U.S. Patent 3,550428, December 29, 1970.
156. **Schomburg, G., Husmann, H., and Weeke, F.**, *J. Chromatogr.*, 112, 205, 1975.
157. **Kaiser, R. E.**, *J. Chromatogr. Sci.*, 12, 36, 1974.
158. **Rijks, J. A., Van den Berg, J. H. M., and Diependaal, J. P.**, *J. Chromatogr.*, 91, 603, 1974.
159. **Shykles, M.**, *Anal. Chem.*, 47, 949, 1975.
160. **Robinson, R. E., Coe, R. H., and O'Neal, M. J.**, *Anal. Chem.*, 43, 591, 1971.
161. **Bertsch, W.**, *J. High Resolut. Chromatogr. Chromatogr. Commun.*, 1, 289, 1978.
162. **Clark, I. R.**, *Chem. N. Z.*, 43, 15, 1979.
163. **Berezkin, V. G.**, *Fresenius' Z. Anal. Chem.*, 296, 1, 1979.
164. **Chizhkov, V. P., Yushina, G. A., Liberman, A. L., and Tyun'kina, N. I.**, *Zavod. Lab.*, 44, 1064, 1978.
165. **Reid, A. M.**, *J. Chromatogr. Sci.*, 14, 203, 1976.
166. **Houghton, E.**, *J. Chromatogr.*, 90, 57, 1974.
167. **Deur-Siftar, D. and Svob, V.**, *Nafta (Zareb)*, 23, 163, 1972.
168. **Mieure, J. P.**, *Anal. Chem.*, 45, 1981, 1973.
169. **Mitooka, M.**, *Bunseki Kagaku*, 21, 1437, 1972; *Chem. Abstr.*, 79, 26901a, 1973.
170. **Evrard, E. and Guiochon, G.**, *Chromatographia*, 5, 587, 1972.
171. **Dorokhov, A. P., Emel'yanova, G. V., Ioffe, I. I., Mel'nikova, N., and Shefter, V. E.**, *Khim. Tekhnol. Topl. Masel.*, 15, 53, 1970.
172. **Boer, H. and Van Arkel, P.**, *Chromatographia*, 4, 300, 1971.
173. **Boer, H. and Van Arkel, P.**, *Hydrocarbon Process*, 51, 80, 1972.
174. **Bagirov, R. A., Kuliev, A. M., and Farkhadov, T. S.**, *Khim. Tekhnol. Topl. Masel.*, 17, 51, 1972.
175. **Burgett, C. A. and Green, L. E.**, *Am. Lab.*, 8, 79, 1976.
176. **Schirrmeister, H.**, *J. Chromatogr.*, 137, 13, 1977.
177. **Rosie, D. M. and Barry, E. F.**, *J. Chromatogr. Sci.*, 11, 237, 1973.
178. **Pecsar, R. E., DeLew, R. B., and Iwao, K. R.**, *Anal. Chem.*, 45, 2191, 1973.
179. **Chowdhury, B. and Karasek, F. W.**, *J. Chromatogr. Sci.*, 8, 199, 1970.
180. **Olacsi, I., Balazs, A., Hoffer, F., and Egri, L.**, *Pr. Vyzk. Ustavu Geol. Inz.*, 35, 154, 1978; *Chem. Abstr.*, 90, 41064A, 1979.
181. **Proksch, E., Gehringer, P., and Szinovatz, W.**, *J. Chromatogr. Sci.*, 17, 568, 1979.
182. **Driscoll, J. N., Ford, J., Jaramillo, L. F., and Gruber, E. T.**, *J. Chromatogr.*, 158, 171, 1978.
183. **Williams, F. W. and Eaton, H. G.**, *Anal. Chem.*, 46, 179, 1974.
184. **Blades, A. T.**, *J. Chromatogr. Sci.*, 11, 251, 1973.
185. **Wojdala, T. and Guichard, N.**, *Analusis*, 2, 432, 1973.
186. **Guiffrida, L.**, in *Recent Advances in Gas Chromatography*, Domsky, I. I. and Perry, J. A., Eds., Marcel Dekker, New York, 1971, 125.
187. **Bocek, P. and Janak, J.**, *Chromatogr. Rev.*, 15, 111, 1971.
188. **McWilliam, I. G.**, *J. Chromatogr.*, 51, 391, 1970.
189. **Rossiter, V.**, *J. Chromatogr. Sci.*, 8, 164, 1970.
190. **Folmer, O. F., Jr. and Haase, D. J.**, *Anal. Chim. Acta*, 48, 63, 1969.
191. **Grant, D. W. and Clarke, A.**, *Anal. Chem.*, 43, 1951, 1971.
192. **Schwarz, F. P., Braun, W., and Wasik, S. P.**, *Anal. Chem.*, 50, 1903, 1978.
193. **Thiede, P. W. and Ehrlich, B. J.**, *Chromatographia*, 8, 709, 1975.
194. **Carson, J. W., Lege, G., and Young, J. D.**, *J. Chromatogr. Sci.*, 11, 503, 1973.
195. **Lasa, J. and Bros, E.**, *Chem. Anal. (Warsaw)*, 18, 825, 1973.
196. **Janghorbani, M. and Freund, H.**, *Anal. Chem.*, 45, 325, 1973.
197. **Rezl, V., Kaplanova, B., and Janak, J.**, *J. Chromatogr.*, 65, 47, 1972.
198. **Umstead, M. E., Woods, F. J., and Johnson, J. E.**, *J. Chromatogr. Sci.*, 8, 375, 1970.
199. **Yokomaya, S., Bodily, D. M., and Wiser, W. H.**, *Fuels*, 58, 162, 1979.
200. **Griffiths, P. R.**, EPA 600/4-76-061, Off. Res. Dev., U.S. Environmental Protection Agency, Washington, D.C., 1976.
201. **Gallaher, K. L. and Grasselli, J. G.**, *Appl. Spectrosc.*, 31, 456, 1977.
202. **De Haseth, J. A. and Isenhour, T. L.**, *Anal. Chem.*, 49, 1977, 1977.
203. **Louw, C. W. and Richards, J. F.**, *Appl. Spectrosc.*, 29, 15, 1975.
204. **Crowley, R. J., Siggia, S., and Uden, P. C.**, *Anal. Chem.*, 52, 1224, 1980.

205. **DiSanzo, F. P.**, *Univ. Mass. Diss.*, 1979.
206. **Sapozhnikova, E. A., Abidova, A. K., Sokol'nikova, M. D., Khasanova, M. N., Cherdakova, G. N., and Nisarbaeva, K. S.**, *Deposited Doc.*, 1976, *VINITI* 3332-76; *Chem. Abstr.*, 90, 8605e, 1979.
207. **Arich, G., Kikio, I., and Longo, V.**, *Chem. Ind. (Milan)*, 52, 433, 1970; *Chem. Abstr.*, 73, 89823y, 1970.
208. **Torradas, J. M., Riviera, J., and Albaiges, J.**, *Afinidad*, 35 (353), 13, 1978.
209. **Greinke, R. A. and Lewis, I. C.**, *Am. Chem. Soc. Div. Pet. Chem. Prepr.*, 20, 787, 1975.
210. **Fujita, M.**, *Sekiyu Gakkai Shi*, 20, 920, 1977.
211. **Machida, T.**, *Sekiyu To Sekiyu Kagaku*, 18, 49, 1974.
212. **Agrawal, K. M. and Joshi, G. C.**, *J. Appl. Chem. Biotechnol.*, 28, 718, 1978.
213. **Carson, J. W., Lege, G., and Gilbertson, R.**, *J. Chromatogr. Sci.*, 16, 507, 1978.
214. **Folmer, O. F., Jr.**, *Anal. Chim. Acta*, 60, 37, 1972.
215. **Stuckey, C. L.**, *Anal. Chim. Acta*, 60, 47, 1972.
216. **Carruth, G. F. and Kobayashi, R.**, *Anal. Chem.*, 44, 1047, 1972.
217. **Gawlowski, J., Niedjielski, J., and Weickowski, A.**, *J. Chromatogr.*, 151, 370, 1978.
218. **Loyd, R. J., Henderson, D. E., and Uden, P. C.**, *Anal. Chem.*, 48, 1645, 1976.
219. **Ilkova, E. L. and Mistryukov, E. A.**, *J. Chromatogr.*, 54, 422, 1971.
220. **Poznyak, T. I., Lisitsyn, D. M., and Dyachkovskii, F. S.**, *Zh. Anal. Khim.*, 34, 2028, 1979.
221. **Guglya, V. G. and Korobeinik, G. S.**, *Neftekhimiya*, 18, 318, 1978.
222. **Nakajima, F. and Sakai, K.**, *Bunseki Kagaku*, 25, 378, 1976.
223. **Bjorseth, A. and Eklund, G.**, *J. High Resolut. Chromatogr. Chromatogr. Commun.*, 2, 22, 1979.
224. **Selucky, M. L., Ruo, T. C. S., and Strausz, O. P.**, *Fuel*, 57, 585, 1978.
225. **D'Orazio, M.**, *Appl. Spectrosc.*, 33, 278, 1979.
226. **Houpt, P. M. and Baalhuis, G. H. W.**, *Appl. Spectrosc.*, 31, 473, 1977.
227. **Albert, D. K. and Kaplan, R. D.**, *Am. Chem. Soc. Div. Pet. Chem. Prepr.*, 18, 546, 1973.
228. **Brown,, R. A. and Searl, T. D.**, *J. Chromatogr. Sci.*, 17, 367, 1979.
229. **Cooney, R. P. and Winefordner, J. D.**, *Anal. Chem.*, 49, 1057, 1977.
230. **Greinke, R. A. and Lewis, I. C.**, *Anal. Chem.*, 47, 2151, 1975.
231. **Robinson, J. W. and Goodbread, J. P.**, *Anal. Chim. Acta*, 66, 239, 1973.
232. **Beroza, M. and Bowman, M. C.**, U.S. Patent 3,506,824, April 14, 1970.
233. **Burchfield, H. P., Wheeler, R. J., and Bernos, J. B.**, *Anal. Chem.*, 43, 1976, 1971.
234. **Gupta, P. L. and Kumar, P.**, *Pet. Hydrocarbons*, 6, 140, 1971.
235. **Searl, T. D., Cassidy, F. J., King, W. H., and Brown, R. A.**, *Anal. Chem.*, 42, 954, 1970.
236. **Katlafsky, B. and Dietrich, M. W.**, *Appl. Spectrosc.*, 29, 24, 1975.
237. **Mulik, J., Cooke, M., Guyer, M. F., Semenink, G. M., and Sawicki, E.**, *Anal. Lett.*, 8, 511, 1975.
238. **Burchfield, H. P., Green, E. E., Wheeler, R. J., and Billedeau, S. M.**, *J. Chromatogr.*, 99, 697, 1974.
239. **Grinberg, A. A., Bigdash, T. V., and Leont'eva, S. A.**, *Fiz. Khim. Metody Issled. Nefteproduktor, M.*, 23, 1979; *Zh. Khim.*, Abstract No. 8P 265, 1980.
240. **Panalaks, T.**, *J. Environ. Sci. Health, Part B*, B11 (4), 299, 1976.
241. **Hedfjaell, B. and Ryhage, R.**, *Anal. Chem.*, 47, 666, 1975.
242. **Heller, S. R.**, *Anal. Chem.*, 44, 1951, 1972.
243. **Justice, J. B., Jr., and Isenhour, T.**, *Anal. Chem.*, 44, 2087, 1972.
244. **Karasek, F. W.**, *Anal. Chem.*, 44, 32A, 1972.
245. **Karasek, F. W.**, *Res. Dev.*, 24 (10), 40, 1973.
246. **Karasek, F. W.**, *Res. Dev.*, 25 (9), 42, 1974.
247. **Thome, F. A. and Young, G. W.**, *Anal. Chem.*, 48, 1423, 1976.
248. **Henderson, W. and Steel, G.**, *Anal. Chem.*, 44, 2302, 1972.
249. **Heeneberg, D., Hendricks, U., and Schomburg, G.**, *Chromatographia*, 8, 449, 1975.
250. **Heeneberg, D., Hendricks, U., and Schomburg, G.**, *J. Chromatogr.*, 112, 343, 1975.
251. **Flath, R. A.**, Guide to Modern Methods on Instrumental Analysis, Gouw, T. H., Ed., Interscience, New York, 1972, 323.
252. **Larskaya, E. S. and Khramova, E. V.**, *Neftekhimiya*, 19, 679, 1979.
253. **Bruner, F., Ciccioli, P., and Zelli, S.**, *Anal. Chem.*, 45, 1003, 1973.
254. **Aczel, T.**, *Erdoel Kohle, Erdgas, Petrochem. Brennst.-Chem.*, 26, 27, 1973.
255. **Aleksandrov, A. N., Rabinovich, A. S., and Skop, S. L.**, *Zh. Anal. Khim.*, 24, 762, 1969.
256. **Schuy, K. D. and Hunneman, D. H.**, *Chem. Ztg.*, 95, 633, 1971.
257. **Staszewski, R.**, *Tluszcze Jadaine*, 16, 165, 1972.
258. **Swansiger, J. T. and Dickson, F. E.**, *Anal. Chem.*, 45, 811, 1973.
259. **Kagler, S. H.**, New Mineral Oil Analysis. Spectroscopy and Chromatography: Background, Apparatus and Applications, Verlag GmbH, Heidelberg, Germany, 1969.
260. **Proctic, G., Svob, V., and Deur-Siftar, D.**, *Sb. Vys. Sk. Chem. Technol. Praze, Technol. Paliv*, D42, 139, 1980.

261. **Termonia, M., Monseur, X., Alaerts, G., and Dourte, P.,** *Pergamon Ser. Environ. Sci.,* 3 (*Anal. Tech. Environ. Chem.*), 135, 1980.

262. **Khadzhieva, P., Sandra, P., Stoyanova-Ivanova, B., and Verzele, M.,** *11th IUPAC Int. Symp. Chem. Nat. Prod.,* 2, 464, 1978.

263. **Ciccioli, P., Hayes, J. M., Rinaldi, G., Denson, K. B., and Meinschein, W. G.,** *Anal. Chem.,* 51, 400, 1979.

264. **Hatch, F. and Munson, B.,** *J. Phys. Chem.,* 82, 2362, 1978.

265. **Kubelka, V., Mitera, J., and Zachar, P.,** *Sb. Vys. Sk. Chem. Technol. Praze, Technol. Paliv,* D36, 49, 1977.

266. **Kuras, M. and Hala, S.,** *J. Chromatogr.,* 51, 45, 1970.

267. **Gibert, J. M. and Oro, J.,** *J. Chromatogr. Sci.,* 8, 295, 1970.

268. **Mitera, J.,** *Sb. Vys. Sk. Chem. Technol. Praze, Technol. Paliv,* D30, 175, 1974; *Chem. Abstr.,* 83, 118130, 1975.

269. **Kovachev, G.,** *Acta Chim. Acad. Sci. Hung.,* 104, 415, 1980.

270. **Albaiges, J., Borbon, J., and Gassiot, M.,** *J. Chromatogr.,* 204, 491, 1981.

271. **Spivakovskii, G. I., Tishchenko, A. I., Zaslavskii, I. I., and Wulfson, N. S.,** *J. Chromatogr.,* 144, 1, 1977.

272. **Raverdino, V. and Sassetti, P.,** *J. Chromatogr.,* 169, 223, 1979.

273. **Doering, C. E. and Estel, D.,** *Chem. Tech. (Leipzig),* 29, 280, 1977.

274. **Gallegos, E. J., Whittemore, I. M., and Klaver, R. F.,** *Anal. Chem.,,* 46, 157, 1974.

275. **DiCorcia, A., Samperi, R., and Capponi, G.,** *J. Chromatogr.,* 160, 147, 1978.

276. **Otvos, I., Iglewski, S., Hunneman, D. H., Bartha, B., Balthazar, Z., and Palvi, G.,** *J. Chromatogr.,* 78, 309, 1973.

277. **Vyknrestyuk, N. I., Polyakova, A. A., Lizogub, A. P., Zhurba, A. S., Bryanshkaya, E. K., and Levenets, V. F.,** *Neftepeerab. Neftekhim. (Moscow),* 1973 (9), 27; *Chem. Abstr.,* 80, 135613z, 1974.

278. **Henneberg, D., Ziegler, E., et al.,** *Angew. Chem., Int. Ed. Engl.,* 11, 357, 1972.

279. **Iida, Y. and Okada, S.,** *Bunseki Kagaku,* 26, 30, 1977.

280. **Kubelka, V.,** *Sb. Vys. Sk. Chem. Technol. Praze, Technol. Paliv,* D42, 49, 1980.

281. *Anon., Ber. Dtsch. Ges. Mineraloelwiss. Kohlechem.,* p. 123, 1977.

282. **Lindeman, L. P.,** *Am. Chem. Soc. Div. Pet. Chem. Prepr.,* 16, A85, 1971.

283. **Gallegos, E. J.,** *Anal. Chem.,* 45, 1399, 1973.

284. **Gallegos, E. J.,** *Anal. Chem.,* 43, 1151, 1971.

285. **Arpino, P., Schmitter, J. M., and Selves, J. L.,** *Rev. Inst. Fr. Pet.,* 33, 467, 1978.

286. **Vorob'eva, N. S., Zernskova, Z. K., and Petrov, Al. A.,** *Neftekhimiya,* 18, 855, 1978.

287. **Nishishita, T., Yoshihara, M., and Oshima, S.,** *Maruzen Sekiyu Giho,* 1971 (16), 99; *Chem. Abstr.,*) 80, 61765b, 1974.

288. **Liao, J. C. and Browner, R. F.,** *Anal. Chem.,* 50, 1683, 1978.

289. **Schultz, R. V., Jorgenson, J. W., Maskarinec, M. P., and Novotny, M.,** *Fuel,* 58, 783, 1979.

290. **Schultz, R. V.,** *Diss. Abstr. Int. B,* 40 (2), 744, 1979.

291. **Lee, M. L., Vassilaros, D. L., Pipkin, W. S., and Sorensen, W. L.,** *NBS Spec. Publ.* 519 (*Trace Org. Anal.: New Front. Anal. Chem.,*), 731, 1979.

292. **Janini, G. M., Muschik, G. M., Schroer, J. A., and Zielinski, W. L., Jr.,** *Anal. Chem.,* 48, 1879, 1976.

293. **Glajch, J. L., Lubkowitz, J. A., and Rogers, L. B.,** *J. Chromatogr.,* 168, 355, 1978.

294. **Gaspar, G., Olivo, J., and Guiochon, G.,** *Chromatographia,* 11, 321, 1978.

295. **Inoue, T., Inoike, Y., Schimogaki, K., and Miyauchi, H.,** *Aromatikkusu,* 28, 209, 1976.

296. **Norman, P. W. and Fells, I.,** *J. Chromatogr.,* 132, 533, 1977.

297. **Bordet, J. and Gourlia, J. P.,** German (W) Patent 2,518,026, October 30, 1975.

298. **Bartoli, B., Mobilio, S., Spinelli, N., and Vanoli, F.,** *J. Chromatogr.,* 107, 51, 1975.

299. **Lekova, K. and Gerasimov, M.,** *Chromatographia,* 7, 69, 1974.

300. **Lekova, K. and Gerasimov, M.,** *Chromatographia,* 7, 595, 1974.

301. **Macnaughtan, D., Jr., Rogers, L. B., and Wernimont, G.,** *Anal. Chem.,* 44, 1421, 1972.

302. **Badinska, K. and Gerasimov, M.,** *Khim. Ind. (Sofia),* 1972, 250.

303. **Follain, G.,** *Analusis,* 2, 377, 1973.

304. **Chastrette, M., Lenfant, G., Remy, A., and Cohen-Makabeh, M.,** *J. Chromatogr.,* 84, 275, 1973.

305. **Yakimenko, L. V. and Chebonenko, N. D.,** *Neftepererab. Neftekhim. (Moscow),* 1972 (4), 13; *Chem. Abstr.,* 77, 50950v, 1972.

306. **Guichard-Loudet, N.,** *Analusis,* 2, 247, 1973.

307. **Malan, E. and Brink, B.,** *Chromatographia,* 5, 182, 1972.

308. **Schomburg, G., Weeke, F., Weimann, B., and Ziegler, E.,** *Gas Chromatogr. Proc. Int. Symp. (Europe),* 8, 280, 1970 (published 1971).

309. **McGarry, R. J. and Bickford, G.,** *Oil Gas J.,* 71 (36), 44, 1973.
310. *Anon., Oil Gas J.,* 71 (19), 85, 1973; *Chem. Week,* 112 (19), 36, 1973; *Anal. Chem.,* 45, 726A, 1973.
311. **Haynes, R. M.,** *Can. Res. Dev.,* 5 (5), 30, 1972.
312. **Guichard, N.,** *Chim. Ind. Genie Chim.,* 105, 1409, 1972.
313. **Uetani, A., Takada, S., and Yamamoto, A.,** *Keiso,* 15 (2), 33, 1972; *Chem. Abstr.,* 77, 7769t, 1972.
314. **Castello, G. and Parodi, P.,** *Chromatographia,* 4, 147, 1971.
315. **Derge, K.,** *Fette, Seifen, Anstrichm.,* 75, 353, 1973.
316. **Hettinger, J. and Hubbard, J.,** *Am. Lab.,* 6 (2), 99, 1974.
317. **Kahn, H. L. and Gill, J. M.,** *Am. Lab.,* 6 (9), 49, 1974.
318. **Leung, A. T. and Gill, J. M.,** *Res. Dev.,* 25 (10), 36, 1974.
319. **Sullivan, J. J. and Silverman, H. A.,** *Am. Lab.,* 7 (5), 110, 1975.
320. **Caesar, F.,** *Chromatographia,* 5, 173, 1972.
321. **Kulikov, V. I.,** *Prakt. Gazov. Khromatogr., (Mater. Semin.),* 1977, 32.
322. **Lekova, K. and Gerasimovf, M.,** *Chromatographia,* 7, 595, 1974.
323. **Badinska, K. and Gerasimov, M.,** *Khim. Ind. (Sofia),* 44 (6), 250, 1972.
324. **Lauer, K. H.,** *Gas-Wasserfach, Gas-Erdgas,* 115, 108, 1974.
325. **Terauchi, O. and Kiwata, Y.,** *Aromatikkusu,* 25, 48, 1973.
326. **Alabuzhev, Yu, A.,** *Tr. Inst. Katal. Sib. Otd. Akad. Nauk. S.S.S.R.,* 3, 18, 1974; *Chem. Abstr.,* 85, 28300b, 1976.
327. **Holderith, J., Toth, T., and Varadi, A.,** *J. Chromatogr.,* 119, 215, 1976.
328. **Schomburg, G., Ziegler, E., et al.,** *Angew. Chem. Int. Ed. Engl.,* 11, 366, 1972.
329. **Landowne, R. A., Morosani, R. W., et al.,** *Anal. Chem.,* 44, 1961, 1972.

Chapter 5

GAS CHROMATOGRAPHY OF HYDROCARBONS — SPECIAL TECHNIQUES AND THERMODYNAMIC STUDIES

This chapter provides literature references concerning a variety of special techniques for hydrocarbon analysis and characterization (Section I) and on thermodynamic studies of hydrocarbons (Section II).

I. SPECIAL TECHNIQUES

Literature references are provided in the following tables of this section on: (1) Sampling Techniques and Systems (Table 1); (2) Reaction Gas Chromatography (Table 2); (3) Pyrolysis Gas Chromatography (Table 3); (4) Hydrocarbon Boiling Point Correlations with GC Retention Data (Table 4); (5) Preparative-Scale Gas Chromatography (Table 5); and (6) Supercritical Chromatography and Other Miscellaneous Special Techniques (Table 6). Sampling techniques for hydrocarbon pollutants are also included under Chapter 6 (Environmental Studies on Hydrocarbons).

II. THERMODYNAMIC STUDIES OF HYDROCARBONS

Studies are summarized in Table 7 in which GC has been used for the determination of various thermodynamic properties of hydrocarbons. The citations for the studies are categorized under: (1) General Applications; (2) Studies on Alkanes; (3) Studies on Unsaturated Hydrocarbons; and (4) Studies on Cyclic and Aromatic Hydrocarbons. The cited studies include the use of GC for the determination of activity coefficients; free energies (G), enthalpies or heats (H), and entropies (S) of solution, mixing, or sorption of the hydrocarbon solutes with the column packing material; and other properties (e.g., solubilities, vapor pressures, equilibrium and partition coefficients, sorption and desorption isotherms).

Table 1
SAMPLING TECHNIQUES AND SYSTEMS

Comments	Ref.	Comments	Ref.
Hydrocarbons, General		**Light Hydrocarbons**	
Up to C_8	1	Ethylene in air (porous polymers)	16
Preconcentration of C_{1-20} hydrocarbons (freeze-out loop)	2	C_3 to C_5 hydrocarbons in stabilizer bottoms (sampling system)	17
Petroleum hydrocarbons (in arson; head-space sampling)	3	Benzene and alkylbenzenes in air	18,19
Headspace sampling and analysis	4,5	C_1 to C_5 hydrocarbons (sampling system)	20
On-line device for pipeline sampling	6	Graphitized carbon black for sample con-centration and GC inlet system	21,22
C_9 to C_{17} alkanes in kerosene (molecular sieve adsorption)	7	Low mol wt hydrocarbons (activated char-coal or alumina for sample concentration and GC inlet system)	23,24
Porous polymer sorbents (Tenax)	8	Innovative sampling system	25
Air hydrocarbon preconcentration (cry-ogenic gradient tube)	9		
Petroleum hydrocarbons from aqueous sur-face films (sampling device)	10	**Polycyclic Aromatic Hydrocarbons**	
Aromatic hydrocarbons (minimizing con-tamination in field sampling techniques)	11	PAH Sampling of combustion effluents	26
Crude oil distillates (GPC for sample preparation)	12	Sample isolation (TLC)	27
		PAH in flames	28
Aromatic hydrocarbons (Tenax sample concentration, precolumn desorption, and GC inlet system)	13	Breakthrough capacity and elution rates on polyurethane and XAD-2	29
Use of C-22 firebrick for sample concen-tration and GC inlet system	14	Tenax for sample concentration and cleanup	30
Hydrocarbons in light petroleum crude (disposable cartridge precolumns)	15		

Table 2
REACTION GAS CHROMATOGRAPHY

Comments	Ref.	Comments	Ref.
Hydrocarbons, General		Microgram-level ozonation and hydrogena-tion of unsaturated and aromatic hydro-carbons for carbon-skeleton and substractive GC analysis	41
Subtractive GC analysis	31		
Hydrogen/carbon ratios	32		
Complex hydrocarbons (postsubstractive column)	33	Reactive GC methods for the identification of hydrocarbon classes	42
Analysis of saturated, aromatic, and ole-finic hydrocarbons	34	Rapid analysis of complex hydrocarbon mixtures using reaction GC	43
Analysis of saturates, aromatics, and ole-finics by type (special GC arrangement)	35		
Determination of alkyl groups by destruc-tive cleavage	36	**Unsaturated Hydrocarbons**	
Analysis of shale oil hydrocarbons	37	Gasoline olefins	44—46
Selective reactions for determining alkanes, conjugated dienes, and alkynes	38	Dienes in hydrocarbon mixtures	47—51
		Unsaturated hydrocarbons	52
Microreactions and GC (hydrocarbon oxi-dations; cracking kinetics)	39	Olefinic hydrocarbons	53
		Saturated/unsaturated hydrocarbon mixtures	54
Subtractive GC analysis for trace hydrocar-bon impurities	40	Analysis of 14 acetylenes and dienes	55

Table 2
REACTION GAS CHROMATOGRAPHY

Comments	Ref.	Comments	Ref.
Unsaturated Hydrocarbons		**Aromatic Hydrocarbons**	
Hydrogenator for analysis of unsaturated hydrocarbons	56	Aromatic hydrocarbons in gasoline	67
Acetylenes and conjugated dienes	57	Conversion of cyclic hydrocarbons to aromatic hydrocarboms	68—70
Determination of cis/trans isomers of olefins	58	A microreactor for characterizing aromatic hydrocarbons	71
Ozonation of terpenes	59	Hydrogenation of naphtha aromatic hydrocarbons to cyclic hydrocarbons	72
Butane/butadiene mixtures (postcolumn reaction)	60	Hydrogenation of acenaphthylene and acenaphthene	73
Characterization of alkenes	61	Hydrogenation of alkyl benzenes	74
Dissociation kinetics of methylcyclopentadiene dimers	62	**Saturated Hydrocarbons**	
Cyclization of ethylene and of propylene or propane to C_6	63		
Location of double bond positions using ozonation	64	Alkanes and cycloalkanes in gasoline	75
		Cyclohexane and C_6H_{14}	76
Quantitative determination of hydrocarbons in hydrocarbon-olefin mixtures	65	Alkanes (subtraction GC)	77
		Alkanes (reaction GC)	78
Microcatalytic method for identification of diene hydrocarbons	66	Saturated hydrocarbons (dehydration technique)	79
		Alkanes (hydrocracking technique)	80

Table 3
PYROLYSIS GAS CHROMATOGRAPHY

Comments	Ref.	Comments	Ref.
Hydrocarbons, General		Identification of alkanes in polymers via thermal pyrolysis	97
Hydrocarbons; bitumens	81—83	Analysis of pyrolysis products of cycloalkanes	98
Polymers; heavy petroleum products	84	Aliphatic hydrocarbons (pyrolysis and hydrogenation)	99
Gasoline hydrocarbons (reactor furnace vs. ribbon pyrolysis probe)	85		
Laser pyrolysis of biphenyl, durene, paraffin wax, terphenyl isomers, and other hydrocarbons	86—87	**Unsaturated (Including Polymers) and Aromatic Hydrocarbons**	
Catalyst for thermal decomposition of C_6 hydrocarbons	88	Unsaturated and saturated hydrocarbon mixtures	100
Oil shales	89	C-1—C-4 hydrocarbons from cis-polybutadiene	101
Pyrolytic GC of hydrocarbons	90	Polyolefins	102
Pyrolysis GC of low-molecular-weight hydrocarbons (Ph.D. thesis)	91	Styrenated alkyd (pyrolysis units)	103
		Copolymers (pyrolysis units)	104
Saturated Hydrocarbons		Styrene copolymers	105
Hexane; 2,2- and 2,3-dimethylbutane	92	Alkenes from pyrolysis of alkanes and polyethylene	106
Alkanes (C_6 to C_8)	93		
C-1—C-10 hydrocarbons from gas-phase radiolysis of n-pentane	94	C-1—C-6 hydrocarbons from laser pyrolysis of ethylene	107
Analysis of products from pyrolysis of n-decane	95	Thermal degradation of polyolefins	108
Conversion of methane to diprene and limonene (Curie point pyrolysis; hydrocarbon characterization)	96	Determination of polybutadiene in polystyrene	109

Table 3 (continued)
PYROLYSIS GAS CHROMATOGRAPHY

Comments	Ref.	Comments	Ref.
Pyrolysis GC of low hydrocarbon compounds and polyethylene	110	Pyrolysis of over 40 alkyl benzenes	113
Copolymers of vinylaromatic hydrocarbons with aliphatic dienes	111	Aromatic hydrocarbons and polyethylene (determination of pyrolysis starting temperature for pyrolysis GC)	114
Alkyl benzenes	112		

Table 4
HYDROCARBON BOILING POINT VS. GC RETENTION (SIMULATED DISTILLATION)

Comments	Ref.
Techniques	
Discussion of retention index and GC columns for alkanes, alkenes, and aromatic hydrocarbons	115,116
Chromadistillation	117,118
Simulated distillation	119—125
Computerized simulated distillation	126
Mathematical models for predicting Reid vapor pressure and the distillation points of gasolines	127
Integrator model for simulated distillation data	128
Distillation curves	129,130
GC retention vs. boiling points	131,132
Continuous feed for fractional distillation column to GC column	133
ASTM distillation curve	134
Correlation of "retention-boiling point" concept with GC retention	135
Simulated distillation of wide boiling range hydrocarbon mixtures	136
Calculated boiling points from GC retention data (saturated and aromatic hydrocarbons)	137
Calculated boiling points from GC retention data (cyclic hydrocarbons)	138
Simulated boiling point curves	139
Gasoline and Petroleum	
Crudes; refined petroleum	140
Gasoline octane number	141
Gasoline, shale oil, hydrocarbon fuels	142—145
Whole crudes (simulated distillation)	146

Table 4 (continued)
HYDROCARBON BOILING POINT VS. GC RETENTION (SIMULATED DISTILLATION)

Comments	Ref.
Narrow distillation cuts	147
Petroleum boiling point range	148,149
Determination of boiling points and distillation range of 34 gasolines	150
Gasoline (boiling points vs. GC retention)	151

Miscellaneous Applications

Chromadistillation of C-6—C-9 alkanes	152
C-6—C-8 hydrocarbons	153
Hydrocarbons isomers (boiling point vs. GC retention)	154
Isomeric decanes (boiling point prediction from GC retention data)	155
Aromatics (determination of saturation vapor pressures)	156

Table 5
PREPARATIVE SCALE GAS CHROMATOGRAPHY

Comments	Ref.

General

Naphtha	157
Preparative GC model design	158
Increasing through-put in preparative GC	159
Text on preparative GC	160
Industrial applications to preparative GC in Hungary	161
Production rate expression for preparative GC	162
Hupe equation for preparative GC	163
Study of optimum designs for preparative GC	164
Use of five parallel columns	165
Temperature programing in preparative GC	166
Column packing techniques and establishing flows	167
Text on preparative GC of light hydrocarbons	168
Applications of continuous GC for hydrocarbon mixtures	169

Saturated and Unsaturated Hydrocarbons

Steranes and Triterpanes (isolation of milligram quantities from oil shale)	170,171
C-6—C-8 alkanes	172
Separation of kyclohexane from dodecadeuterocyclohexane	173
C-6—C-19 alkanes using a carrier gas splitter	174
Cyclic hydrocarbons	175
Cyclopentene hydrocarbons	176

Table 5 (continued)
PREPARATIVE SCALE GAS CHROMATOGRAPHY

Comments	Ref.
Aromatic Hydrocarbons	
p-Xylene	177
Purification of *m*-xylene	178
Aromatic hydrocarbons from gasoline fractions	179

Table 6
SUPERCRITICAL GAS CHROMATOGRAPHY AND OTHER MISCELLANEOUS SPECIAL TECHNIQUES

Comments	Ref.
Supercritical GC	
Polycyclic aromatic hydrocarbons	180,181
System for C_{14} to C_{20} alkanes	182
Mobile phases for supercritical fluid chromatography	183
C5—C22 *n*-alkanes; aromatic and cyclic hydrocarbons	184
Review of applications (benzo [*a*]- and benzo [*e*]pyrene in polycyclic aromatic hydrocarbons; C16, C18 olefins; C10 saturates	185
Fractionation of monodisperse polystyrenes	186
Programable pneumatic amplifier pump (control device)	187
Specific polycyclic aromatic hydrocarbons; CO_2 mobile phase	188
Other Miscellaneous Special Techniques	
Flow programing (C10—C18 alkanes)	189
Patent on apparatus and method for temperature programing (*n*-alkanes)	190
Backflushing (aromatics)	191
Addition of *n*-C_5 to carrier (polycyclic aromatic hydrocarbons)	192
Twofold improvement in GC separation speed for *n*-paraffins (use of ammonia as carrier in place of nitrogen)	193
Temperature and flow programing; hydrocarbon analyses	194,195

Table 7
THERMODYNAMIC DATA ON
HYDROCARBONS FROM GC ANALYSIS

Comments	Ref.

General Applications

Thermodynamically defined retention	196
Vapor pressure predictions	197, 198
Correlation between retention indices, boiling points, and activity coefficients	199
Activity coefficients: 21 compounds in *n*-alkane stationary phases	200
Activity coefficients: *n*-heptane, benzene, substituted benzenes in *n*-C-18 and *n*-C-16 halide stationary phases	201
Activity coefficients: 23 hydrocarbons in three *n*-alkane stationary phases	202
Activity coefficients: hydrocarbons in sulfoxides	203
Activity coefficients: gaseous hydrocarbons in squalane	204
Activity coefficients, H, S: 42 alkanes, alkenes, and aromatics in two liquid crystal stationary phases	205
Activity coefficients; excess H	206
Activity coefficients, H: hydrocarbons in benzoxanthiin derivative (MDDB)	207
H of solution	208
H of solution: alkanes, cyclics, olefins in squalane	209
Reliability of H of evaporation: 900 retention volumes	210
H of solution and adsorption: *n*-nonane, benzene, and cyclohexane in 11 stationary phases	211
H and S of solution: branched alkanes and alkyl benzenes in squalane	212
H and G: various functional groups in 75 stationary phases	213
G: Rohrschneider solutes in 16 stationary phases	214
G, H, and S of mixing: heptane and toluene in polymers and oligomers	215
H of adsorption: benzene and *n*-hexane on glass beads	216
G of adsorption: methylene groups on alumina, Chromosorbs, Porasil	217
G of adsorption: methylene groups on Amberlite XAD, Chromosorbs, Porapaks, Spheron, Polymer-1	218
H of adsorption: C-4—C-6 hydrocarbons on $CoCl_2$	219
H of adsorption and Henry constants: C-1—C-6 alkanes, alkenes, alkadienes, alkynes, cyclanes, cyclenes on graphitized carbon black	220
G, H, and S of solution: C-5—C-9 hydrocarbons in dinonylphthalate	221
H of adsorption: cyclic olefins and aromatics on $BaSO_4$	222
Adsorption thermodynamics: hydrocarbons on zeolites	223
H-controlled adsorption: hydrocarbons on 1,3,5-trinitrobenzene	224

Table 7 (continued)
THERMODYNAMIC DATA ON
HYDROCARBONS FROM GC ANALYSIS

Comments	Ref.
H of adsorption and effective diffusivity: C-5— C-8 hydrocarbons on Pt-alumina catalyst	225
Adsorption thermodynamics: hydrocarbons on organoclays	226
Adsorption thermodynamics: hydrocarbons on organomonillonite	227
H of adsorption: C-4—C-6 hydrocarbons on cobalt oxide	228
H of adsorption: hydrocarbon gases on Fe-Al silicate	229
Thermodynamic properties of liquid crystals using alkanes and aromatics as test solutes	230
Characterization of liquid crystal stationary phases via partition coefficients and solution thermodynamics	231
Solubilities of ethylene, n-hexane, n-octane, benzene, and toluene in low-density polyethylene	232
Solubilities of n-alkanes, cyclic alkanes and aromatics in polystyrene	233
Henry constants; benzene, n-hexane, and cyclohexane on porous glass beads	234
Adsorption isotherms: hydrocarbons on alumina	235
Diffusion coefficients; methane and benzene in polyethylene films	236
Gaseous diffusion coefficients: ethylene and n- and iso-butane in nitrogen	237
Determination of interaction parameters	238
Interaction parameters for methyl and methylene groups	239
Effect of π-electron densities and H of adsorption on separations of alkanes, alkenes, and alkynes on VCl_2, $MnCl_2$, and $CoCl_2$	240
Solution and adsorption of hydrocarbons on diethylene and triethylene glycols	241
Correlation of retention and thermodynamic functions with chemical structure	242, 243
Interactions of hydrocarbons with polyoxyethylene nonylphenyl ether stationary phases	244
Partition coefficients of light hydrocarbons in solar oil	245

Alkanes

Comments	Ref.
Thermodynamic properties, n-hexane, cyclohexane, cyclohexene, isooctane, benzene	246
Thermodynamic properties: n-alkanes	247
Vapor pressures and activity coefficients: nonanes	248
Thermodynamic significance of the retention index: n-alkanes	249
Activity coefficients: C-6—C-9 alkanes, benzene	250
Molar heats of solution: n-alkanes in squalane	251
Activity coefficients: ar-alkanes, binary alkane mixtures	252

Table 7 (continued)
THERMODYNAMIC DATA ON
HYDROCARBONS FROM GC ANALYSIS

Comments	Ref.
Activity coefficient vs. solute structure: branched nonanes	253
Interaction parameters, heats of solution, excess heat capacities: normal and branched alkanes in OV-101	254
Interaction parameters: *n*-alkanes, aromatic and cyclic hydrocarbons in polyisobutylene	255
Heats of mixing: normal and branched alkanes in Apiezon M, squalane, and OV-101	256
Enthalpy and entropy of mixing: alkanes in *n*-$C_{36}H_{74}$	257
Heats of sorption: alkanes on lithium stearate and oxystearate	258
Heats of adsorption, effective diffusivities: *n*-alkanes on Pt-alumina catalyst	259
Desorption isotherms: *n*-butane on activated carbon	260
Adsorption isotherms: *n*-pentane on silica gels	261
Diffusion coefficients: long-chain *n*-alkanes in GC silicone phases	262
Diffusion coefficients, partition coefficients, ethalphies and entropies of evaporation: *n*-alkanes on silicone polymers	263
Activity coefficients and vapor pressures: branched alkanes on SF-96 at four temperatures	264
Interaction parameters: *n*-alkanes in *n*-alkane stationary phases	265
Activity coefficients: 36 binary *n*-alkane systems at 80, 100, and 120°C	266
Thermodynamic stability: monoethylalkanes over Al halides	267
Isomerization equilibria: methyl pentanes and methyl heptanes over Al halides	268

Unsaturated Hydrocarbons

Comments	Ref.
Activity coefficients: unsaturated C-5 and C-6 *n*-hydrocarbons in normal chain hexadecane, hexadecene, octadecane, and octadecene at 35 and 50°C	269
Equilibrium constants in π-complexing systems: hexenes with $PdCl_2$ and Ag^+	270
Enthalpies and entropies of adsorption, chemical potentials: C-6—C-10 alkenes on graphitized carbon black	271
Enthalpy and entropy of isomerization, isomerization equilibria: positional and stereo isomers of hexene	272
Complex formation: unsaturated hydrocarbons with copper ion	273
Partition coefficients: unsaturated hydrocarbons and deuterated hydrocarbons	274

Table 7 (continued)
THERMODYNAMIC DATA ON
HYDROCARBONS FROM GC ANALYSIS

Comments	Ref.
Aromatic and Cyclic Hydrocarbons	
Heats of adsorption: 31 model dialkyl benzenes on GSC packings	275
Adsorption energies: aromatic hydrocarbons on graphitized carbon black	276
Sorption of benzene on Porasil C	277
Henry's constants, heats of adsorption: C-6—C-14 aromatic hydrocarbons on graphitized carbon black	278
Enthalpies, entropies, free energies: alkyl benzenes on $CoCl_2$ and $MnCl_2$	279
Linear free energies: alkyl benzenes	280
Linear free energies: identification of alkyl aromatic hydrocarbons on capillary GC	281
Enthalpies, entropies, free energies of mixing: technical divinylbenzene components on a liquid crystal stationary phase	282
Thermodynamic properties: cycloalkanes	283
Adsorption isotherm: cyclohexane on silica gel	284

REFERENCES

1. **Schaefer, R. G., Leythaeuser, D., and Weiner, B.,** *J. Chromatogr.,* 167, 355, 1978.
2. **Leinster, P., Perry, P., and Young, R. J.,** *Talanta,* 24, 205, 1977.
3. **Midkiff, C. R., Jr. and Washington, W. D.,** *J. Assoc. Offic. Anal. Chem.,* 55, 840, 1972.
4. **Kolb, B.,** *Appl. Gas Chromatogr.,* 15E, 9, 1972.
5. **Vitenberg, A. G., Ioffe, B. V., and Borisov, V. N.,** *Chromatographia,* 7, 610, 1974.
6. **Allen, P. V. and McAllister, E. W.,** U.S. Patent 3,681,997, August 8, 1972.
7. **Altamirano, I. J. J., Cortes, R. A., and Cuevas, P. A.,** *Rev. Inst. Mex. Pet.,* 11, 34, 1979.
8. **Berkley, R. E. and Pellizzari, E. D.,** *Anal. Lett.,* A11, 327, 1978.
9. **Ullrich, D. and Seifert, B.,** *Fresenius' Z. Anal. Chem.,* 291 (4), 299, 1978.
10. **Miget, R. and Kator, H.,** *Anal. Chem.,* 46, 1154, 1974.
11. **Bruce, H. E. and Cram, S. P.,** Marine Pollution Monitoring (Petroleum), NBS Spec. Publ. No. 409, Gaithersburg, Md., 1974, 181.
12. **Coleman, H. J., Dooley, J. E., Hirsch, D. E., and Thompson, C. J.,** *Anal. Chem.,* 45, 1724, 1973.
13. **May, W. E., Chesler, S. N., Cram, S. P., Gump, B. H., Hertz, H. S., Enagonio, D. P., and Dyszel, S. M.,** *J. Chromatogr. Sci.,* 13, 535, 1975.
14. **Anon.,** *Kagaku Kogyo (Chem. Fact.),* 17 (10), 66, 1973.
15. **Dogra, P. V. and Mallik, K. L.,** *Res. Ind.,* 24 (3), 177, 1979.
16. **De Greef, J., De Proft, M., and De Winter, F.,** *Anal. Chem.,* 48, 38, 1976.
17. **Roof, L. B. and DeFord, D. D.,** U.S. Patent 4,007,626, April 18, 1975.
18. **Saha, N. C.,** *Chromatogr. Polym. Pet. Petrochem. (Pap. Symp. Workshop), 1978,* 1979, 126; *Chem. Abstr.,* 92, 173956u, 1980.
19. **Mindrup, R.,** *J. Chromatogr. Sci.,* 16, 380, 1978.
20. **Malan, E., Loubser, N. H., Boshoff, A. S. J., and Eps, O.,** *Chromatographia,* 4, 475, 1971.
21. **Raymond, A. and Guiochon, G.,** *Analusis,* 2 (5), 357, 1973.
22. **Raymond, A. and Guiochon, G.,** *J. Chromatogr. Sci.,* 13, 173, 1975.
23. **Okhotnikov, B. P., Zhukhovitskii, A. A., and Krasnova, G. V.,** U.S.S.R. Patent 393,672, December 27, 1973.

24. **Swinnerton, J. W. and Lamontagne, R. A.,** *Environ. Sci. Technol.,* 8, 657, 1974.
25. **Gaspar, G., Arpino, P., and Guiochon, G.,** *J. Chromatogr. Sci.,* 15, 256, 1977.
26. **Jones, P. W., Glammer, R. D., Strup, P. E., and Stanford, T. B.,** *Environ. Sci. Technol.,* 10, 806, 1976.
27. **Daisey, J. M. and Leyko, M. A.,** *Anal. Chem.,* 51, 24, 1979.
28. **Ambrosio, M. and Di Lorenzo, A.,** *Riv. Combust.,* 29, 337, 1975.
29. **Navratil, J. D., Sievers, R. E., and Walter, H. F.,** *Anal. Chem.,* 49, 2260, 1977.
30. **Leoni, V., Puccetti, G., and Grella, A.,** *J. Chromatogr.,* 106, 119, 1975.
31. **Berezkin, V. G.,** *Anal. Chem. (Fresenius'),* 296, 1, 1979.
32. **Scheil, G. W. and Harris, W. E.,** *J. Chromatogr. Sci.,* 14, 412, 1976.
33. **Mitra, G. D., Mohan, G., and Sinha, A.,** *J. Chromatogr.,* 91, 633, 1974.
34. **Soulages, N. L. and Brieva, A. M.,** *J. Chromatogr. Sci.,* 9, 492, 1971.
35. **Diskina, D. E. and Vigdergauz, M. S.,** *Sb. Nekot. Probl. Org. Khim., Mater. Nauch. Sess., Inst. Org. Fiz. Khim., Akad. Nauk. S.S.S.R.,* 1972, 123; *Chem. Abstr.,* 78, 100017q, 1973.
36. **Franc, J. C.,** Czech. Patent 139,431, December 15, 1970.
37. **DiSanzo, G. P., Uden, P. C., and Siggia, S.,** *Anal. Chem.,* 51, 1529, 1979.
38. **Kugucheva, Ye. Ye., and Alekseyeva, A. V.,** *Sb. Tr. Nauchn. Issled. Inst. Sintet. Spirtov Organ. Prod.,* 5, 165, 1974.
39. **Nand, S., Desai, B. K., and Sarkar, M. K.,** *J. Chromatogr.,* 133, 359, 1977.
40. **Nigam, R. N. and Moochandra, R.,** *Anal. Chem.,* 43, 1683, 1971.
41. **Beroza, M.,** in *Column Chromatography,* Kovats, E. sz., Ed., Swiss Chemists' Association, 1970, 92.
42. **Kugucheva, E. E. and Alekseeva, A. V.,** *Usp. Khim.,* 42, 2247, 1973.
43. **Ivanov, A. and Eisen, O.,** *J. Chromatogr.,* 69, 53, 1972.
44. **Schulz, H., Sedighi, N., Gregor, H. B., Van, T. D., and Min, S. S.,** *Dtsch. Ges. Mineraloelwiss. Kohlechem. e. V., Compend.,* 1976, 4509; *Chem. Abstr.,* 90, 124186t, 1979.
45. German Society for Petroleum Sciences and Coal Chemistry, *Ber. Dtsch. Ges. Mineraloelwiss. Kohlechem.,* 1978, 139; *Chem. Abstr.,* 91, 41653n, 1979.
46. **Schulz, H.,** *Erdoel Kohle, Erdgas, Petrochem. Brennst. Chem.,* 32, 391, 1979.
47. **Simon, A., Palagyi, J., Speier, G., and Furedi, Z.,** *J. Chromatogr.,* 150, 135, 1978.
48. **Ten Noever de Brauw, M. C.,** *J. Chromatogr.,* 165, 3, 1979.
49. **Simon, A., Palagyi, J., Speiger, G., and Furedi, Z.,** *J. Chromatogr.,* 150, 135, 1978.
50. **Simon, A., Palagyi, J., Speier, G., and Furedi, Z.,** *J. Chromatogr.,* 150, 135, 1978.
51. **Nametkin, N. S., Kolesnikova, L. P., Tyurin, V. D., Nekhaev, A. I., and Potapova, L. G.,** *Neftekhimiya,* 15 (5), 763, 1975.
52. **Matsuoka, S., Takano, T., and Tamura, T.,** *Bunseki Kagaku,* 27, 777, 1978.
53. **Schulz, H.,** *Erdoel Kohle, Erdgas, Petrochem. Brennst. Chem.,* 30, 182, 1977.
54. **Poznyak, T. I., Lisitsyn, D. M., D'yachkovskii, F. S., and Razumovskii, S. D.,** *Zh. Anal. Khim.,* 32, 783, 1977.
55. **Kuklinskii, A. Ya. and Pushkina, R. A.,** *Neftekhimiya,* 13, 467, 1973.
56. **Simmonds, P. G. and Smith, C. F.,** *Anal. Chem.,* 44, 1548, 1972.
57. **Kugucheva, E. E. and Alekseeva, A. V.,** *Khim. Tekhnol. Topl. Masel,* 17 (3), 50, 1972.
58. **McDonough, L. M. and George, D. A.,** *J. Chromatogr. Sci.,* 8, 158, 1970.
59. **Moore, B. P. and Brown, W. V.,** *J. Chromatogr.,* 60, 157, 1971.
60. **Harbourn, C. L. A. and Kirby, F. B.,** British Patent 1,190,591, May 6, 1970.
61. **Francis, G. W. and Tande, T.,** *J. Chromatogr.* 150, 139, 1978.
62. **Langer, S. H., Melton, H. R., and Griffith, T. D.,** *J. Chromatogr.,* 122, 487, 1976.
63. **Braghin, O. V., Preobrazhenskii, A. V., and Liberman, A. L.,** *Izv. Akad. Nauk. S.S.S.R., Ser. Khim.,* 1974, 2751.
64. **Johnston, A. E. and Dutton, H. J.,** *J. Am. Oil Chem. Soc.,* 49, 98, 1972; **Anon.,** *Ind. Res.,* 14 (6), C1, C4, 1972.
65. **Ohtaki, T., Ozaki, H., Yata, N., Itoh, M., and Suzuki, A.,** *Nippon Kagaku Kaishi,* 1972 (5), 934.
66. **Kugucheva, E. E. and Alekseeva, A. V.,** *Neftekhimiya,* 10 (5), 778, 1970.
67. **Schulz, H., Gregor, B., et al.,** *Dtsch. Ges. Mineraloelwiss. Kohlechem. e. V., Compend.,* 2, 1407, 1978— 79.
68. Texaco Development Corp., French Patent 2,137,088, February 2, 1973.
69. **Goryaeva, E. M., Azerbaeva, R. G., and Shomenkova, V. V.,** *Khim. Tekhnol. (Alma-Ata),* 1972, 128; *Chem. Abstr.,* 80, 135587u, 1974.
70. **Ivanov, A. and Eisen, O.,** *J. Chromatogr.,* 69, 53, 1972.
71. **Stamm, R. E.,** E. German Patent 2,118,538, October 26, 1972; *Chem. Abstr.,* 78, 32382c, 1973.
72. **Diskina, D. E., Lazareva, I. S., Burmistrov, G. G., and Kurbskii, G.,** *Neftepererab. Neftekhim. (Moscow),* 8, 3, 1971; *Chem. Abstr.,* 75, 119706y, 1971.

73. **Takeuchi, T., Seno, H., Tsuge, S., and Ishii, Y.,** *Bunseki Kagaku,* 19, 1424, 1970; *Chem. Abstr.,* 74, 94137c, 1971.

74. **Franc, J. and Pour, J.,** *J. Chromatogr.,* 131, 291, 1977.

75. **Koci, K.,** *Bul. Shkencave Nat.,* 32, 59, 1978; *Chem. Abstr.,* 90, 171106b, 1979.

76. **Nigam, R. N.,** *J. Chromatogr.,* 119, 620, 1976.

77. **Nigam, R. N. and Moolchandra, R.,** *Anal. Chem.,* 43, 1683, 1971.

78. **Ohtaki, T., Ozaki, H., Yata, N., Itoh, M., and Suzuki, A.,** *Nippon Kagaku Kaishi,* 1972, 934; *Chem. Abstr.,* 77, 42894q, 1972.

79. **Antonucci, P. and Giordana, N.,** *J. Chromatogr.,* 150, 309, 1978.

80. **Scott, K. F. and Phillips, C. S. G.,** *J. Chromatogr.,* 112, 61, 1975.

81. **Dubonska, V., Dubanska, V., and Hejl, V.,** *Rudy,* 23, 383, 1975.

82. **Ramljak, Z., Deur-Siftar, D., and Solc, A.,** *J. Chromatogr.,* 119, 445, 1976.

83. **Uden, P., Henderson, D. E., and Lloyd, R. J.,** *J. Chromatogr.,* 126, 225, 1976.

84. **Afanas'ev, M. I., Kozlov, S. P., Lozovskii, V. M., and Datskevich, A. A.,** *Tr. Vses Nauchno-Issled. Proektno-Konstr. Inst. Kompleksn. Avtom. Neft. Gazov. Prom-sti.,* 1973 (5), 289; *Chem. Abstr.,* 81, 155429t, 1974.

85. **Herian, J. and Mokra, M.,** *Ropa Uhlie,* 15, 547, 1973; *Chem. Abstr.,* 80, 61803n, 1974.

86. **Ristau, W. T. and Vanderborgh, N. E.,** *Anal. Chem.,* 43, 702, 1971.

87. **Folmer, O. F., Jr. and Azarraga, L. V.,** *J. Chromatogr. Sci.,* 7, 665, 1969.

88. **Packakova, V. and Kozlik, V.,** *Chromatographia,* 11, 266, 1978.

89. **Hanson, R. L., Brookins, D., and Vanderborgh, N. E.,** *Anal. Chem.,* 48, 2210, 1976.

90. **Herian, J. and Mokra, M.,** *Ropa Uhlie,* 15 (10), 547, 1973.

91. **Topping, J. J.,** *Diss. Abstr. Int. B,* 30 (9), 4017, 1970.

92. **Higgins, T. N. and Harris, W. E.,** *J. Chromatogr. Sci.,* 11, 588, 1973.

93. **Brown, R. A.,** *Anal. Chem.,* 43, 900, 1971.

94. **Lukas, S.,** *Chromatographia,* 12, 17, 1979.

95. **Braekman-Danheux, C. and Bredael, P.,** *J. Chromatogr.,* 106, 395, 1975.

96. **Krishen, A. and Tucker, R. G.,** *Anal. Chem.,* 46, 29, 1974.

97. **Brown, R. A.,** *Anal. Chem.,* 43, 900, 1971.

98. **Bajus, M. and Vesely, V.,** *Ropa Uhlie,* 21 (7), 375, 1979.

99. **Gough, T. A. and Walker, E. A.,** *J. Chromatogr. Sci.,* 8, 134, 1970.

100. **Ramljak, Z., Solc, A., Arpino, P., Schmitter, J. M., and Guichon, G.,** *Anal. Chem.,* 49, 1222, 1977.

101. **Tsuge, S. and Takeuchi, T.,** *Anal. Chem.,* 49, 348, 1977.

102. **Seeger, M., Cantow, H.-J., and Marti, S.,** *Z. Anal. Chem.,* 276, 267, 1975.

103. **Gough, T. A. and Jones, C. E. R.,** *Chromatographia,* 8, 696, 1975.

104. **Walker, J. Q.,** *J. Chromatogr. Sci.,* 15, 267, 1977.

105. **Zizin, V. G., Grigorieva, L. A., and Berdina, L. Kh.,** *Plast. Massy,* 1, 55, 1976.

106. **Voorhees, K. J., Hileman, F. D., and Einhorm, I. N.,** *Anal. Chem.,* 47, 2385, 1975.

107. **Bell, J. P., Edwards, R. V., Nott, B. R., and Angus, J. C.,** *Ind. Eng. Chem. Fundam.,* 13, 89, 1974.

108. **Buniyat-Zade, A. A., Androsova, V. M., Bulatnikova, E. L., Ryabova, T. M., and Efendieva, T. Z.,** *Vysokomol. Soedin. Ser. A,* 12, 2494, 1970.

109. **Armitage, F.,** *J. Chromatogr. Sci.,* 9, 245, 1971.

110. **Abo, T. and Watanabe, T.,** *Kogyo Kagaku Zasshi,* 74 (5), 885, 1971.

111. **Alekseeva, K. V., Khramova, L. P., and Strel'nikova, I. A.,** *Zavod. Lab.,* 36 (11), 1304, 1970.

112. **Svob, V. and Deur-Siftar, D.,** *J. Chromatogr.,* 135, 85, 1977.

113. **Svob, V., Deur-Siftar, D., and Cremers, C. A.,** *J. Chromatogr.,* 91, 659, 1974.

114. **Korshak, V. V., Mozgova, K. K., Val'kovskii, D. G., and Khomutov, V. A.,** *Vysokomol. Soedin. Ser. B,* 13 (9), 695, 1971.

115. **Gonzalez, D. V., Gonzalez, A. A., and Garcia, M. F.,** *An. Quim,* 75, 865, 1979.

116. **Dahlmann, G., Koeser, H. J. K., and Oelert, H. H.,** *Chromatographia,* 12, 665, 1979.

117. **Zhukhovitskii, A. A., Yanovskii, S. M., and Shvarzman, V. P.,** *J. Chromatogr.,* 119, 591, 1976.

118. **Zhukhovitskii, A. A., Yanovskii, S. M., and Shavrzman, V. P.,** *J. Chromatogr.,* 119, 591, 1976.

119. **Green, L. E.,** *Hydrocarbon Process,* 55 (5), 205, 1976.

120. **Ford, D. C., Miller, W. H., Thren, R. C., and Wertzler, R.,** Spec. Tech. Publ. No. 557, *American Society Testing and Materials,* Philadelphia, 1975.

121. **Jackson, B. W., Judges, R. W., and Powell, J. L.,** *J. Chromatogr. Sci.,* 14 (2), 49, 1976.

122. **Hickerson, J. F.,** Spec. Tech. Publ. No. 557, *American Society Testing and Materials,* Philadelphia, 1975.

123. **Mikkelsen, L. and Green, L. E.,** *J. Chromatogr. Sci.,* 14 (4), 190, 1976.

124. **Crawford, W. J. and Overton, R. A.,** Paper 424, in *Pitt. Conf. Anal. Chem. Appl. Spectrosc.,* 1975.

125. **Geary, S., Hoberecht, H. D., Klimowski, R. J., and March, E. W.,** Paper 425, in *Pitt. Conf. Anal. Chem. Appl. Spectrosc.,* 1975.

126. **Gol'dshtein, K. R., Frenkel, B. A., Lerman, A. G., Lipavskii, V. N., Zhuravleva, M. A., and Vasin, L. S.,** *Khim. Tekhnol. Topl. Massel,* 1976 (9), 17.

127. **Bird, W. L. and Kimball, J. L.,** Spec. Tech. Publ. No. 557, American Society Testing Materials, Philadelphia 1975.

128. **Rion, J. C. and Wallaert, B.,** *Analusis,* 1, 224, 1972.

129. **Rion, J. C. and Wallaert, B.,** *Analusis,* 1, 224, 1972.

130. **Sokolova, V. I., Berg, G. A., Shklovskii, Ya. A., and Shokurova, N. I.,** *Neftekhimiya,* 11, 931, 1971.

131. **O'Donnell, R. J.,** *Ind. Eng. Chem. Process Des. Develop.,* 12, 208, 1973.

132. **Sojak, L., Krupcik, J., and Rijks, J.,** *Chromatographia,* 7, 1974.

133. **Klement, I.,** *J. Chromatogr.,* 87, 401, 1973.

134. **Green, L. E.,** *Riv. Combust.,* 14, 536, 1970.

135. **Bach, R. W., Doetsch, E., Friedrichs, H. A., and Marx, L.,** *Chromatographia,* 4 (12), 561, 1971.

136. **Gouw, T. H., Wittemore, I. M., and Jentoft, I. M., Jr.,** *Anal. Chem.,* 42, 1394, 1970.

137. **Weingaertner, E., Guer, T., and Bayunus, O.,** *Z. Anal. Chem.,* 254, 28, 1971.

138. **Makushina, V. M., Aref'ev, O. A., et al.,** *Neftekhimiya,* 10, 165, 1970.

139. **Stuckey, C. L.,** *J. Chromatogr. Sci.,* 16, 482, 1978.

140. **Stuckey, C. L.,** *J. Chromatogr. Sci.,* 16, 482, 1978.

141. **Dracheva, S. I., Bryanskaya, E. K., Sabirova, G. V., and Zhurba, A. S.,** *Rasshir. Utochnenie Programmy Issled. Neftei, Sb. Kzbr. Dokl. Mater. Vses. Nauchno-Tekh. Konf.,* p. 139, 1976; *Chem. Abstr.,* 84, 186858k, 1978.

142. **Otozai, K.,** *Z. Anal. Chem.,* 268, 257, 1974.

143. **Sojak, L., Krupcik, J., and Rijks, J.,** *Chromatographia,* 7, 26, 1974.

144. **Kimball, J. L. and McCracken, E. A.,** *Am. Soc. Test. Mater., Annu. Meet.,* Montreal, 1975.

145. **Klesment, I.,** *J. Chromatogr.,* 87, 401, 1973.

146. **Trusell, F. C. and Mikulas, L. J.,** *Am. Chem. Soc. Div. Petrol. Chem. Prepr.,* 18, 517, 1973.

147. **O'Donnell, R. J.,** *Ind. Eng. Chem. Process Des. Develop.,* 12, 208, 1973.

148. **Green, L. E.,** *Anal. Instrum.,* 8 (VI-2), 1, 1970.

149. **Jackson, B. W., Judges, R. W., and Powell, J. L.,** *J. Chromatogr. Sci.,* 14, 49, 1976.

150. **de Bruine, W. and Ellison, R. J.,** *Am. Soc. Test. Mater.,* Comm. D-2 Symp., Prepr., Dallas, 1973.

151. **Yakimenko, L. V. and Chebonenko, N. D.,** *Neftepererab. Neftekhim. (Moscow),* 1972 (4), 13.

152. **Zhukhovitskii, A. A., Yanovskii, S. M., and Shvartsman, V. P.,** *Zh. Fiz. Khim.,* 52, 1442, 1978; *Chem. Abstr.,* 89, 239965Z, 1978.

153. **Bender-Ogly, A. O., Arustamova, L. G., Sultanov, N. T., and Babaev, F. R.,** *Zerb. Khim. Zh.,* 1977 (1), 129.

154. **Sojak, L., Druscova, A., and Janak, J.,** *Ropa Uhlie,* 14, 238, 1972.

155. **Sultanov, N. T. and Arustamova, L. G.,** *J. Chromatogr.,* 115, 553, 1975.

156. **Voitkevish, S. A., Shchedrina, M. M., Prilepskaya, K. K., and Rudolfi, T. A.,** *Maslob—Zh. Prom.,* 40 (9), 29, 1974.

157. **Saha, N. C. and Nitra, G. D.,** *J. Chromatogr. Sci.,* 11, 419, 1973.

158. **Valentin, P. and Guiochon, G.,** *Sep. Sci.,* 10, 289, 1975.

159. **Roz, B., Bonmati, R., Hagenbach, G., and Valentin, P.,** *J. Chromatogr. Sci.,* 14, 367, 1976.

160. **Sakodynskii, K. I. and Volkov, S. A.,** *Preparative Gas Chromatography,* Khimiya, Moscow, 1972.

161. **Szepesy, L.,** *Magy. Kem. Kapja,* 25, 376, 1970.

162. **DeClerk, K. and Pretorius, V.,** *Sep. Sci.,* 6, 401, 1971.

163. **Hupe, K. P.,** *J. Chromatogr. Sci.,* 9, 11, 1971.

164. **Musser, W. N. and Sparks, R. E.,** *J. Chromatogr. Sci.,* 9, 116, 1971.

165. **Landis, W. R., Perrine, T. D., and Thomas, G.,** *J. Chromatogr.,* 53, 125, 1970.

166. **Weinreich, P.,** *Chromatographia,* 3, 1971.

167. **Albrecht, J. and Verzele, M.,** *J. Chromatogr. Sci.,* 8, 586, 1970.

168. **Kolesnikova, R. D. and Egel'skaya, L. P.,** *Preparative Gas Chromatography of Light Hydrocarbons,* Khimiya, Moscow, 1970.

169. **Carter, D. E. and Esterson, G. L.,** *Ind. Eng. Chem. Fundam.,* 9, (4), 661, 1970.

170. **Gelpi, E., Wszolek, P. C., Yang, E., and Burlingame, A. L.,** *Anal. Chem.,* 43, 864, 1971.

171. **Gelpi, E., Wszolek, P. C., Yang, E., and Burlingame, A. L.,** *J. Chromatogr. Sci.,* 9, 147, 1971.

172. **Carel, A. B., Clement, R. E., and Perkins, G.,** *J. Chromatogr. Sci.,* 7, 218, 1969.

173. **Zabokritskii, M. P., Rudenko, B. A., and Chizhkov, V. P.,** *Izv. Akad. Nauk, SSSR, Ser, Khim.,* 1977, 1627.

174. **Hopf, S., Ecknig, W., and Thurmer, H.,** *J. Chromatogr.,* 139, 249, 1977.

175. **Borisov, S. B., Datskevich, N. A., and Tolmachev, A. M.,** *Vestn. Mosk. Univ. Khim.,* 17 (1), 122, 1976.

176. **Raude, H., Eisen, O., Saks, T., and Talvari, A.,** *Eesti NSV Tead. Akad. Toim. Keem. Geol.,* 21 (3), 224, 1972.

177. Mobil Oil Corp., British Patent 1,334,302, April 22, 1971.
178. **Maryakhin, R. Kh. and Vigdergauz, M. S.,** *Zavod. Lab.,* 42, 1314, 1976.
179. **Smol'skii, A. M., Kulikov, V. I., and Egiazarov, Yu. G.,** *Vestsi Akad. Navuk. B.S.S.R. Ser. Khim. Navuk,* 1976 (6), 51.
180. **Semonian, B. P. and Rogers, L. B.,** *J. Chromatogr. Sci.,* 16, 49, 1978.
181. **Fuzita, K., Shimokobe, I., and Nakazima, F.,** *Nippon Kagaku Kaishi,* 8, 1348, 1975.
182. **Ecknig, W. and Polster, H.-J.,** *Chem. Tech. (Leipzig),* 31, 245, 1979.
183. **Asche, W.,** *Chromatographia,* 11, 411, 1978.
184. **Bartmann, D.,** *Ber. Bunsenges. Phys. Chem.,* 76, 336, 1972.
185. **Gouw, T. H. and Jentoft, R. E.,** *J. Chromatogr.,* 68, 303, 1972.
186. **Conaway, J. E., Graham, J. A., and Rogers, L. B.,** *J. Chromatogr. Sci.,* 16, 102, 1978.
187. **Conaway, J. E., Smith, E. F., and Rogers, L. B.,** *Chem. Instrum.,* 8, 167, 1978.
188. **Jentoft, R. E. and Gouw, T. H.,** *Anal. Chem.,* 48, 2195, 1976.
189. **Nygren, S.,** *J. High Resolut. Chromatogr. Chromatogr. Commun.,* 2, 319, 1979.
190. **Hermann, K. and Sasse, H.,** E. German Patent 109,265, October 20, 1974; *Chem. Abstr.,* 83, 150226f, 1975.
191. Antar Petroles de L'Atlantique, French Patent 2,177,142, December 7, 1973.
192. **Semonian, B. P. and Rogers, L. B.,** *J. Chromatogr. Sci.,* 16, 49, 1978.
193. **Ilkova, E. L. and Mistryukov, E. A.,** *J. Chromatogr.,* 54, 422, 1971.
194. **Grant, D. W. and Hollis, M. G.,** *J. Chromatogr.,* 158, 3, 1978.
195. **Struppe, H. G.,** *Zh. Fiz. Khim.,* 46 (7), 1782, 1972.
196. **Grant, D. W. and Hollis, M. G.,** *J. Chromatogr.,* 158, 3, 1978.
197. **De Bruine, W. and Ellison, R. J.,** Spec. Tech. Publ. No. 557, American Society Testing and Materials, Philadelphia, 1975.
198. **Luskin, M. M. and Morris, W. E.,** Spec. Tech. Publ. No. 557, American Society Testing and Materials, Philadelphia, 1975.
199. **Sojak, L., Krupcik, J., and Rijks, J.,** *Chromatographia,* 7, 26, 1974.
200. **Parcher, J. F., Weiner, P. H., et al.,** *J. Chem. Eng. Data,* 20, 145, 1975.
201. **Janini, G. M. and Martire, D. E.,** *J. Phys. Chem.,* 78, 1644, 1974.
202. **Tewari, Y. B., Martire, D. E., and Sheridan, J. P.,** *J. Phys. Chem.,* 74, 2345, 1970.
203. **Hradetzky, G., Hauthal, H. G., and Bittrich, H. J.,** *Z. Chem.,* 19 (6), 224, 1979.
204. **Guermouche, M. H. and Vergnaud, J. M.,** *J. Chromatogr.,* 94, 25, 1974.
205. **Chow, L. E. and Martire, D. E.,** *J. Phys. Chem.,* 75, 2005, 1971.
206. **Wicarova, O., Novak, J., and Janak, J.,** *J. Chromatogr.,* 65, 241, 1972.
207. **Bittrich, H. J., Sarius, A., Muehlstaedt, M., and Seifert, A.,** *Z. Chem.,* 20 (8), 304, 1980.
208. **Pacakova, V. and Ullmannova, H.,** *Chromatographia,* 7, 75, 1974.
209. **Shevchuk, I. M., Bogoslovskii, Y. N., and Sakharov, V. M.,** *Neftekhimiya,* 10, 913, 1970.
210. **Dondi, F., Betti, A., and Bighi, C.,** *J. Chromatogr.,* 66, 191, 1972.
211. **Castells, R. C.,** *J. Chromatogr.,* 111, 1, 1976.
212. **Korol, A. N.,** *Chromatographia,* 8, 385, 1975.
213. **Figgins, C. E., Reinbold, B. L., and Risby, T. H.,** *J. Chromatogr. Sci.,* 15, 208, 1977.
214. **Figgins, C. E., Risby, T. H., and Jurs, P. C.,** *J. Chromatogr. Sci.,* 14, 453, 1976.
215. **Nesterov, A. Ye. and Lipatov, Yu. S.,** *Vysokomol. Soedin.,* 16, 1919, 1974.
216. **Gawdzik, J. and Jaroniec, M.,** *J. Chromatogr.,* 131, 1, 1977.
217. **Mathur, D. S. and Saha, N. C.,** *J. Chromatogr.,* 138, 33, 1977.
218. **Hradil, J. and Stamberg, J.,** Part 1, 4th Natl. Conf., Pardubice, 1976, 115.
219. **Moro-oka, Y.,** *Trans. Faraday Soc.,* 67, 3381, 1971.
220. **Kalaschnikova, E. V., Kiselev, A. V., Petrova, R. S., and Shcherbakova, K. D.,** *Chromatographia,* 4, 495, 1971.
221. **Vigdergauz, M. S., Pomazanov, V. V., Bogdanchikov, A. I., and Gunchenko, E. I.,** *Izv. Akad. Nauk S.S.S.R. Ser.Khim.,* 1972, 646.
222. **Belyakova, L. D., Kiselev, A. V., and Soloyan, G. A.,** *Chromatographia,* 3, 254, 1970.
223. **Artemova, A. A., Keibal, V. L., Nikitin, Yu. S., and Shcherbakova, A. A.,** *Vestn. Mosk. Univ. Ser. 2. Khim.,* 22 (2), 172, 1981.
224. **Akimoto, M. and Echigoya, E.,** *Bull. Chem. Soc. Jpn.,* 49 (12), 3687, 1976.
225. **Choudhary, V. R. and Menon, P. G.,** *J. Chromatogr.,* 116, 431, 1976.
226. **Gavrichev, V. S. and Berezkin, V. G.,** *Zh. Fiz. Khim.,* 49 (1), 269, 1975.
227. **Bondarenko, S. V., Zhukova, A. I., Vdovenko, N. V., and Tarasevich, Yu. I.,** *Kolloidn. Zh.,* 36 (5), 845, 1974.
228. **Morooka, Y.,** *Trans. Faraday Soc.,* 67 (Pt. 11), 3381, 1971.
229. **Gornak, A. I. and Skurko, O. F.,** *Org. Katal.,* 1970, 65.

230. **Jeknavorian, A. A. and Barry, E. F.,** *J. Chromatogr.*, 101, 299, 1974.
231. **Jeknavorian, A. A., Barrett, P., Watterson, A. C., and Barry, E. F.,** *J. Chromatogr.*, 107, 317, 1975; **Kelker, H.,** *J. Chromatogr.*, 112, 165, 1975; **Bocquet, J. F. and Pommier, C.,** *J. Chromatogr.*, 117, 315, 1976.
232. **Maloney, D. P. and Prausnitz, J. M.,** *Am. Inst. Chem. Eng. J.*, 22, 74, 1976.
233. **Steel, L. I. and Harnish, D. G.,** *Am. Inst. Chem. Eng. J.*, 22, 117, 1976.
234. **Gawdzik, J., Suprynowicz, Z., and Jaroniec, M.,** *J. Chromatogr.*, 121, 185, 1976.
235. **Neumann, M. G.,** *J. Chem. Educ.*, 53, 708, 1976.
236. **Rosolovskaya, Ye. N. and Salvinski, Ya.,** *Vysokomol. Soedin*, 18, 1428, 1976.
237. **Garusov, A. V. and Vigdergauz, M. S.,** *Zasvod. Lab.*, 41, 1481, 1975.
238. **Marcille, P., Audebert, R., and Quivoron, C.,** *J. Chim. Phys.*, 72, 72, 1975.
239. **Sugiyama, T., Takeuchi, T., and Suzuki, Y.,** *J. Chromatogr.*, 105, 273, 1975.
240. **Grob, R. L. and McGonigle, E. J.,** *J. Chromatogr.*, 59, 13, 1971.
241. **Arancibia, E. L. and Catoggio, J. A.,** *J. Chromatogr.*, 197, 135, 1980.
242. **Zielinski, W. L., Jr.,** Ph.D. thesis, Georgetown University, Washington, D.C., 1972.
243. **Nabivach, V. M. and Kirilenko, A. V.,** *Chromatographia*, 13, 29, 1980; 13, 93, 1980.
244. **Konno, K., Sakiyama, M., and Kithara, A.,** *Yukagaku*, 23 (8), 501, 1974.
245. **Simon, J.,** *Chromatographia*, 4, 98, 1971.
246. **Kuchhal, R. K. and Mallik, K. L.,** *Anal. Chem.*, 51, 392, 1979.
247. **Karaiskakis, G., Lycourghiotis, A., and Katsanos, N. A.,** *Z. Phys. Chem. (Wiesbaden)*, 111, 207, 1978.
248. **Castello, G. and D'Amato, G.,** *J. Chromatogr.*, 116, 249, 1976.
249. **Kalashnikova, E. V., Kiselev, A. V., Poshkus, D. P., and Sh'cherbakova, K. D.,** *J. Chromatogr.*, 119, 233, 1976.
250. **Martire, D. E.,** *Anal. Chem.*, 46, 626, 1974.
251. **Packakova, V., Suejda, J., and Smolkova, E.,** *Czech. Chem. Commun.*, 42, 2850, 1977.
252. **Laub, R. J., Martire, D. E., and Purnell, J. H.,** *J. Chem. Soc., Faraday Trans.*, 73, 1686, 1977; 74, 213, 1978.
253. **Castello, G. and D'Amato, G.,** *J. Chromatogr.*, 116, 249, 1976.
254. **Hammers, W. E. and De Ligny, C. L.,** *J. Polym. Sci. Polym. Phys. Ed.*, 12, 2065, 1974.
255. **Leung, Y. and Eichinger, B. E.,** *J. Phys. Chem.*, 78, 60, 1974.
256. **Manners, W. E., Bos, B. C., Vaas, L. H., Loomans, Y. J. W. A., and De Ligny, C. L.,** *J. Polym. Sci. Polym. Phys. Ed.*, 13, 401, 1975.
257. **Meyer, E. F. and Baiocchi, F. A.,** *Anal. Chem.*, 49, 1029, 1977.
258. **Mysak, A. E., Nikitina, N. S., Mankovskaya, N. K., and Ermolenko, I. V.,** *Kolloid. Zh.*, 36, 1175, 1974.
259. **Choudhary, V. R. and Menon, P. G.,** *J. Chromatogr.*, 116, 431, 1976.
260. **Semonian, B. P. and Manes, M.,** *Anal. Chem.*, 49, 991, 1977.
261. **Aleksnis, O. N. and Yanovskii, S. M.,** *Zavod. Lab.*, 41, 976, 1975.
262. **Kong, J. M and Hawkes, S. J.,** *J. Chromatogr. Sci.*, 14, 279, 1976.
263. **Millen, W. and Hawkes, S.,** *J. Chromatogr. Sci.*, 15, 148, 1977.
264. **Castello, G. and D'Amato, G.,** *J. Chromatogr.*, 107, 1, 1975.
265. **Sugiyama, T., Takeuchi, T., and Suzuki, Y.,** *J. Chromatogr.*, 105, 265, 1975.
266. **Parcher, J. F. and Yun, K. W.,** *J. Chromatogr.*, 99, 193, 1974.
267. **Morozova, O. Y., Zemskova, Z. K., and Petrov, A. A.,** *Neftekhimiya*, 12, 635, 1972.
268. **Roganov, C. N., Kabot, G. Y., and Andreyevskii, D. N.,** *Neftekhimiya*, 12, 495, 1972.
269. **Letcher, T. M. and Marcicano, F.,** *J. Chem. Thermodyn.*, 6, 501, 1974.
270. **Kraitr, M., Komers, R., and Cuta, F.,** *Anal. Chem.*, 46, 974, 1974.
271. **Eisen, O. G., Kiselev, A. V., Pilt, A. E., Rang, S. A., and Shcherbakova, K. D.,** *Chromatographia*, 4, 448, 1971.
272. **Radyuk, Z. A., Kabo, G. Y., and Andreyevskii, D. N.,** *Neftekhimiya*, 12, 679, 1972.
273. **Doering, C. E., Geyer, R., and Burkhardt, G.,** *Z. Chem.*, 15 (8), 319, 1975.
274. **Wasik, S. P. and Tsang, W.,** *J. Phys. Chem.*, 74 (15), 2970, 1970.
275. **Bezukhanova, T. and Dimitrov, K.,** *J. Catal.*, 22, 145, 1971.
276. **Hofmann, H. F., Engewald, W., Heidrich, D., Porschmann, J., Thieroff, K., and Uhlmann, P.,** *J. Chromatogr.*, 115, 299, 1975.
277. **Deininger, G., Asshauer, J., and Halasz, I.,** *Chromatographia*, 8, 143, 1975.
278. **Kalashnikova, E. V., Kiselev, A. V., and Shcherbakova, K. D.,** *Chromatographia*, 7, 2, 1974.
279. **McGonigle, E. J. and Grob, R. L.,** *J. Chromatogr.*, 101, 39, 1974.
280. **Ono, A.,** *J. Chromatogr.*, 110, 233, 1975.
281. **Krupcik, J., Liska, O., and Sojak, L.,** *J. Chromatogr.*, 51, 119, 1970.
282. **Zielinski, W. L., Jr., Freeman, D. H., Martire, D. E., and Chow, L. C.,** *Anal. Chem.*, 42, 176, 1970.

283. **Karaiskakis, G., Lychourghiotis, A., and Katsanos, N. A.,** *Z. Phys. Chem. (Wiesbaden),* 111, 207, 1978.
284. **Waksmundzki, A., Jaroniec, M., and Suprynowicz, Z.,** *J. Chromatogr.,* 110, 381, 1975.

Chapter 6

ENVIRONMENTAL STUDIES ON HYDROCARBONS

This chapter provides literature references concerning air pollution studies on hydrocarbons (Section I) and on hydrocarbons in oil spills and other water pollution studies (Section (II).

I. AIR POLLUTION

Literature references are provided in Table 1 on: (1) Polycyclic Aromatic Hydrocarbons and on (2) Other Hydrocarbons.

II. WATER POLLUTION

Literature references are provided in Table 2 on: (1) Oil Spills and on (2) Other Water Pollution Studies.

Table 1
AIR POLLUTION

Comments	Ref.

Polycyclic Aromatic Hydrocarbons

Comments	Ref.
Auto exhaust analysis	1—8
Analysis in various air pollution samples	9—26
Coke ovens	27,28
Air particulates	29—40
Atmospheric dust: review article (150 references)	41
Urban air	42
Tokyo air particulates	43
PAH in air pollution: review article (1055 references)	44
Identification of over 100 unsubstituted and alkylated PAH	45
Identification of PAH in coal, wood, and kerosene combustion	46
Analysis of over 40 PAH	47
Polyvinylchloride smoke particulates	48
Cigarette and tobacco smoke	49—51
PAH in soot	52,53
Photodecomposition of atmospheric PAH	54
Analysis of PAH in air using high efficiency packed columns	55—57
PAH and methylated PAH in petroleum pitch volatiles	58
PAH in combustion emissions using Tenax preconcentration	59
Diesel exhaust particulates	60
PAH from high vol filter collections	61,62
PAH in working atmospheres	63
PAH and aliphatic hydrocarbons in air particulates	64
Use of GC-MS in the analysis of PAH in air pollution studies	17,20,21,26,31—33,35,42,47,51,52,60—63

Other Hydrocarbons

Comments	Ref.
Auto exhaust	65—69
Aromatic hydrocarbons in ambient air	70—74
Jet engine exhaust hydrocarbons: substractive GC	75
GC profiles in routine analysis of exhaust hydrocarbons	76
C-1—C-2 and C-6—C-20 hydrocarbons: freeze-out loop collection	77
Atmospheric C-1—C-3 hydrocarbons	78
C-1—C-5 hydrocarbons in air: automatic GC measurement	79
C-2—C-5 hydrocarbons in air	80,81
C-1—C-5 hydrocarbons in air	82,83
C-1—C-6 alkanes and alkenes	84,85
C-2—C-6 alkanes and alkenes	86,87
C-2—C-11 olefins and aromatics; capillary GC	88
C-2—C-5 alkanes	89
C-5—C-12 hydrocarbons	90
C-6—C-20 organics: GC-MS ultratrace analysis	91
C-8—C-18 hydrocarbons in Paris air	92
C-3 and C-5—C-8 *n*-alkanes, butadiene, ketene, methylcyclohexane, cumene, cyclohexane, ethylbenzene, styrene, terphenyl, vinyl toluene	93
C-15—C-36 *n*-alkanes in air particulates	94

Table 1 (continued)
AIR POLLUTION

Comments	Ref.
Hydrocarbons: capillary GC	95
Light hydrocarbons	96
Alkanes	97
Nonreactive hydrocarbons	98
Sampling of hydrocarbons	99
Unsaturated hydrocarbons: modification of commercial hydrocarbon analyzer	100
Benzene derivatives	101
Hydrocarbons: personal sampler	102
Hydrocarbons in air: HPLC used for sample preparation	103
Ethylene in air: use of porous polymers for sample preconcentration	104
Nonmethane hydrocarbon analysis using backflush GC method	105
Stack gas GC analyzer	106
Benzene and alkyl benzenes	107
Analysis of benzene in air	108
EPA reports on benzene in air analysis	109
Hydrocarbons from coal and oil shale combustion: GC-MS analysis	110
EPA method for nonmethane hydrocarbons	111
Styrene analysis	112
Pentane, hexane, benzene, toluene	113
Acetylene in air	114
Ethylene analysis	115,116
Benzene, toluene, xylene	117—119
Cold trap sampling for hydrocarbons in air	120—122
Aromatic hydrocarbons: piezoelectric detector	123
Alkanes: TLC cleanup	124
Review of GC methods for hydrocarbon analysis	125
High-molecular-weight alkanes and alkenes	126
Automated GC system for analysis of sub-ppb level hydrocarbons in air	127
Aromatic hydrocarbons: quantitative analysis in city air	128
Aromatic hydrocarbons: head-space analysis	129
Volatile aromatic hydrocarbons in atmospheric air: head-space analysis	130
Aromatic hydrocarbons in atmosphere: equilibrium preconcentration	131
Aliphatic hydrocarbons in air particulates: capillary GC analysis	132
Hydrocarbons in ambient air: cryogenic sampling	133
Analysis of environmental hydrocarbons: capillary GC application	134
Hydrocarbon air pollutants by automated GC	135
Direct GC analysis of photochemically nonreactive hydrocarbons	136
Aromatic hydrocarbons in air: semiautomatic preconcentration	137
Sampling and analysis of hydrocarbon emissions	138
Hydrocarbon emissions from combusion processes: three-dimensional GC analysis	139

Table 1 (continued)
AIR POLLUTION

Comments	Ref.
Light hydrocarbons in air: use of a chemically bonded stationary phase	140
Standard GC methods for atmospheric hydrocarbon analysis	141
Hydrocarbons in cigarette smoke condensate	142

Table 2
WATER POLLUTION

Comments	Ref.
Oil Spills	
Reviews concerning GC analysis of oil spills	143—147
Specific GC analyses of oil spills	148—163
Petroleum hydrocarbons in surface water: sampling device	164
Analysis of crude oils in water	165
Benzene, toluene, xylenes	166
Hydrocarbons in water	167
Sampling hydrocarbons in water	168
Petroleum hydrocarbons in marine sediments: capillary GC	169
Oil spill "fingerprints": capillary GC	170
Petroleum and synthetic fuel "fingerprints"	171
Aromatics in river water: GC-MS	172
Hydrocarbons in water: GC, IR	173
Identification of spilled residual fuel oils	174
Identification of gasoline contamination in ground waters	175
Nonvolatile hydrocarbons in open ocean water: sampling and analysis	176
C-3—C-6 alkanes in sea water: portable GC method	177
Volatile hydrocarbons in marine sediments and seawater: head-space sampling; GC-MS	178
Hydrocarbons in seawater: degassing, head-space sampling, GC analysis	179
Oil pollution "fingerprinting": two-stage GC	180
Trace GC analysis of gasoline in water	181
GC analysis of oil and oil product volatiles in water	182
Gasoline in water: GC data system readout	183
Identification of petroleum products in water	184
Identification of petroleum and volatile hydrocarbons in estuarine waters	185
Analysis of hydrocarbons in Amoco oil spill: capillary GC analysis	186
Analysis of petroleum hydrocarbons in Tallium Bay, U.S.S.R.: IR, GC	187
Petroleum hydrocarbons in seawater: capillary GC, IR	188
Hydrocarbons in sediments at ship wreck site: UV fluoresence vs. GC analysis	189,190
Polycyclic aromatic hydrocarbons in sediments: GC-MS	191

Table 2 (continued)
WATER POLLUTION

Comments	Ref.
Polycyclic aromatic hydrocarbons in sediments and mussels: capillary GC	192
n-Alkanes in oils on sea surface: capillary GC	193
Polycyclic aromatic hydrocarbons in lake water and sediments: GC-MS	194
Gasoline hydrocarbons in water: capillary GC	195
Polycyclic aromatic hydrocarbons in shellfish: short capillary column	196
Hydrocarbons in beach sands: automatic solids injector	197
Identification of source of marine hydrocarbons	198
Hydrocarbons on seas and beaches	199

Other Water Pollution Studies

Comments	Ref.
Polycyclic aromatic hydrocarbons in water: literature review	200
Quantitative analysis of polycyclic aromatic hydrocarbons in water	201
Petroleum hydrocarbons in wastewater	202
Benzene in water	203
Hydrocarbons in marine biota: HPLC for sample preparation	204
Polycyclic aromatic hydrocarbons in water: HPLC for sample preparation	205
Polycyclic aromatic hydrocarbons in water: extraction procedure for GC analysis	206
Determination of hydrocarbons in water at the ppb level: head-space analysis	207
GC monitoring methods for EPA Consent Decree priority pollutants in water and wastewater	208
Polycyclic aromatic hydrocarbons in water near coal coking plant	209
GC profile analysis of over 200 polycyclic aromatic hydrocarbons in water	210
GC analysis of aromatic hydrocarbons in natural water and wastewater	211
Hydrocarbons as marine pollutants: GC-MS	212
Ultra-trace (ppt) measurement of terphenyls in water and sediment: GC-MS	213
Standard addition method for head-space and extraction analysis for hydrocarbons in water	214
Hydrocarbons in water: head-space analysis	215
Determination of hydrocarbons and other organics in natural and sewage waters: review article (187 references)	216
Isolation method for hydrocarbons and other organics from water	217
Low-molecular-weight hydrocarbons and other organics in drinking water: GC for automated monitoring	218
Analysis of trace volatile hydrocarbons and other hydrocarbons in water: GC-MS	219
Low-molecular-weight hydrocarbon gases in water: isolation and GC analysis	220

Table 2 (continued)
WATER POLLUTION

Comments	Ref.
Sampling and identification of pollutant oil in industrial water	221
Trace analysis (ppm) of naphthalenes in water and marine sediment	222
Polycyclic aromatic hydrocarbons in water at ppm and ppb levels: HPLC isolation and GC analysis using a photoionization detector	223
Polycyclic aromatic hydrocarbons in water: isolation and preconcentration using three different sorbents	224
Polycyclic aromatic hydrocarbons in lake and river sediments and in river particulates: capillary GC	225
Trace analysis of C-6—C-13 aromatic hydrocarbons in water at the 0.1 ppb level: method for isolation and analysis	226
Aliphatic, olefinic, and polycyclic aromatic hydrocarbons in anoxic sediment: GC-MS	227
Isolation and trace analysis of aliphatic and polycyclic aromatic hydrocarbons: GC, GC-MS	228
Xylene and toluene in wastewater: isolation and GC-MS	229
Identification of normal hydrocarbons and polycyclic aromatic hydrocarbons in drinking (tap) water: resin cleanup and GC-MS analysis at the ppb level	230
Volatile hydrocarbon solvents in water and wastewater: isolation and quantitative analysis	231
Polycyclic aromatic hydrocarbons in water: isolation and GC-MS analysis	232
Extraction and quantitative GC analysis of polycyclic aromatic hydrocarbons in water and other environmental samples	233
GC identification of petroleum products in natural water down to 0.25 ppm	234
Identification of high-boiling (at or above 140°) hydrocarbons in polluted drinking water: isolation and GC-MS analysis	235
Analysis of hydrocarbons above C-15 in San Francisco Bay water: GC-MS	236
C-1—C-14 hydrocarbons in water	237
Methane and trace amounts of C-2—C-5 hydrocarbons in water: GC method	238
Hydrocarbons in potable waters: GC method	239
Sampling and GC analysis of volatile hydrocarbons and other organics in water	240
Hydrocarbons in water: analysis by GC and by hydrocarbon analyzer	241
Identification of organic pollutants in water: review article (99 references)	242
Organics in water: GC-MS	243
Analysis of aqueous solutions of crude oils and refinery products: GC vs. IR analysis	245
Isolation methods of hydrocarbons from aqueous solutions for GC analysis	246
Trace volatiles in water: assessment of reversion GC and low-temperature GC analysis	244

Table 2 (continued)
WATER POLLUTION

Comments	Ref.
Analysis of aqueous solutions of crude oils and refinery products: GC vs. IR analysis	245
Isolation methods of hydrocarbons from aqueous solutions for GC analysis	246
Determination of benzene in demineralized water	247
Light aromatic hydrocarbons: new GC method	248
Hydrocarbons in ground and surface water: stripping procedure and capillary GC analysis	249
Petroleum hydrocarbons in tap water	250
Polycyclic aromatic hydrocarbons in water and wastewater: GC-UV method	251
Aromatic hydrocarbons in natural and wastewaters: equilibrium vapor analysis	252
Trace (ppb-level) hydrocarbons in water: capillary GC, head-space analysis	253
Trace polycyclic aromatic hydrocarbons in aqueous systems: XAD-2 cleanup and capillary GC-MS analysis	254
GC profiles of polycyclic aromatic hydrocarbons in sewage	255
Aromatic hydrocarbons in wastewater from styrene and cumene production	256
Trace analysis of hydrocarbons in water: vapor phase extraction	257
Aromatic hydrocarbons in drinking water: GC-MS	258
Improved GC method for "environmental" hydrocarbons	259

REFERENCES

1. **Grimmer, G., Boehnke, H., and Glaser, A.,** *Erdoel Kohle, Erdgas, Petrochem. Brennst Chem.,* 30, 411, 1977.
2. **Giger, W. and Blumer, M.,** *Anal. Chem.,* 46, 1663, 1974.
3. **Grimmer, G., Hildebrandt, A., and Boehnke, H.,** Publ. No. International Agency for Research on Cancer, Lyon, 29, 1979, 141.
4. **Argirova, M. and Lundsjo, A.,** *Khig. Zdraveopaz,* 20, 166, 1977.
5. **Grimmer, G., Hildebrandt, A., and Boehnke, H.,** *Erdol Kohle,* 25, 442, 1972.
6. **Grimmer, G., Hildebrandt, A., and Boehnke, H.,** *Erdol Kohle,* 25, 531, 1972.
7. **Doran, T. and McTaggart, N. G.,** *J. Chromatogr. Sci.,* 12, 715, 1974.
8. **Jentoft, R. E. and Gouw, T. H.,** *Anal. Chem.,* 48, 2195, 1976.
9. **Bartle, K. D., Lee, M. L., and Novotny, M.,** *Proc. Anal. Div. Chem. Soc.,* 13 (10), 304, 1976.
10. **Van Vaeck, L. and Van Cauwenberghe, K.,** *Anal. Lett.,* 10, 467, 1977.
11. **Georgescu, M., Iordache, I., Lipkovics, R., and Neagoe, M.,** *Rev. Chim.,* 25, 1030, 1974; *Anal. Abstr.,* 29, 5C62, 1975.
12. **Deaconeasa, V., Constantinesau, T., and Trestinau, S.,** *Rev. Chim.,* 28, 777, 1977.
13. **Daisey, J. M. and Leyko, M. A.,** *Anal. Chem.,* 51, 24, 1979.
14. **Becher, G.,** *VDI-Ber.,* 358, 95, 1980.
15. **Bjoerseth, A.,** *Carcinog. Compr. Surv. Polynucl. Aromat. Hydrocarbons,* 3, 75, 1978.
16. **Yavorovskaya, S. F.,** *Gig. Sanit.,* 10, 72, 1974.

17. **Lao, R. C., Thomas, R. S., Oja, H., and Dubois, L.,** *Anal. Chem.,* 45, 908, 1973.
18. **Pellizzari, E. D.,** No. EPA-650/2-74-121, U. S. Environmental Protection Agency, Washington, D.C., July 1974.
19. **Greinke, R. A. and Lewis, I. C.,** *Anal. Chem.,* 47, 2151, 1975.
20. **Hites, R. A.,** *Am. Chem. Soc. Petrol. Chem. Prepr.,* 20 (4), 824, 1975.
21. **Lao, R. C., Thomas, R. S., and Monkman, J. L.,** *J. Chromatogr.,* 112, 681, 1975.
22. **Bhatia, K.,** *Anal. Chem.,* 43, 609, 1971.
23. **Grob, K. and Grob, G.,** *J. Chromatogr.,* 62, 1, 1971.
24. **Searl, T. et al.,** *Anal. Chem.,* 42, 954, 1970.
25. **Szepesy, L., Lakszner, K., Ackermann, L., Podmaniczky, L., and Literathy, P.,** *J. Chromatogr.,* 206, 611, 1981.
26. **Lao, R. C., Thomas, R. S., Oja, H., and Dubois, L.,** *Anal. Chem.,* 45, 908, 1973.
27. **Searl, T. D., Cassidy, F. J., King, W. H., and Brown, R. A.,** *Anal. Chem.,* 42, 954, 1970.
28. **Broddin, G., Van Vaeck, L., and Van Cauwenberghe, K.,** *Atmos. Environ.,* 11, 1061, 1977.
29. **Bjorseth, A. and Eklund, G.,** *J. High Resol. Chromatogr. Chromatogr. Commun.,* 2, 22, 1979; **Giger, W. and Schaffner, C.,** *Anal. Chem.,* 50, 243, 1978.
30. **Hill, H. H., Jr., Chan, K. W., and Karasek, F. S.,** *J. Chromatogr.,* 131, 245, 1977.
31. **Thomas, R. S. and Lao, R. C.,** *Proc. 70th Annu. Meet. Air Pollut. Control Assoc.,* 4, 77, 1977.
32. **Tabata, T., Sueta, S., and Shigemori, N.,** *Taiki Osen Gakkaishi,* 13, (2), 58, 1978.
33. **Thomas, R. S., Lao, R. C., Wang, D. T., Robinson, D., and Sakuma, T.,** *Carcinog. Compr. Surv. Polynucl. Aromat. Hydrocarbons,* 3, 9, 1978.
34. **Bjoerseth, A.,** *Anal. Chim. Acta,* 94 (1), 21, 1977.
35. **Lee, M. L., Novotny, M., and Bartle, K. D.,** *Anal. Chem.,* 48, 1566, 1976.
36. **Liberti, A. and Zoccolillo, L.,** *Spec. Environ. Rep. W.M.O., Obs. Meas. Atmos. Pollut., Proc. 1973 Tech. Conf,* 3, 79, 1974.
37. **Grimmer, G. and Boehnke, H.,** *Fresenius' Z. Anal. Chem.,* 261 (4—5), 310, 1972.
38. **Brocco, D., Cantuti, V., and Cartoni, G. P.,** *J. Chromatogr.,* 49, 66, 1970.
39. **Bjoerseth, A.,** *Anal. Chim. Acta,* 94 (1), 21, 1977.
40. **Giger, W. and Schaffner, C.,** *Anal. Chem.,* 50, 243, 1978.
41. **Korol, A. N. and Lvsyuk, L. S.,** *Zh. Anal. Chim.,* 34 (3), 577, 1979.
42. **Lee, M. L., Prado, G. P., Howard, J. B., and Hites, R. A.,** *Biomed. Mass Spectrom.,* 4 (3), 182, 1977.
43. **Akiyama, K., Nagashima, C., Takahashi, R., Fujitani, K., Yagyu, H., and Nagasaki, M.,** *Tokyo Toritsu Eisei Kenkyusho Kenkyu Nempo,* 25, 337, 1974.
44. **Heins, C. F., Johnson, F. D., and Mangold, E. C.,** *Environ. Sci. Technol.,* 9 (8), 720, 1975.
45. **Lee, M. L., Novotny, M., and Bartle, K. D.,** *Anal. Chem.,* 48, 1566, 1976.
46. **Lee, M. L., Prado, G. P., Howard, J. B., and Hites, R. A.,** *Biomed. Mass Spectrom.,* 4 (3), 182, 1977.
47. **Bjorseth, A. and Lunde, G.,** *J. Am. Ind. Hyg. Assoc.,* 38, 224, 1977.
48. **Liao, J. C. and Browner, R. R.,** *Anal. Chem.,* 50, 1683, 1978.
49. **Severson, R. F., Snook, M. E., Arrendale, R. F., and Chortyk, O. T.,** *Anal. Chem.,* 48, 1866, 1976.
50. **Severson, R. F., Snook, M. E., Arrendale, R. F., and Chortyk, O. T.,** *Anal. Chem.,* 48, 1866, 1976.
51. **Lee, M. L., Novotny, M., and Bartle, K. D.,** *Anal. Chem.,* 48, 405, 1976.
52. **Tausch, H. and Stehlik, G.,** *Chromatographia,* 10 (7), 350, 1977.
53. **Lorenzo, A.,** *Chim. Ind. (Milan),* 55 (7), 573, 1973.
54. **Lane, D. A.,** *Diss. Abstr. Int. B,* 37 (1), 181, 1976.
55. **Zoccolillo, L., Liberti, A., and Brocco, D.,** *Atmos. Environ.,* 6 (100), 715, 1972.
56. **Sawicki, E.,** *Atmos. Environ.,* 7 (2), 233, 1973.
57. **Liberti, A.,** *Atmos. Environ.,* 7 (2), 233, 1973.
58. **Greinke, R. A. and Lewis, I. C.,** *Anal. Chem.,* 47, 2151, 1975.
59. **Jones, P. W., Glammer, R. D., Strup, P. E., and Stanford, T. B.,** *Environ. Sci. Technol.,* 10, 806, 1976.
60. **Hanson, R. L., Royer, R. E., and Brooks, A. L.,** *Annu. Rep. Inhalation Toxicol. Res. Inst.,* 1979, 203.
61. **Van Vaeck, L. and Van Cauwenberghe, K.,** *Quant. Mass Spectrom. Life Sci.,* 2, 459, 1978.
62. **Moller, M. and Alfheim, I.,** *Atmos. Environ.,* 14, 83, 1980.
63. **Bjoerseth, A. and Eklund, G.,** *Anal. Chim. Acta,* 105, 119, 1979.
64. **Daisey, J. M. and Leyko, M. A.,** *Anal. Chem.,* 51, 24, 1979.
65. **de Ruwe, H. J. J. M., Schulting, F. L., and van Grondelle, J.,** *High Resolut. Chromatogr. Chromatogr. Commun.,* 1, 211, 1978.
66. **Anghelach, I.,** *Rev. Chim. (Bucharest),* 26, 962, 1975.
67. **Schofield, K.,** *Environ. Sci. Technol.,* 8, 826, 1974.

68. **Oshima, S., Iwamiya, Y., Machida, C., and Takagi, Y.,** *Maruzen Sekiyu Giho,* 1973 (18), 57.
69. **Spengler, G., Maier, K., Woerle, R., and Schriewer, M.,** *Erdoel, Kohle, Erdgas, Petrochem. Brennst. Chem.,* 27 (5), 246, 1974.
70. **Imamura, K. and Fujii, T.,** *Bunseki, Kagaku,* 28, 549, 1979.
71. **Esposito, G. G. and Jacobs, B. W.,** *Ind. Hyg. Assoc. J.,* 38 (8), 401, 1977.
72. **Burghardt, E. and Jeltes, R.,** *Atmos. Environ.,* 9, 935, 1975.
73. **Volfson, V. Y., Zanievskaya, O. S., Korol, A. N., Ruban, P. P., and Chugayeva, O. T.,** *Sb. Nauchn. Tr. Gazov. Khromatogr., N. I. Fiz. Khim. Inst.,* 18, 87, 1972.
74. **Esposito, G. G. and Jacobs, B. W.,** *J. Am. Ind. Hyg. Assoc.,* 38 (8), 401, 1977.
75. **Black, M. S., Rehg, W. R., Sievers, R. E., and Brooks, J. J.,** *J. Chromatogr.,* 142, 809, 1977.
76. **Dimitriades, B., Raible, C. J., and Wilson, C. A.,** *Gov. Rep. Announce. (U.S.),* 73 (6), 64, 1973.
77. **Leinster, P., Perry, P., and Young, R. J.,** *Talanta,* 24, 205, 1977.
78. **Mrose, H.,** *Z. Meteorol.,* 23 (3—4), 112, 1972.
79. **Jeltes, R. and Burghardt, E.,** *Atmos. Environ.,* 6 (11), 793, 1972.
80. **Westberg, H. H., Rasmussen, R. A., and Holdres, M.,** *Anal. Chem.,* 46, 1852, 1974.
81. **Brocco, D., Cantuti, V., and Moscatelli, F.,** *Riv. Combust.,* 25 (4), 164, 1971.
82. **Sawicki, E.,** *Health Lab. Sci.,* 7, 23, 1970.
83. **Yuan, R.-Y., Yu, C.-C., Li, C.-L., Chang, C.-K., Wang, M.-Y., and Tang, H.-Y.,** *Huan Ching K'o Hsueh,* 1979 (3), 22
84. **Ball, H.,** *Fresenius' Z. Anal. Chem.,* 282, 301, 1976.
85. **Saha, N. C., Jain, S. K., and Dua, R. K.,** *J. Chromatogr. Sci.,* 16, 323, 1978; **Cudney, R. A., Walther, E. G., and Malm, W. C.,** *J. Air Pollut. Control Assoc.,* 27 (5), 468, 1977.
86. **Cudney, R. A., Walther, E. G., and Malm, W. C.,** *J. Air Pollut. Control Assoc.,* 27 (5), 468, 1977.
87. **Brocco, D., Cantuti, V., and Moscatelli, F.,** *Riv. Combust.,* 25 (4), 164, 1971.
88. **Tsujino, Y.,** *Taiki Osen Gakkaishi,* 14 (6), 231, 1979.
89. **Burghardt, E.,** *Sci. Tech. Aerosp. Rep.,* 18 (14), Abstr. No. N80-23409, 1980.
90. **Charlton, J., Sarteur, R., and Sharkey, J. M.,** *Oil Gas J.,* 73 (28), 96, 1975.
91. **Greinke, R. A. and Lewis, I. C.,** *Anal. Chem.,* 47, 2151, 1975.
92. **Raymond, A. and Guiochon, G.,** *Environ. Sci. Technol.,* 8 (2), 143, 1974.
93. **Saltzman, B. E. and Burg, W. R.,** *Anal. Chem,* 49 (5), 8R, 1978.
94. **Hauser, T. R. and Pattison, J. N.,** *Environ. Sci. Technol.,* 6, 549, 1972.
95. **Schneider, W., Frohne, J. C., and Brudereck, H.,** *J. Chromatogr.,* 155, 311, 1978.
96. **Guillemin, G. L., Martinez, F., and Thiault, S.,** *J. Chromatogr. Sci.,* 17, 677, 1979.
97. **Middleditch, B. S. and Basile, B.,** *Anal. Lett.,* 9 (11), 1031, 1976.
98. **Black, F. M., High, L. E., and Sigsby, J. E.,** *J. Chromatogr. Sci.,* 14 (5), 257, 1976.
99. **Aue, W. A., Vogt, C. R., and Younker, D. R.,** *J. Chromatogr.,* 114, 184, 1975.
100. **Saltzman, B. E., Burg, W. R., and Cuddleback, J. E.,** *Anal. Chem.,* 47, 2234, 1975.
101. **Iordache, I., Lipkovics, R., Vasilescu, E., and Georgescu, M.,** *Rev. Chem. (Bucharest),* 26, 338, 1975.
102. *Anon., Proc. Anal. Div. Chem. Soc.,* 14, 108, 1977.
103. **Leontieva, S. A., Drugov, Yu. S., Lulova, V. I., and Koroleva, N. M.,** *Zh. Anal. Khim.,* 32, 1638, 1977.
104. **De Greef, J., De Proft, M., and De Winter, F.,** *Anal. Chem.,* 48, 38, 1976.
105. **Stephens, E. R. and Hellrich, O. P.,** *Environ. Sci. Technol.,* 14 (7), 836, 1980.
106. **Miller, D. L., Woods, J. S., Grubaugh, K. W., and Jordon, L. M.,** *Environ. Sci. Technol.,* 14 (1), 97, 1980.
107. **Hester, N. E. and Meyer, R. A.,** *Environ. Sci. Technol.,* 13, 107, 1979.
108. **Jonsson, A. and Berg, S.,** *J. Chromatogr.,* 190, 97, 1980.
109. Report No. EPA-600/4-80-024, U. S. Environmental Protection Agency, Washington, D.C., 1980; Report No. EPA-600/4-78-057, 1978.
110. **Hanson, R. L., Carpenter, R. L., Clark, C. R., and Brooks, A. L.,** in Annu. Rep. Inhalation Toxicol. Res. Inst., 1978.
111. Report No. EPA-600/4-77-003, U. S. Environmental Protection Agency, Washington, D.C., 1977.
112. **Kuwada, K., Yamazaki, Y., Eguchi, M., Uwabori, M., and Matsuo, K.,** *J. Jpn. Soc. Air Pollut.,* 10 (4), 318, 1975.
113. **Hayashi, T.,** *Chugoku Denryoku Gijutsu Kenkyujo Giken Jiho,* 47, 153, 1975.
114. **Dencker, W., Robinson, M., and Villalobos, R.,** Report No. EPA-650/2-74-056, U. S. Environmental Protection Agency, Washington, D.C., August 1974.
115. **DeGreef, J., DeProft, M., and DeWinter, F.,** *Anal. Chem.,* 48, 38, 1976.
116. **Harbourn, C. L. A., McCambley, W. H., and Trollope, T. J.,** *Proc. Int. Clean Air Congr.,* 1973, C38.
117. **Burghardt, E. and Jeltes, R.,** *Atmos. Environ.,* 9, 935, 1975.

118. **Shirayama, H. and Okumura, H.,** *J. Jpn. Soc. Air Pollut.,* 9, (2), 198, 1974.
119. **Suzuki, R., Nakaya, T., Ito, M., and Oda, H.,** *J. Jpn. Soc. Air Pollut.,* 10 (4), 311, 1975.
120. **Kikuchi, T. and Goto, G.,** *Miyagi Prefect. Pollut. Contr. Tech. Center Rep.,* 3, 23, 1975.
121. **Tyson, B. J. and Carle, G. C.,** *Anal. Chem.,* 46, 610, 1974.
122. **Wang, I. C., Swofford, H. S., Jr., Price, P. C., Martinsen, D. P., and Buttrill, S. E., Jr.,** *Anal. Chem.,* 48, 491, 1976.
123. **Karmarkar, K. H. and Guilbault, G. G.,** *Environ. Lett.,* 10 (3), 237, 1975.
124. **Brocco, D., Dipalo, V., and Possanzini, M.,** *J. Chromatogr.,* 86, 234, 1973.
125. **Rasmussen, R. A., Westberg, H. H., and Holdren, M.,** *J. Chromatogr. Sci.,* 12, 80, 1974.
126. **Hauser, T. R. and Pattison, J. N.,** *Environ. Sci. Technol.,* 6, 549, 1972.
127. **Ferman, M. A.,** Paper 80-39.2, in Proc. 73rd Annu. Meet., Air Pollut. Control Assoc., 3, 1980.
128. **Halket, J. M. and Angerer, J.,** *Anal. Chem. Symp. Ser., Recent Dev. Chromatogr. Electrophor.,* 3, 211, 1980.
129. **Ioffe, B. V., Vitenberg, A. G., and Tsibul'skaya, I. A.,** *J. Chromatogr.,* 186, 851, 1979.
130. **Ioffe, B. V., Vitenberg, A. G., and Tsibul'skaya, I. A.,** *14th Adv. Chromatogr. (Houston),* 1979, 929.
131. **Vitenberg, A. G. and Tsibul'skaya, I. A.,** *Zh. Anal. Khim.,* 34 (9), 1830, 1979.
132. **Sueta, S., Tabata, T., and Shigemori, N.,** *Eisei Kagaku,* 25 (2), 67, 1979.
133. **Ullrich, D. and Seifert, B.,** *Fresenius' Z. Anal. Chem.,* 291 (4), 299, 1978.
134. **Jeltes, R., Burghardt, E., Thijsse, Th. R., and Den Tonkelaar, W. A. M.,** *Chromatographia,* 10, 430, 1977.
135. **Ball, H.,** *Fresenius' Z. Anal. Chem.,* 282 (4), 301, 1976.
136. **Black, F. M., High, L. E., and Sigsby, J. E.,** *J. Chromatogr. Sci.,* 14 (5), 257, 1976.
137. **Burghardt, E. and Jeltes, R.,** *Atmos. Environ.,* 9 (10), 935, 1975.
138. **Schneider, W. and Frohne, J. C.,** *Staub Reinhalt. Luft,* 35 (7), 275, 1975.
139. **Siegert, H., Delert, H. H., and Zajontz, J.,** *Chromatographia,* 7, 599, 1974.
140. **Westberg, H. H., Rasmussen, R. A., and Holdren, M.,** *Anal. Chem.,* 46, 1852, 1974.
141. **Rasmussen, R. A., Westberg, H. H., and Holdren, M.,** *J. Chromatogr. Sci.,* 12, 80, 1974.
142. **Gelpi, E. and Oro, J.,** *J. Chromatogr. Sci.,* 8, 210, 1970.
143. **Higashi, K. and Hagiwara, K.,** *PPM,* 9, 33, 1978; *Chem. Abstr.,* 90, 161707j, 1979.
144. **Frame, G. M., Carmody, D. C., and Flanigan, G. A.,** *Gov. Rep. Announce. Index (U.S.),* 78 (18), 209, 1978.
145. **Adlard, E. R.,** *J. Chromatogr. Sci.,* 11, 137, 1979.
146. **Adlard, E. R.,** *J. Inst. Petrol. (London),* 58, 63, 1972.
147. **Jeltes, R. and den Tonkelaar, W. A. M.,** H₂O, 5, 288, 1972.
148. **Higashi, K. and Hagiwara, K.,** *Bunseki Kagaku,* 27, 772, 1978.
149. **Chesler, S. N., Hertz, H. S., May, W. E., Wise, S. A., and Guenther, F. R.,** *Int. J. Environ. Anal. Chem.,* 5, 259, 1978.
150. **Clark, H. A. and Jurs, P. C.,** *Anal. Chem.,* 51, 616, 1979.
151. **Utashiro, S. and Matsua, H.,** *Kaijo Hoan Daigakko Kenkyu Hokoku, Dai-2-Bu,* 24 (1), 23, 1978; *Chem. Abstr.,* 90, 124146e, 1979.
152. **Hertz, H. S., May, W. E., Chesler, S. N., and Gump, B. H.,** *Environ. Sci. Technol.,* 10, 900, 1976.
153. **Petrovic, K. and Vitorovic, D.,** *J. Chromatogr.,* 119, 413, 1976.
154. **Hilbert, L. R., May, W. E., Wise, S. A., Chesler, S. N., and Hertz, H. S.,** *Anal. Chem.,* 50, 458, 1978.
155. **Snowdon, L. R. and Peake, E.,** *Anal. Chem.,* 50, 379, 1978.
156. **Davis, C. E., Kro, A. E., Szakasits, J. J., and Hodgson, R. L.,** *Proc. Conf. Prev. Control Oil Pollut.,* 1975, 93.
157. **Jackson, B. W., Judges, R. W., and Powell, J. L.,** *Environ. Sci. Technol.,* 9, 656, 1975.
158. **Clark, H. A. and Jurs, P. C.,** *Anal. Chem.,* 47, 374, 1975.
159. **May, W. E., Chesler, S. N., Cram, S. P., Gump, B. H., Hertz, H. S., Enagonio, D. P., and Dyszel, S. M.,** *J. Chromatogr. Sci.,* 13, 535, 1975.
160. **Ehrhardt, M. and Blumer, M.,** *Environ. Pollut.,* 3, 179, 1972.
161. **Zafiriou, O. C.,** *Anal. Chem.,* 45, 952, 1973.
162. **Kawahara, F. K.,** *J. Chromatogr. Sci.,* 10, 629, 1972.
163. **Novak, J., Zluticky, J., Kubelka, V., and Mostecky, J.,** *J. Chromatogr.,* 76, 45, 1973.
164. **Miget, R. and Kator, H.,** *Anal. Chem.,* 46, 1154, 1974.
165. **Cole, R. E.,** *Nature (London),* 233, 546, 1971.
166. **Mel'kanovitskaya, S. G.,** *Gidrokhim. Mater.,* 53, 153, 1972; *Chem. Abstr.,* 77, 22466y, 1972.
167. **Drozd, J., Novak, J., and Rijks, J. A.,** *J. Chromatogr.,* 158, 471, 1978.
168. **Voznakova, Z., Popl, M., and Berka, M.,** *J. Chromatogr. Sci.,* 16, 123, 1978.
169. **Overton, E. B., Bracken, J., and Laseter, J. L.,** *J. Chromatogr. Sci.,* 15, 169, 1977.

170. **Rasmussen, D. V.**, *Anal. Chem.*, 48, 1562, 1976.
171. **Petrovic, K. and Vitorovic, D.**, *J. Chromatogr.*, 119, 413, 1976.
172. **Hites, R. A. and Biemann, K.**, *Science*, 178, 160, 1972.
173. **Jeltes, R. and Tonkelaar, W. D.**, *Water Res.*, 6, 271, 1972.
174. **Kawahara, F. K.**, *J. Chromatogr. Sci.*, 10, 629, 1972.
175. **Dell'Acqua, R., Bush, B., and Egan, J.**, *J. Chromatogr.*, 128, 271, 1976.
176. **Brown, R. A., Elliott, J. J., Kelliher, J. M., and Searl, T. D.**, *Adv. Chem. Ser.*, 147, 172, 1975.
177. **Perras, J. C.**, AD Rep. No. 786583/5GA, *U.S. National Technical Information Service, Springfield, Va.*,
178. **May, W. E., Chesler, S. N., Cram, S. P., Gump, B. H., Hertz, H. S., Enagonio, D. P., and Dyszel, S. M.**, *J. Chromatogr. Sci.*, 13, 535, 1975.
179. **Wasik, S. P.**, *J. Chromatogr. Sci.*, 12, 845, 1974.
180. **George, A. E., Smilery, G. T., Montgomery, D. S., and Sawatzky, H.**, *Can. Mines Br. Res. Rep.*, R267, 1973; *Chem. Abstr.*, 80, 19276t, 1974.
181. **Dell'Acqua, R. and Bush, B.**, *Int. J. Environ. Anal. Chem.*, 3, 141, 1973.
182. **Polak, J. and Lu, B. C. Y.**, *Anal. Chim. Acta*, 68, 231, 1974.
183. **DiFilippo, M. and Maljevac, J.**, *Chromatogr. Newsl.*, 2, 20, 1973.
184. **Lure, Yu. Yu., Panova, V. A., and Nikolaeva, Z. V.**, *Gidrokhim. Mater.*, 55, 108, 1971; *Chem. Abstr.*, 75, 121164p, 1971.
185. **Maehler, C. Z. and Greenberg, A. E.**, *J. Sanit. Eng. Div. (Amer. Soc. Civil Eng.)*, 94, (SA5), 969, 1968.
186. **Berthou, F., Gourmelun, T., Dreano, Y., and Friocourt, M. P.**, *J. Chromatogr.*, 203, 279, 1981.
187. **Tiks, K., Talvari, A., and Yankovskii, Kh. I.**, Tezisy Dokl. Resp. 2nd Konf. Molodykh Uch., Khim., 1977, 2.
188. **Tiks, K.**, Sint. Issled. Biol. Soedin., Tezisy Dokl. 6th Konf. Molodykh Uch., 1978, 91.
189. **Hoffman, E. J., Quinn, J. G., Jadamec, J. R., and Fortier, S. H.**, *Bull. Environ. Contam. Toxicol.*, 23 (4—5), 536, 1979.
190. **Hoffman, E. J. and Quinn, J. G.**, *Mar. Pollut. Bull.*, 10 (1), 20, 1979.
191. **Tan, Y. L.**, *J. Chromatogr.*, 176, 319, 1979.
192. **Bjoerseth, A., Knutzen, J., and Skei, J.**, *Sci. Total Environ.*, 13 (1), 71, 1979.
193. **Utashiro, S. and Matsuo, H.**, *Kaijo Hoan Daigakko Kenkyu Hokoku, Dai-2-Bu*, 24 (1), 23, 1978.
194. **Strosher, M. T. and Hodgson, G. W.**, Water Qual. Parameters, Spec. Tech. Publ. American Society for Testing and Materials, Philadelphia, 573, 259, 1975.
195. **Hrivnak, J. and Hassler, J.**, *Vodni Hospod. B*, 26 (7), 193, 1976.
196. **Onuska, F. I., Wolkoff, A. W., Comba, M. E., Larose, R. H., Novotny, M., and Lee, M. L.**, *Anal. Lett.*, 9 (5), 451, 1976.
197. **McKay, T. R.**, *Gas Chromatogr., Proc. Int. Symp. (Eur.)*, 9, 33, 1972.
198. **Ehrhardt, M. and Blumer, M.**, *Environ. Pollut.*, 3 (3), 179, 1972.
199. **Adlard, E. R., Creaser, L. F., and Matthews, P. H. D.**, *Anal. Chem.*, 44, 64, 1971.
200. **Anal. Div. of the Chem. Soc.**, *Proc. Anal. Div. Chem. Soc.*, 15, 148, 1978.
201. **Nonaka, A.**, *Jpn. Anal.*, 20, 422, 1971.
202. **Goeke, G.**, *Fette, Seifen, Anstrichm.*, 74 (3), 168, 1972.
203. **Drozd, J. and Novak, J.**, *J. Chromatogr.*, 152, 55, 1978.
204. **Chesler, S. N., Gump, B. H., Hertz, H. S., May, W. E., and Wise, S. A.**, *Anal. Chem.*, 50, 805, 1978.
205. **Oyler, A. R., Bodenner, D. L., Welch, K. J., Liukkonen, R. J., Carlson, R. M., Kopperman, H. L., and Caplo, R.**, *Anal. Chem.*, 50, 837, 1978.
206. **Acheson, M. A., Harrison, R. M., Perry, R., and Wellings, R. A.**, *Water Res.*, 10, 207, 1976.
207. **Drozd, J., Novak, J., and Rijks, J. A.**, *J. Chromatogr.*, 158, 471, 1978.
208. **Keith, L. H., Lee, K. W., Provost, L. P., and Present, D. L.**, Spec. Publ. STP 686, *American Society for Testing and Materials*, Philadelphia, 1979, 85.
209. **Griest, W. H., Maskarinec, M. P., Herbes, S. E., and Southworth, G. R.**, Report, CONF-790663-1, *National Technical Information Service*, Springfield, Va., 1979.
210. **Grimmer, G. and Naujack, K. W.**, *Vom Wasser*, 53, 1, 1979.
211. **Ioffe, B. V., Stolyarov, B. V., and Smirnova, S. A.**, *Zh. Anal. Khim.*, 33, 2196, 1978.
212. **Albaiges, J. and Albrecht, P.**, *Int. J. Environ. Anal. Chem.*, 6, 171, 1979.
213. **Shinohara, R., Hori, T., and Koga, M.**, *Bunseki Kagaku*, 27, 400, 1978.
214. **Khazal, W. J., Vejrosta, J., and Novak, J.**, *J. Chromatogr.*, 157, 125, 1978.
215. **Friant, S. L. and Suffet, I. H.**, *Anal. Chem.*, 51, 2167, 1979.
216. **Yavorovskoya, S. F. and Anvaer, B. I.**, *Zh. Anal. Khim.*, 32, 2044, 1977; *Chem. Abstr.*, 88, 65709z, 1978.
217. **Sievers, R. E., Barkley, R. M., Eiceman, G. A., Shapiro, R. H., Walton, H. F., Kolonko, K. J., and Field, L. R.**, *J. Chromatogr.*, 142, 745, 1977.

218. **Dowty, B., Green, L. E., and Laseter, J. L.,** *J. Chromatogr. Sci.,* 14, 187, 1976.
219. **Lingg, R. D., Melton, R. G., Kopfler, F. C., Coleman, W. E., and Mitchell, D. E.,** *J. Am. Water Works Assoc.,* 69, 605, 1977.
220. **Sultz, D. J., Pankow, J. F., Tai, D. Y., Stephens, D. W., and Rathbun, R. E.,** *J. Res. U.S. Geol. Surv.,* 4, 247, 1976.
221. **Baier, R. E., Gasiecki, E. A., Leonard, R. P., and Mack, E. J.,** Rep. No. 249362, National Technical Information Service, Springfield, Va., 1976.
222. **Chesler, S. N., Gump, B. H., Hertz, H. S., May, W. E., and Wise, S. A.,** NBS Spec. Publ. 464, 1977, 81.
223. **Oyler, A. R., Bodenner, D. L., Welch, K. H., Liukkonen, R. J., Carlson, R. M., Kopperman, H. L., and Caple, R.,** *Anal. Chem.,* 50, 837, 1978.
224. **Navratil, J. D., Sievers, R. E., and Walton, H. F.,** *Anal. Chem.,* 49, 2260, 1977.
225. **Giger, W. and Schaffner, C.,** *Anal. Chem.,* 50, 243, 1978.
226. **Voznakova, Z., Popl, M., and Berka, M.,** *J. Chromatogr. Sci.,* 16, 123, 1978.
227. **Hites, R. A. and Biemann, W. G.,** *Adv. Chem. Ser.,* 147, 188, 1975.
228. **Shinohara, R., Koga, M., Shinohara, J., and Hori, T.,** *Bunseki Kagaku,* 26, 856, 1977.
229. **Viden, I., Kubelka, V., and Mostecky, J.,** *Fresenius' Z. Anal. Chem.,* 280, 369, 1976.
230. **Van Rossum, P. and Webb, R. G.,** *J. Chromatogr.,* 150, 381, 1978.
231. **Bellar, T. A. and Lichtenberg, J. J.,** *J. Am. Water Works Assoc.,* 66, 739, 1974.
232. **Matsushima, H. and Hanya, T.,** *Bunseki Kagaku,* 24, 505, 1975.
233. **Hellmann, H.,** *Fresenius' Z. Anal. Chem.,* 275, 109, 1975; 278, 263, 1976.
234. **Dell'Acqua, R., Egan, J. A., and Bush, B.,** *Environ. Sci. Technol.,* 9, 38, 1975.
235. **Novak, J., Zluticky, J., Kubelka, V., and Mostecky, J.,** *J. Chromatogr.,* 76, 45, 1973.
236. **Simoneit, B. R., Smith, D. H., Eglinton, G., and Burlingame, A. L.,** *Arch. Environ. Contam. Toxicol.,* 1, 193, 1973.
237. **Kaiser, R.,** *Haus Tech. Essen Vortragsveroeff,* p. 13, 1972.
238. **Kim, A. G. and Douglas, L. J.,** *J. Chromatogr. Sci.,* 11, 615, 1973.
239. **Grob, K.,** *J. Chromatogr.,* 84, 255, 1973.
240. **Zlatkis, A., Lichtenstein, H. A., and Tishbee, A.,** *Chromatographia,* 6, 67, 1973.
241. **Desbaumes, E. and Imhoff, C.,** *Water Res.,* 6, 885, 1972.
242. **Croll, B. T.,** *Water Treat. Exam.,* 21, 213, 1972.
243. **Van Cauwenberghe, K.,** *Bull. Inform. Assoc. Nat. Serv. Eau Belg.,* p. 119, 1973.
244. **Josefsson, B. O.,** *Anal. Chim. Acta,* 52, 65, 1970.
245. **Jeltes, R. and Den Tonkelaar, W. A. M.,** *Water Res.,* 6, 271, 1972.
246. **McAuliffe, C.,** *Chem. Geol.,* 4, 225, 1969.
247. **Kontsova, V. V.,** *Org. Veshchestvo Podzemn. Vod. Ego Znachenie Neft. Geol.,* p. 266, 1967; *Chem. Abstr.,* 70, 90661p, 1969.
248. **Kozlov, V. F.,** *Porod i Vod, L.,* p. 171, 1980.
249. **Hassler, J.,** *Proc. Int. Symp. Ground Water Pollut. Oil Hydrocarbons,* 1978, 267.
250. **Durdovic, V.,** *Acta Hydrochim. Hydrobiol.,* 7 (5), 527, 1979.
251. **Searl, T. D., Robbins, W. K., and Brown, R. A.,** Meas. Org. Pollut. Water Wastewater, Spec. Tech. Publ. 686, American Society for Testing and Materials, Philadelphia, 1979, 164.
252. **Ioffe, B. V., Stolyarov, B. V., and Smirnova, S. A.,** *Zh. Anal. Khim.,* 33 (11), 2196, 1978.
253. **Drozd, J., Novak, J., and Rijks, J. A.,** *J. Chromatogr.,* 158, 471, 1978.
254. **Strup, P. E., Wilkinson, J. E., and Jones, P. W.,** *Carcinog., Compr. Surv. Polynucl. Aromat. Hydrocarbons,* 3, 131, 1978.
255. **Grimmer, G., Boehnke, H., and Borwitzky, H.,** *Fresenius' Z. Anal. Chem.,* 289 (2), 91, 1978.
256. **Sokolowska, J. and Maciejowski, F.,** *Gaz Woda Tech. Sanit.,* 51 (2), 54, 1977.
257. **Mackay, D., Shiu, W. Y., and Wolkoff, A. W.,** Water Qual. Parameters, Spec. Tech. Publ. 1973 Symp., American Society for Testing and Materials, Philadelphia, 573, 251, 1975.
258. **Dowty, B. J., Carlisle, D. R., and Laseter, J. L.,** *Environ. Sci. Technol.,* 9 (8), 762, 1975.
259. **Zafiriou, O. C.,** *Anal. Chem.,* 45, 952, 1973.

Chapter 7

LIQUID PHASES AND ADSORBENTS

Studies reporting the use of specific stationary liquid phases (gas-liquid chromatography) and solid sorbents (gas-solid chromatogarphy) for hydrocarbon separations and analyses are cited in detailed tables in Section I (Table 1) and Section II (Table 2), respectively, of this chapter. Specific information on liquid phases, solid supports for preparing gas-liquid chromatography (GLC) packings, and solid sorbents used for gas-solid chromatography (GSC) separations is given in Sections III, IV, and V, respectively.

I. STUDIES USING SPECIFIC LIQUID PHASES (GLC)

Studies using various stationary liquid phases for GLC analyses of hydrocarbons are summarized in Table 1. The literature citations for these studies are categorized under: (1) General Hydrocarbon Analyses; (2) Studies on Alkanes; (3) Studies on Unsaturated Hydrocarbons; (4) Studies on Cyclic Hydrocarbons; (5) Studies on Aromatic Hydrocarbons; and (6) Studies on Polycyclic Aromatic Hydrocarbons.

II. STUDIES USING SPECIFIC SOLID ADSORBENTS (GSC)

Studies using various solid adsorbents for GSC analyses of hydrocarbons are summarized in Table 2. The literature citations for these studies are categorized under: (1) General Hydrocarbon Analyses; (2) Studies on Alkanes; (3) Studies on Unsaturated Hydrocarbons; and (4) Studies on Cyclic, Aromatic, and Polycyclic Aromatic Hydrocarbons.

III. PROPERTIES AND CHARACTERISTICS OF LIQUID PHASES

Various properties of liquid phases used for hydrocarbon analyses (composition; minimum and maximum recommended operating temperatures; appropriate solvent used for dissolving the liquid phases for preparing columns for gas-liquid chromatographic analysis; and proximate polarity) are given in Table 3, while the polar selectivity of a number of these liquid phases relative to squalane and of silicone OV phases relative to squalane is given in Tables 4 and 5, respectively. The data in Table 5 for OV phases are arranged relative to increasing retention of benzene. For the convenience of the reader, the liquid phases listed in Tables 3 and 4 are arranged alphabetically, with all silicone phase types listed under "Silicone". Literature citations for a wide variety of applications using specific liquid phases are given in Table 1.

IV. SOLID SUPPORTS USED FOR GLC PACKINGS

While a number of different types of solid support materials have been used to prepare GLC packings, the silaceous diatomaceous earth supports continue to be the most popular. A description of diatomaceous earth supports for GC packings is available in Analabs-Foxboro Catalog K23-B. These supports tend to be white or pink in character. The white materials tend to have lower loading capacities than pink materials, but have lower surface energies and a higher degree of inertness, and are the more popular of the two types. Diatomaceous earths in general consist of highly porous skeletons of fossilized diatoms from monocellular algae and contain roughly about 90% silica.

Table 1
LIQUID PHASES (GAS-LIQUID CHROMATOGRAPHY)

Stationary phase: solutes separated	Ref.
General Hydrocarbon Analyses	
Squalane: light gasoline hydrocarbons	1
1,2,3-tris(2-cyanoethoxy)propane: gasoline hydrocarbons	2
NaOH-treated 13X molecular sieve: gasoline alkanes and cycloalkanes	3
OPN and phenyl isocyanate on Porasil C: 16 C-1—C-5 alkanes and alkenes	4
Urea-hexadecane: alkanes, alkenes, aromatics	5
Mixed bis-lactams: 26 hydrocarbons	6
Phenylmethyl silicones: hydrocarbons	7
Squalane: many hydrocarbons	8
Squalane: naphtha hydrocarbons	9
Hexadecane: light naphtha hydrocarbons	10
Dipropionitriles: hexane, cyclohexane, cyclohexene, isooctane, benzene	11
Apiezon L, phenylisocyanate, and squalane: C-1—C-10 hydrocarbons	12
Al$_2$O$_3$: hydrocarbons in air	13
Factor analysis on 49 stationary phases: hydrocarbons	14
SP-2100: petroleum hydrocarbons	15
C-87-Hydrocarbon phase (Kovats): hydrocarbons	16
Three isoalkoxypropionitriles: light hydrocarbons	17
Dibutyltetrachlorophthalate: 70 hydrocarbons	18
Polyimides: hydrocarbons	19
C-22—C-36 alkane stationary phases: 21 compounds	20
Liquid crystals: alkanes and aromatics	21
Poly(methylphenyl)siloxane: high-molecular-weight hydrocarbons	22
Preferred liquid phases (6 listed; 18 others given): hydrocarbons	23
C-24—C-36 alkane stationary phases: C-6—C-9 alkanes and benzene	24
Effect of stationary phase structure on retention of hydrocarbons	25
Computer selection of stationary phases	26
Influence of stationary phase loading	27
Stationary phases on modified attapulgite: hydrocarbons	28
Thiodipropionitrile: alkanes and cyclics	29
Isodecylphthalate on Bentone-34: styrenes, indenes, cyclenes	30
Squalane and tetracyanoethylated-pentaerythritol: naphtha hydrocarbons	31
Squalane (capillary column): dodecanes and alkylaromatics	32
Squalane (capillary column): 200 gasoline hydrocarbons	33
MS-550 (capillary column): cyclic olefins and alkanes	34
Dinonylphthalate: C-8—C-9 naphtha hydrocarbons	35
Squalane and polyethylene glycol: alkenes and C-6—C-11 alkyl benzenes	36
Polyimides: C-14—C-36 hydrocarbons	37
Two liquid crystals: 42 solutes including alkanes, alkenes, and aromatic isomers	38
13 liquid phases with 4 aluminas (best results obtained with 1,2,3-tris-(2-cyanoethoxy)propane): C-1—C-5 hydrocarbons	39

Table 1 (continued)
LIQUID PHASES (GAS-LIQUID CHROMATOGRAPHY)

Stationary phase: solutes separated	Ref.
Dexsil-300 GC and two carborane polymers: high-boiling hydrocarbons up to C-60	40
N,N'-bis(2-cyanoethyl)formamide: saturated, olefinic, and aromatic hydrocarbons in gasoline	41
OV-101: C-1—C-20 hydrocarbons	42
Characterization of stationary phases: hydrocarbons	43
"Master column": all 32 C-1—C-4 hydrocarbons	44
Furan derivatives: close-boiling naphtha hydrocarbon fractions	45
Squalane: hydrocarbons	46
Hydrogenated Apiezon M (general evaluation compared to squalane and Kovats C-87 hydrocarbon phase): hydrocarbons	47
Sebacic acid dinitrile and 1,10-decanediol: aromatics, alkanes, alkadienes	48
Stationary phase polarity scale based on "retention polarities" (161 phases): hydrocarbons	49
Modified graphatized carbon black: aliphatic and aromatic hydrocarbons	50
Density measurement of liquid phases	51
Monomer polymerized in GC column containing support (Phillips Petroleum patent)	52
2,2-Diethyl-1,3-bis(2-cyanoethoxy)propane: low- and high-boiling hydrocarbons	53
Characterization of 81 stationary phases: partition coefficients of octane, toluene, etc.	54
n-Heptylhypophosphorus acid: high-boiling paraffins and aromatics	55
Lukopren G 1000 (Czechoslovakian stationary phase): retention data on 195 nonolefinic hydrocarbons	56
Standardization of stationary phases (International conference)	57
Rohrschneider constants to characterize liquid phases	58
200 liquid phases characterized relative to squalane	59
Polyphenylether (capillary columns): tar aromatics and aliphatics	60
Liquid crystals (capillary columns): hydrocarbon isomers	61
Squalane: prediction of hydrocarbon separations	62
Immobile stationary phase (U.S.S.R. patent): hydrocarbon mixtures	63
Kovats "squalane-like" C-87 hydrocarbon phase on chemically modified supports	64
C-87 hydrocarbon phase	65
Di-n-butyl-tetrachlorophthalate: identification of hydrocarbons	66
2,4,5,7-Tetranitrofluorenone on modified graphatized carbon black: aliphatic and aromatic hydrocarbons	67
Computer selection of suitable stationary phases for separation of hydrocarbon mixtures	68
Mixed stationary phases (capillary GC): C-5—C-8 hydrocarbons	69
1,2,3-Tris(2-cyanoethoxy)propane: C-1—C-5 hydrocarbons	70

Table 1 (continued)
LIQUID PHASES (GAS-LIQUID CHROMATOGRAPHY)

Stationary phase: solutes separated	Ref.
New stationary phase (patent): hydrocarbons	71
Squalane: prediction of hydrocarbon separations at different temperatures	72

Studies on Alkanes

Apiezon L: alkanes	73
Hydrocarbon and methylsilicone phases: iso-alkanes	74
Poly (*m*-phenylether) and methylsilicones: higher alkanes	75
Squalane: isoalkanes	76
Alkane-urea adducts: *n*-alkanes	77
Urea inclusion compounds containing *n*-octane and *n*-decane: branched methyl heptane isomers	78
Polyphenylsiloxane: alkanes	79
Squalane on carbon black: C-2—C-4 alkanes and deuterated and tritiated analogs	80
Apiezon L: 300 C-7—C-24 alkanes	81
N,N'-dimethyl- and *N,N'*-diethyl-1-naphthylamine: hexane and heptane isomers in a 14-component mixture of C-5—C-7 normal, iso-, and cyclic hydrocarbons	82
Apiezon L: C-9—C-14 *n*-alkanes	83
Squalane: 16 C-8—C-10 isomeric alkanes	84
Urea: C-8—C-16 *n*-alkanes	85
Apiezon L: tri- and tetra-methylated C-13, C-14, and C-16 alkanes	86
SF-96: C-6—C-8 branched alkanes	87
Specific retention volumes predicted for *n*-octane on 65 stationary phases	88
5A Molecular sieve: *n*-paraffins	89

Studies on Unsaturated Hydrocarbons

Ethyl 2-oxocyclopentanoate: isomeric butenes	90
ODPN and dimethylsulfolane containing aqueous Na$_2$CO$_3$: alkynes and alkadienes in C-1—C-5 hydrocarbon fractions	91
Phenylglycidyl ether: unsaturated C-5—C-6 hydrocarbons from parent *n*-alkanes	92
Nitro-terephthalic acid on Carbowax 20M: alpha-olefins	93
1,2,3-Tris(2-cyanoethoxy)propane: alkenes	94
Five stationary phases: alkenes	95
Polyphenylether: C-6—C-14 *n*-alkenes	96
Silver nitrate and alpha-cyanotoluene: cis/trans pentene-2	97
Ion-exchanger: cis/trans pentene-2 and butene-2	98
Squalene: *n*-penta- and *n*-hexa-decenes	99
Charge-transfer complexing phases: olefins	100
Propylene carbonate: butenes, including cis/trans isomers	101
Squalane (capillary column): all possible 85 *n*-alkenes up to C-14	102
Tetrachlorophthalates (capillary GC): C-7—C-10 *n*-alkenes	103

Table 1 (continued)
LIQUID PHASES (GAS-LIQUID
CHROMATOGRAPHY)

Stationary phase: solutes separated	Ref.
7,8- and 5,6-Benzoquinoline (capillary GC): *n*-C-8—C-11 alkenes	104
1,5-Dicyanopentane: 2-methylpropene and butene-1	105
Silver nitrate and polyols: *n*-2- and *n*-3-alkenes	106
Silver nitrate and carbowax: C-8 and C-10 cyclic diolefins	107
Formamide: specific C-3—C-4 impurities in propylene	108
Ucon LB and dibutylphthalate: alkynes	109
1,2-Bis(2-cyanoethoxy)ethane containing silver nitrate: alkenes, including cis/trans isomers	110
Squalane: all 9 *n*-undecenes	111
Phenylglycid ether: C-5—C-6 unsaturated hydrocarbons	112
Squalane; Ucon LB 550X and 50HB 280X; carbowax 1000 (capillary GC): isomeric unsaturated octanes	113
Thallium(I)-tetraphenylborate: unsaturated hydrocarbons	114
Bis(2-ethylhexyl)- and di-*n*-butyl-tetrachloro-phthalates: C-7—C-10 olefins	115
Squalane containing dicarbonyl-rhodium-beta-diketonates: selective separation of olefins	116
Silver ion charge-transfer phases: olefins	117
Squalane; Apiezon L; polyethylene glycol 4000: *n*-alkenes	118
Silver ion stationary phase: unsaturated hydrocarbon isomers	119
Tri-isobutyl-aluminum: conjugated dienes	120
Aqueous silver nitrate on Chromosorb P: unsaturated hydrocarbons and deuterated isomers	121

Studies on Cyclic Hydrocarbons

1,1-Diphenylethane on Porolith: cyclohexane, cyclohexene, benzene	122
Squalane: cyclic hydrocarbons	123
Evaluation of 60 stationary phases (best found were vacuum oil VM-4, tricresyl phosphate, and polyethylene glycol 2000): terpenes	124
Carbowax 20M and SE-30: tricyclic saturated hydrocarbons	125
Squalane; OV-1; Ucons LB550X/HB280X; Carbowax 20M: dicyclopentadienes	126
Squalane: 30 C-5—C-9 cyclalkenes	127
Polyphenylethers: monoterpene hydrocarbons	128
Squilane: retention data for 156 saturated and unsaturated cyclopropanes	129
Squalane: C-6—C-13 cyclo-alkanes, -alkenes, and -alkadienes	130
Squalane; Apiezon L; polyethylene glycol 4000: cyclopentanes and -hexenes	131

Studies on Aromatic Hydrocarbons

OV-1 and TCEP (2 columns in series): benzene and toluene	132
2,4,5,7-Tetranitrofluorenone: alkyl benzenes	133
Multiple liquid phases: *meta*- and *para*-xylenes	134

Table 1 (continued)
LIQUID PHASES (GAS-LIQUID CHROMATOGRAPHY)

Stationary phase: solutes separated	Ref.
Squalane and Ucon 50HB-208X: alkyl-aromatics	135
Squalane: *n*-alkane hydrocarbon phases; liquid crystals: aromatics and alkanes	136
ES-52: aromatics	137
Poly-*m*-phenylether (capillary GC): aromatic hydrocarbons	138
Squalane: aromatics	139
Poly[Cu(II)-di-*n*-hexylphosphinate]: aromatic isomers, alpha-methylstyrene	140
Cyanoethylated polyols: aromatics	141
Apiezon L; polyphenyl ether; SP-1000: biphenyls and diphenylalkanes	142
PEGA and 1,2,3-tris(2-cyanoethoxy)propane: pseudo-cumene (1,2,4-trimethylbenzene)	143
Dibutyltetrachlorophthalate: aromatics	144
Tetrachlorophthalate oligomers: *meta*- and *para*-xylenes	145
Polyimides: high-boiling aromatics	146
Clathrate picoline compounds: xylene and diethylbenzene isomers, ethylbenzene	147
PEG-1500: aromatic hydrocarbons in petroleum	148
1,2,3,4,5,6-Hexa-bis(cyanoethoxy)hexane: C-6—C-10 aromatics in C-1—C-12 hydrocarbon mixtures	149
Dinonylphthalate and SF-96 silicone oil on Bentone-34: isomeric xylenes	150
m-Bis(m-phenoxyphenoxy)benzene; Igepal Co-880; Chrom Rb 470-1: benzene, toluene, xylenes	151
Tricresylphosphate (capillary GC): aromatic hydrocarbons in gasoline	152
p-Azoxyanisole: *meta*- and *para*-xylenes	153
1,2,3-Tris(2-cyanoethoxy)propane: alkyl benzenes and other aromatics	154
Apiezon L: dodecylbenzenes	155
4,4'-Azoxydiphenetole and anisole, and their mixtures: benzene, toluene, and *ortho*- and *para*-xylenes	156
Binary liquid crystal mixtures: *meta*- and *para*-xylenes and C-9—C-11 alkanes	157
Naphthalene-1,8-diamine: *meta*- and *para*-xylenes	158
Three liquid crystals: *ortho*-, *meta*-, and *para*-disubstituted benzenes	159
Bentone-34 plus modifying agents: isomeric C-6—C-10 aromatics	160
PEG and squalane: benzene and C-7—C-11 alkyl benzenes	161
DC-550: aromatic hydrocarbons	162
Apiezon L; SE-30; squalane: di- and tri-substituted benzenes	163
Liquid crystals: isomeric substituted benzenes	164—166
Evaluation of 11 stationary phases: benzene	167
SE-30; Citroflex A-4; squalane; carbowax 6000: retention data for alkyl benzenes	168
Nickel-quinone tetragonal clathrate: *ortho*-, *meta*-, and *para*-xylenes	169

Table 1 (continued)
LIQUID PHASES (GAS-LIQUID CHROMATOGRAPHY)

Stationary phase: solutes separated	Ref.
Cobalt-phthalocyanine on graphatized carbon black: aromatics	170
4,4'-dimethoxyazooxybenzene liquid crystal: *meta*- and *para*-xylenes	171
Interlamellar organic complexes prepared from clay minerals: selectivity for aromatic hydrocarbons	172
Liquid crystal: aromatic isomers	173
1,3,5-Trinitrobenzene charge-transfer complexing phase: aromatic hydrocarbons	174
New stationary phase: aromatic hydrocarbons	175

Studies on Polycyclic Aromatic Hydrocarbons (PAH)

Liquid crystals vs. Dexsil 300: separation of benz[*a*]anthracene from chrysene	176
Bentone-34: PAH	177
17:13 Ucon LB-550X:tris(2-cyanoethoxy)-propane: substituted naphthalenes and biphenyls	178
OV-17; OV-275: naphthalene, 10 methylnaphthalenes, biphenyl, fluorene	179
High-transition-temperature liquid crystals:PAH isomers	180—182
Nematic liquid crystals: PAH isomers	183,184
Dexsil 300; Apiezon L: 1—2 ring PAH	185
Various phases: 209 PAH	186
Phenylsilicone: PAH	187
Several phases: PAH	188
OV-101; OV-17: 59 PAH	189
p,p'-Azoxyphenetole plus PEG 2000; binary liquid crystal mixtures (eutectic): dimethylnaphthalenes	190
OV-7: Benzo[*a*]- and benzo[*e*]pyrene, benzo[*k*]fluoranthene, perylene, etc.	191
Liquid crystals: PAH isomers	192
Cadmium chloride modified with Carbowax 20M: naphthalenes	193
OV-101; OV-17: 51 PAH	194
SE-52 (capillary GC): 10 PAH	195
SE-30; OV-101: PAH and PAH derivatives	196
Liquid crystal: naphthalenes	197
Nematic liquid crystals: isomeric substituted naphthalenes and benzenes	198
Carborane: PAH	199
New high temperature liquid crystal	200

Table 2
ADSORBENTS (GAS-SOLID CHROMATOGRAPHY)

Stationary phase: solutes separated	Ref.
General Hydrocarbon Analyses	
Porous polymer and molecular sieve: light hydrocarbons	201
Graphatized carbon black: 65 C-2—C-10 alkanes and alkenes	202
Graphatized carbon black: alkanes and alkenes	203
Graphatized carbon black: C-6—C-9 aromatics and C-4—C-5 alkanes	204
PPE-20 on graphatized carbon black: C-4 hydrocarbons	205
Chromosorb 101: cycloalkanes, *n*-alkenes, aromatics	206
Solid Carbowax 20M: hydrocarbons	207
Ion exchangers (factor analysis): saturated and unsaturated hydrocarbons	208
Inorganic salt eutectics: petroleum hydrocarbons	209
Lithium chloride: petroleum hydrocarbons	210
Cobalt(II) hydroxide: C-1—C-7 alkanes and alkenes	211
Modified porous glass: propane, propylene	212
Urea-*n*-C-16-clathrate: hydrocarbons	213
Carbopack C modified with 1,3,5-trinitrobenzene: 19 C-4—C-5 hydrocarbons	214
Vitreous carbons: light hydrocarbons	215
Carbon black modified with 2,4,5,7-tetranitrofluorenone: alkanes and alkenes	216
Zeolites: hydrocarbons	217
Zirconium potassium phosphate type synthetic inorganic ion exchangers: hydrocarbons	218
Silicas: petroleum hydrocarbons	219
Copper(II) complexes on silica: alkanes and unsaturated and aromatic hydrocarbons	220
Porapak Q: light hydrocarbons	221
Silica-alumina: light hydrocarbons	222
Porapak and molecular sieve: light hydrocarbons	223
Bonded phase: C-2—C-5 hydrocarbons	224
3-Methoxy- or 3-pentoxy-propionitrile: C-1—C-5 hydrocarbons	225
Thallium(I)-tetraphenyl borate: olefins, aromatics	226
Modified porous polymer: light hydrocarbons	227
Barium sulfate: C-5—C-20 *n*-alkanes and C-6—C-8 arenes	228
Modified aluminas and silica gels: light hydrocarbons	229
Silica-containing polymers: gaseous and C-5—C-6 saturated hydrocarbons and light naphtha hydrocarbons	230
Carbon molecular sieves: light hydrocarbons	231
ODPN bonded to silica: C-1—C-4 hydrocarbons	232
Molecular sieves: naphtha hydrocarbons	233
Ion exchanger: hydrocarbons	234
Modified SiO_2/Al_2O_3: C-1—C-4 hydrocarbons	235
Polyethylene glycol, octane, or phenyl isocyanate chemically-bonded to Porasil: specific alkanes and alkenes; gasoline hydrocarbons	236
Chemically bonded stationary phases	237

Table 2 (continued)
ADSORBENTS (GAS-SOLID CHROMATOGRAPHY)

Stationary phase: solutes separated	Ref.
Four nickel complexes: C-5—C-20 aliphatic ali-cyclic, and aromatic hydrocarbons	238
Surface-treated silica: gasoline hydrocarbons	239
Modified silica gel: aliphatic and aromatic hydrocarbons	240
Treated alumina: H_2, CH_4, C_2H_6, C_2H_4	241
Graphite: hydrocarbons	242
Attapulgite: petroleum products	243
Organoclays: hydrocarbons	244
Trinitrofluorenone-modified alumina: saturated and unsaturated hydrocarbons	245
Trinitrofluorenone-modified graphatized carbon black: hydrocarbon isomers	246
Various sorbents: light hydrocarbons	247,248
Various sorbents: alkenes and alkanes	249
Various sorbents: C-5—C-8 hydrocarbons	250
Mixed silica-alumina: light hydrocarbons	251
Molecular sieves: hydrocarbons	252
GSC improvement using steam as carrier gas: high-boiling hydrocarbons	253
Preparation of adsorbent packing material: light hydrocarbons	254
Characterization of Porapak columns: light hydrocarbons	255
Mixed Porapak columns: light hydrocarbons	256
Porous polymers: light hydrocarbons	257
Graphatized carbon (GC-MS): high-boiling hydrocarbons	258
Molecular sieves: hydrocarbons	259
Sorbents containing an organoclay: hydrocarbons	260
Molecular sieve 13X: content of alkane and cyclic hydrocarbons in saturated hydrocarbon distillates	261
Organo-vermiculite: isomeric hydrocarbons	262
Molecular sieve 13X: separation of saturated and naphtha hydrocarbons by carbon number and hydrocarbon type	263
Water-modified soil, silica, and Chromosorb W (sorptive interactions study): *n*-hydrocarbons	264
Macroporous polymer (GPC-3): hydrocarbon gases	265
Preparation of silica gel adsorbent: light hydrocarbons	266
Graphatized carbon black: C-6—C-10 *n*-alkenes	267

Studies on Alkanes

NaOH-treated 13X molecular sieve: gasoline paraffins	268
Active silica (capillary GC): isotopic methanes	269
Potassium-lithium-sodium nitrate mixture: saturated isoprenoids and C-4—C-40 *n*-alkanes	270
Modified silicas: *n*-alkanes	271
PVC powder: *n*-, iso-, and cyclo-alkanes	272
Graphatized carbon black: C-6—C-10 *n*-alkanes	273
Ion-exchanger: methane and *n*-, branched-, and cy-clo-alkanes	274

Table 2 (continued)
ADSORBENTS (GAS-SOLID CHROMATOGRAPHY)

Stationary phase: solutes separated	Ref.
Heat-cured Carbowax 20M on Chromosorb W: *n*-alkanes	275
In situ formed open-pore polyurethane (for GSC, GLC, etc.): *n*-alkanes	276
5A and 13X molecular sieves: *n*-, iso-, and cyclo-alkanes	277
Zeolites: *n*-paraffins	278

Studies on Unsaturated Hydrocarbons

Silver nitrate: unsaturated hydrocarbons	279
Ion-exchanger: cis-trans alkenes	280
Iron(III) oxide: butene-1, isobutene, and cis-trans butene-2	281
Graphatized carbon black: acetylene	282
$PdCl_2/AgNO_3$: hexenes	283
Sulfobenzyl-modified silicas: unsaturated hydrocarbons	284
Rhodium compounds: olefins	285
Graphatized carbon black: C-6—C-11 *n*-alkynes	286
Vanadium-manganese-cobalt chlorides (pi-bonded retention): alkanes, alkenes, alkynes	287
Rhodium diketonate coordination compounds: 27 C-2—C-6 *n*-alkenes and 12 cyclic olefins	288
Zeolites: C-5—C-8 olefins	289
Graphatized carbon black: determination of acetylene in ethylene	290
Various sorbents: *n*-alkynes	291
Molecular sieves: acetylenes and dienes in propylene	292
Type X zeolite cations: olefins	293

Studies on Cyclic, Aromatic, and Polycyclic Aromatic Hydrocarbons (PAH)

NaOH-treated 13X molecular sieve: gasoline cycloparaffins	268
Various sorbents: methylcyclohexanes	294
Graphatized carbon black: C-6—C-13 cycloalkanes, cycloalkenes, and cycloalkadienes	295
Molecular sieves: cyclic hydrocarbons	296,197
Graphatized carbon black: aromatics	298
Graphatized carbon black: C-9—C-14 alkyl benzenes	299
Cation exchangers: benzene, toluene, xylene, and lower aliphatic hydrocarbons	300
Graphatized carbon black: methylbenzenes	301
Specifically modified alumina: alkyl benzenes	302
Graphatized carbon black: C-3—C-8 alkyl benzenes	303
Effect of Na(I)-ion content in zeolite on retention: aromatic hydrocarbons	304
Molecular sieves: isomeric triethylbenzenes	305
Modified aluminas: aromatic hydrocarbons	306
Decationized zeolite: aromatic hydrocarbons	307

Table 2 (continued)
ADSORBENTS (GAS-SOLID CHROMATOGRAPHY)

Stationary phase: solutes separated	Ref.
Bentone-34: nephthalene homologs	308
Lithium chloride on silica: PAH	309
Na-treated alumina: PAH and alkylbenzenes	310
Carbon black on Chromosorb W: PAH	311
Chromosorb P: PAH	312
Silica gels: PAH	313
Graphatized carbon black: naphthalenes	314
Glass beads: PAH	315
Carbon black: PAH	316

Table 3
LIQUID PHASES USED FOR GC ANALYSIS OF HYDROCARBONS

Liquid phase	Temperature (°C) Min.	Max.	Solvent[a]	Polarity[b]
Acetonylacetone (2,5-hexane-dione)		25	A	P
Apiezon L	50	300	H	N
Apiezon M	50	275	H	N
Apiezon N	50	325	H	N
Asphalt		300	B	N
BBBT (*N,N'*-bis[*p*-butoxy-benzylidene]-α,α'-bi-*p*-toluidine)	165	240	C	[c]
BMBT (*N,N'*-bis[*p*-methoxy-benzylidene]-α,α'-bi-*p*-toluidine)	185	225	C	[c]
BPhBT (*N,N'*-bis[*p*-phenyl-benzylidene]-α,α'-bi-*p*-toluidine)	260	275	C	[c]
Bentone 34 (dimethyldioctadecylammonium bentonite)	20	200	T	S
7,8-Benzoquinoline		150	C	I
Benzyl Diphenyl	60	100	A	N
Benzyl ether		50	A	I
Bis[2-(methoxyethoxy)ethyl]-ether		50	C	P
Butanediol succinate (Craig polyester; BDS)	50	225	C	P
Carbowax 400 (polyethylene glycol; av. mol wt 380—420)	20	125	M	P
Carbowax 600 (polyethylene glycol; av. mol wt 570—630)	20	125	M	P
Carbowax 750 (methoxypoly-propylene glycol; av. mol wt 715—785)	25	125	M	P
Carbowax 1000 (polyethylene glycol; av. mol wt 950—1050)	40	125	C	P
Carbowax 1450[a] (polyethylene glycol; av. mol wt 1300—1600) formerly 1540	50	200	C	P
Carbowax 1500 (polyethylene glycol; av. mol wt 1500—1600)	40	200	C	P
Carbowax 3350[a] (polyethylene glycol; av. mol wt 3000—3700) formerly 4000	60	200	C	P

Table 3 (continued)
LIQUID PHASES USED FOR GC ANALYSIS OF HYDROCARBONS

Liquid phase	Temperature (°C) Min.	Max.	Solvent[a]	Polarity[b]
Carbowax 8000[a] (polyethylene glycol; av. mol wt 6000—7500) formerly 6000	60	200	C	P
Carbowax 20M (polyethylene glycol; av. mol wt 15,000—20,000)	60	225	C	P
Carbowax 20M-TPA (terminated with terephthalic acid)	60	250	M	P
Castorwax (hydrogenated castor oil)	90	200	C	P
Dexsil 300 GC (carborane-methyl silicone)	40	450	T	N
Dexsil 400 GC (carborane-phenyl silicone)	30	375	T	N
Dexsil 410 GC (carborane-nitrile silicone)	20	360	T	N
Dibutyl phthalate	20	125	A	P
Diisodecyl phthalate	20	160	A	I
Di-*n*-decyl phthalate	20	160	A	I
Di-*n*-propyl tetrachloro-phthalate		75	M	P
Diethylene glycol succinate (stabilized)	20	260	A	P
Diglycerol	20	120	M	P
Dimethyl sulfolane		50	A	P
Dioctyl phthalate (di(2-ethylhexyl)phthalate)	20	150	M	I
Dioctyl sebacate (di(2-ethylhexyl)sebacate)	20	100	A	I
Emulphor ON-870 (aryloxypolyethyleneoxyethanol)	20	175	M	I
Ethyl benzoate		150	M	P
Ethylene glycol adipate (stabilized)	100	260	A	P
Ethylene glycol isophthalate	100	250	C	P
Ethylene glycol phthalate	100	210	C	P
Ethylene glycol succinate	100	200	C	P
Eutectic (27.3% LiNO$_3$, 18.2% NaNO$_3$, 54.5% KNO$_3$)		400	W	
Hallcomid M-18 (*N,N*-dimethylstearamide)	40	150	M	I
Hallcomid M-18OL (*N,N,*-dimethyloleylamide)	40	150	M	I
Halocarbon oil 13-21 (56 cSt)	0	100	A	I
Halocarbon oil 700 (700 cSt, Kel-F oil No. 10: Halocarbon oil 14-25)	0	100	A	I
n-Hexadecane	20	50	T	N
HMPA (hexamethylphosphoramide)	10	50	M	P
Isoquinoline		50	M	P
LAC-1-R-296 (diethylene glycol adipate (DEGA); Resoflex)	20	190	A	P
LAC-2-R-446 (DEGA crosslinked with pentaerythritol)	50	190	A	P
Lanolin		200	C	I
Lithium chloride		500	W	P
Microcrystalline wax bottoms		250	C	I
Naphthylamine		75	M	P
Nitrobenzene		150	M	I
Nujol (paraffin oil)		200	T	N
n-Octadecane	30	55	T	N
Oxydipropionitrile	20	100	M	P
Oxypropionitrile	20	100	A	P
PDEAS (phenyldiethanolamine succinate)	15	225	A	P
Polyphenylether (5-ring)	20	200	A	I
Polyphenylether (6-ring)	0	225	A	I
Propylene carbonate	0	50	C	P

Table 3 (continued)
LIQUID PHASES USED FOR GC ANALYSIS OF HYDROCARBONS

Liquid phase	Temperature (°C)		Solvent[a]	Polarity[b]
	Min.	Max.		
Quadrol (*N,N,N',N'*-tetrakis-[2-hydroxypropyl]ethylenediamine)	20	150	A	P
Reoplex 100 (poly[propylene glycol sebacate])	0	200	C	P
Reoplex 400 (poly[proplyene glycol adipate])	20	220	C	P
Silar 5CP (silicone)	50	275	C	N
Silar 7CP (silicone)	50	275	C	N
Silar 9CP (silicone)	50	275	C	N
Silar 10C (100% 3-cyanopropyl silicone)	50	275	A	I
Silastic (LS-420; LS-X-3-0295; 1% vinyl silicone gum rubber) 50% 3,3,3-trifluoropropyl plus	100	275	E	I
Silicone AN-600 (25% 2-cyanoethyl silicone)	20	300	MEK	I
Silicone DC-11 compound (silicone grease)	20	300	E	N
Silicone DC-200 oil (100% methyl silicone; 350 cSt)	20	250	T	N
Silicone DC-200 oil (100% methyl silicone; 500 cSt)	0	200	T	N
Silicone DC-200 oil (100% methyl silicone; 1000 cSt)	20	250	T	N
Silicone DC-200 oil (100% methyl silicone; 12,500 cSt)	20	250	T	N
Silicone DC-410 gum (methyl silicone)		325	T	N
Silicone DC-550 oil (25% phenyl silicone)	20	225	A	I
Silicone DC-555 oil (methyl phenyl silicone)		275	C	I
Silicone DC-560 oil (Silicone fluid F-60; chloro-phenyl methyl silicone)		250	A	I
Silicone DC-703 oil (methyl phenyl silicone)		225	C	I
Silicone DC-710 oil (50% phenyl silicone)	20	250	A	I
Silicone DC QF-1 (3,3,3-trifluoropropyl silicone; FS-1265; QF-1-0065)	20	250	E	I
Silicone Versilube F-50 fluid (unknown chloro-phenyl content; chlorophenyl methyl silicone)	20	275	T	I
Silicone fluid F-60 (Silicone DC-560 oil; chloro-phenyl methyl silicone)		250	A	I
Silicone DC-560 fluid (11% chlorophenyl silicone)	20	250	A	I
Silicone SF-96 fluid (100% methyl silicone; 1000 cSt)	0	250	T	N
Silicone SF-96 fluid (100% methyl silicone; 50 cSt)		250	T	N
Silicone SF-96 fluid (100% methyl silicone; 100 cSt)		250	T	N
Silicone fluid (100% methyl silicone; 500,000 cSt)		300	T	N
Silicone W-98 gum (1% vinyl silicone)	100	300	T	N
Silicone W-96 gum (vinyl-methyl silicone)	100	300	T	N
Silicone oil L-525		200	T	N
Sorbitol-Silicone oil X-525 (1:1)	15	150	M	P
Sorbitol	15	150	M	P
Silicone OV-1 gum (100% methyl silicone)	100	350	T	N
Silicone OV-3 fluid (10% phenyl silicone)	20	350	C	I
Silicone OV-7 fluid (20% phenyl silicone)	20	350	C	I
Silicone OV-11 fluid (35% phenyl silicone)	20	350	C	I

Table 3 (continued)
LIQUID PHASES USED FOR GC ANALYSIS OF HYDROCARBONS

Liquid phase	Temperature (°C) Min.	Max.	Solvent[a]	Polarity[b]
Silicone OV-17 fluid (50% phenyl silicone)	20	350	C	I
Silicone OV-22 fluid (65% phenyl silicone)	20	300	C	I
Silicone OV-25 fluid (75% phenyl silicone)	20	300	C	I
Silicone OV-61 fluid (33% phenyl silicone)	20	350	C	I
Silicone OV-73 gum (5.5% phenyl silicone)		350	T	I
Silicone OV-101 fluid (100% methyl silicone)	20	350	T	N
Silicone OV-105 fluid (cyanopropylmethyl, di-methyl silicone)		300	A	I
Silicone OV-202 fluid (trifluoropropyl silicone)		275	E	I
Silicone OV-210 fluid (50% 3,3,3-trifluoropropyl silicone)	20	300	E	I
Silicone OV-215 gum (trifluoropropylmethyl silicone)		275	E	I
Silicone OV-225 fluid (25% 3-cyanopropyl-25% phenyl silicone)	20	275	A	P
Silicone OV-275 fluid (dicyanoallyl silicone)		250	A	P
Silicone OV-330 fluid (phenyl silicone-carbowax copolymer)	20	250	T	I
Silicone OV-351 solid (carbowax-terephthalic acid polymer)	50	270	C	N
Silicone gum rubber SE-30 (100% methyl silicone)	50	300	T	N
Silicone rubber SE-30 (GC Grade; 100% methyl silicone)	50	350	T	N
Silicone rubber SE-31 (vinyl-methyl silicone)	50	300	T	N
Silicone rubber SE-33 (1% vinyl-99% methyl silicone)	50	300	T	N
Silicone rubber SE-52 (5% phenyl silicone)	50	300	T	N
Silicone rubber SE-54 (1% vinyl-5% phenyl silicone)	100	300	T	N
Silicone rubber XE-60 (25% 2-cyanoethyl silicone)	20	275	A	I
Silicone rubber XE-61	20	275	A	I
Squalane	20	140	T	N
Squalene	20	140	T	N
Surfonic N-300 (alcoholic-alkylaryl polyether)		200	T	P
Tergitol NPX (polyethylene glycol-nonylphenyl ether)		200	A	P
Tergitol NP-35 (polyethylene glycol-nonylphenyl ether)		200	A	P
Tetraethylene glycol		70	T	P
Tetraethylene glycol dimethyl ether (Ansul ether)		80	M	P
Tetrahydroxyethyl ethylene-diamine (THEED)	20	135	A	P
1,2,3,4-Tetrakis(2-cyano-ethoxy)butane	110	200	C	I
β,β'-Thiodipropionitrile	25	100	C	P
Tricresyl phosphate (TCP; tritolyl phosphate)	20	125	A	P
Triisobutylene		25	A	N
2,4,7-Trinitro-9-fluorenone			A	P
Triphenylmethane		150	C	N
1,2,3-Tris(2-cyanoethoxy)-propane (TRIS)	20	180	M	P
Triton® X-100 (octylphenoxy-polyethoxy ethanol)	20	190	M	P
Triton® X-305 (octylphenoxy-polyethoxy ethanol)	20	250	M	P

Table 3 (continued)
LIQUID PHASES USED FOR GC ANALYSIS OF HYDROCARBONS

| Liquid phase | Temperature (°C) | | Solvent[a] | Polarity[b] |
	Min.	Max.		
Trixylol phosphate	20	250	A	P
Ucon 50 HB-270X		225	M	P
Ucon 50 HB-280X		200	M	P
Ucon 50 HB-660	20	225	M	P
Ucon 50 HB-2000	20	200	M	P
Ucon 50 HB-5100	20	200	M	P
Ucon 50 LB-300X	20	200	M	P
Ucon 50 LB-550X	20	200	M	P
Ucon 50 LB-1200X	20	200	M	P
Ucon 75-H-90,000		250	M	P
Ucon 1800 XMP-1018		160	A	P
Versamide 900 (polyamide)	190	275	C:Bu (1:1; hot)	P
Versamide 940 (polyamide)	115	200	C:Bu (1:1; hot)	P

[a] A, acetone; B, benzene; Bu, butanol; C, chloroform; E, ethyl acetate; H, hexane; M, methanol; MEK, methylethyl ketone; T, toluene.

[b] N, nonpolar; I, weak to intermediate polarity; P, polar; S, slightly polar.

[c] Liquid crystal; separations related to differences in rod-like geometric shape of solutes.

Table 4
RETENTION INDEX DIFFERENCES (ΔI) FOR SOLUTE PROBES OF DIFFERENT FUNCTIONALITY BETWEEN VARIOUS LIQUID PHASES AND SQUALANE[a]

| Liquid Phase | ΔI for solute probes[b] | | | | | |
	Benz	BuOH	P	Pyr	O	D
Squalane	0	0	0	0	0	0
Apiezon L	32	22	15	42	11	31
Apiezon M	31	22	15	40	10	28
Apiezon N	38	40	28	58	15	43
Butanediol succinate	370	571	448	611	242	533
Carbowax 600	350	631	428	605	240	503
Carbowax 1000	347	607	418	589	240	493
Carbowax 4000	325	551	375	520	224	443
Carbowax 6000	322	540	369	512	222	437
Carbowax 20M	322	536	368	510	221	434
Carbowax 20M-TPA	321	537	367	520	220	435
Castorwax	108	265	175	246	73	196
Diisodecyl phthalate	84	173	137	155	59	130
Di-*n*-decyl phthalate	136	255	213	235	101	202
Diethylene glycol succinate	499	751	593	860	323	725
Diglycerol	371	826	560	854	141	724
Dioctyl phthalate	92	186	150	167	66	140
Dioctyl sebacate	72	168	108	123	49	106
Emulphor ON-870	202	395	251	344	140	289
Ethylene glycol adipate	371	579	454	633	248	550
Ethylene glycol isophthalate	326	508	425	561	213	498

Table 4 (continued)
RETENTION INDEX DIFFERENCES (ΔI) FOR SOLUTE PROBES OF
DIFFERENT FUNCTIONALITY BETWEEN VARIOUS LIQUID PHASES AND
SQUALANE[a]

Liquid Phase	ΔI for solute probes[b]					
	Benz	BuOH	P	Pyr	O	D
Ethylene glycol phthalate	453	697	602	872	306	699
Ethylene glycol succinate	537	787	643	889	348	795
Hallcomid M-18	79	268	130	146	48	106
Hexatriacontane	12	2	−3	11	2	5
LAC-1-R-296 (diethylene glycol adipate (DEGA))	377	601	458	655	253	551
LAC-2-R-446 (DEGA cross-linked with pentaerythritol)	387	616	471	667	257	567
Nujol (paraffin oil)	9	5	2	11	2	6
PDEAS (phenyldiethanolamine succinate)	386	555	472	654	242	562
Reoplex 400 (poly[propylene glycol adipate])	364	619	449	671	245	540
Silicone DC-11	17	86	48	56	23	51
Silicone DC-200	16	57	45	43	23	46
Silicone DC-550	74	116	117	135	72	128
Silicone DC-560	32	72	70	68	35	69
Silicone DC-703	76	123	126	140	78	134
Silicone DC-710	107	149	153	190	98	174
Silicone DC QF-1	144	233	355	305	53	280
Silicone Versilube F-50	19	57	48	47	23	50
Silicone SF-96	12	53	42	37	21	41
Silicone OV-1	16	55	44	42	23	45
Silicone OV-3	44	86	81	88	46	84
Silicone OV-7	69	113	111	128	66	120
Silicone OV-11	102	142	145	178	92	164
Silicone OV-17	119	158	162	202	105	184
Silicone OV-22	160	188	191	253	132	228
Silicone OV-25	178	204	208	280	147	251
Silicone OV-101	17	57	45	43	23	46
Silicone OV-210	146	238	358	310	56	283
Silicone OV-225	228	369	338	386	150	342
Silicone rubber SE-30	15	53	44	41	22	44
Silicone rubber SE-31	16	54	45	43	23	46
Silicone rubber SE-33	17	54	45	42	23	46
Silicone rubber SE-52	32	72	65	67	36	67
Silicone rubber SE-54	33	72	66	67	36	68
Silicone rubber XE-60	204	381	340	367	120	327
Squalene	152	341	238	344	101	265
Surfonic N-300	261	462	313	427	180	364
Tergitol NPX	197	386	258	351	39	293
Tetrahydroxyethyl ethylenediamine (THEED)	463	942	626	893	269	721
Tricresyl phosphate	176	321	250	299	131	254
Triton® X-100	203	399	268	362	145	304
Triton® X-305	262	467	314	430	183	366
Ucon 50 HB-280X	177	362	227	302	130	256
Ucon 50 HB-660	193	380	241	321	141	274
Ucon 50 HB-2000	202	394	253	341	147	289
Ucon 50 HB-5100	214	418	278	375	155	316
Ucon LB-550X	118	271	158	206	91	177

Table 4 (continued)
RETENTION INDEX DIFFERENCES (ΔI) FOR SOLUTE PROBES OF DIFFERENT FUNCTIONALITY BETWEEN VARIOUS LIQUID PHASES AND SQUALANE[a]

	ΔI for solute probes[b]					
Liquid Phase	Benz	BuOH	P	Pyr	O	D
Ucon 75-H-90,000	255	452	299	406	180	348
Versamide 940	109	314	145	209	57	159

[a] The ΔI values listed in the above table represent the actual retention index for a given solute probe on a given liquid phase at 120°C minus the actual retention index for the same solute probe on squalane at 120°C. These ΔI values were reported by McReynolds, W. O., *J. Chromatogr. Sci.*, 8, 685, 1970. The actual retention indices (I) reported for the solute probes on squalane at 120°C were benzene, 653; 1-butanol, 590; 2-pentanone, 627; pyridine, 699; 2-octyne, 841; and 1,4-dioxane, 654.

[b] Solute probes: Benz., benzene; BuOH, 1-butanol; P, 2-pentanone; Pyr, pyridine; O, 2-octyne; D, 1,4-dioxane.

Table 5
RETENTION INDEX DIFFERENCES (ΔI) FOR SOLUTE PROBES OF DIFFERENT FUNCTIONALITY BETWEEN VARIOUS OV LIQUID PHASES AND SQUALANE[a]

	ΔI for Solute Probes[b]				
Liquid Phase	Benz	BuOH	P	Pyr	O
Squalane	0	0	0	0	0
Methyl silicones					
Silicone OV-1	16	55	44	42	23
Silicone OV-101	17	57	45	43	23
Cyano silicones					
Silicone OV-105	36	108	93	86	29
Phenyl silicones					
Silicone OV-73	40	86	76	85	39
Silicone OV-3	44	86	81	88	46
Silicone OV-7	69	113	111	128	66
Silicone OV-61	101	143	142	174	86
Silicone OV-11	102	142	145	178	92
Silicone OV-17	119	158	162	202	105
Trifluoropropyl silicones					
Silicone OV-202	146	238	358	310	56
Silicone OV-210	146	238	358	310	56
Silicone OV-215	149	240	363	315	56
Phenyl silicones					
Silicone OV-22	160	188	191	253	132
Silicone OV-25	178	204	208	280	147
Phenylsilicone-carbowax copolymer					
Silicone OV-330	222	391	273	368	158
Cyano silicones					
Silicone OV-225	228	369	338	386	150

Table 5 (continued)
RETENTION INDEX DIFFERENCES (ΔI) FOR SOLUTE
PROBES OF DIFFERENT FUNCTIONALITY BETWEEN
VARIOUS OV LIQUID PHASES AND SQUALANE[a]

Liquid Phase	ΔI for Solute Probes[b]				
	Benz	BuOH	P	Pyr	O
Carbowax-nitroterephthalic acid polymer					
Silicone OV-351	335	552	382	540	—
Cyano silicones					
Silicone OV-275	629	872	763	849	318

Note: Arranged according to increasing benzene polarity.

[a] The ΔI values listed in the above table represent the actual retention index for a given solute probe on a given liquid phase at 120°C minus the actual retention index for the same solute probe on squalane at 120°C. These ΔI values were reported by McReynolds, W. O., *J. Chromatogr. Sci.*, 8, 685, 1970. The actual retention indices (I) reported for the solute probes on squalane at 120°C were benzene, 653; 1-butanol, 590; 2-pentanone, 627; pyridine, 699; and 2-octyne, 841. Data arrangement taken from Analabs Guide to Chromatographic Phases, The Foxboro Company.

[b] Solute probes: Benz, benzene; BuOH, 1-butanol; P, 2-pentanone; Pyr, pyridine; O, 2-octyne.

The white diatomaceous earth supports have been heat-fluxed with Na_2CO_3 at temperatures in excess of 900°C. The diatoms are fused in this process via a silicate glass bond in which the silica is partially converted to crystalline cristobalite. Iron present in the diatoms is converted to a colorless sodium iron silicate, affording a white material such as Chromosorb W. These white flux-calcined supports are somewhat fragile and fine particles can be produced in the course of preparing GLC packings and packing the GC column. The production of excessive fines during column packing will produce a wider particle size range and increased solute band broadening as discussed in Chapter 2, Section II.

The pink diatomaceous earth supports are produced from crushed firebrick that is obtained by calcining diatoms with a clay binder at temperatures in excess of 1000°C. Metals present in the diatoms (principally, iron) form complex oxides that produce the pink coloration of these supports and tend to make them less inert than white diatomaceous earth supports, resulting in peak tailing of polar solutes when low-loaded GLC column packings are used. These supports have been used successfully, however, for separations of volatile nonpolar hydrocarbons. They have a higher density than the white diatomaceous earth supports due to the breakdown of the diatom structure during calcining and are harder, not as friable, and have greater liquid phase loading capacities (up to 30 wt%, vs. about 25 wt% for white diatomaceous earth). The density of pink vs. white diatomaceous earth is in the order of 0.47 g/mℓ vs. 0.3 g/mℓ, respectively. The surface area for pink diatomaceous earth is fourfold that of white diatomaceous earth (4.0 vs. 1.0 m²/g, respectively).

Both types of diatomaceous earth supports are usually treated further (e.g., by acid washing and/or silanization) to reduce solute sorption and peak tailing. Acid washing (HCl) treatment will remove metal oxides (Fe, Al) that offer degradative adsorption sites, while silanization (generally by reaction with dimethyldichlorosilane [DMCS]) will chemically deactivate hydoxyl groups attached to silica (i.e., silanol groups) that can hydrogen bond to polar solutes and produce excessive peak tailing. Silanized supports, however, have a low surface energy and are not as useful for loading polar liquid phases (e.g., polyesters, silicone phases having a high cyano content) since such phases do not "wet" the support very well. In addition, the analysis of highly polar solutes (e.g., free acids) or the presence of water in injected samples can hydrolyze the support silyl ether groups produced during silanization to produce a partially unsilanized support having a higher surface energy that may produce changes in chromatographic behavior. Silanized supports are recommended, however, for preparing nonpolar GLC packings that are frequently used in hydrocarbon analyses. Base washing is also sometimes used for treating supports that will be used in the analysis of basic solutes. Examples of treated diatomaceous earth supports are given in the following chart:

Treatment[a,b]	Manville	Analabs	Supelco	Alltech/Applied Science
AW	AW Chromosorb W	Anakrom A		GasChrom A
AW-S	AW-DMCS Chromosorb W	Anakrom AS		GasChrom Z
AW-S	HP Chromosorb W	Anakrom Q	Supelcoport	GasChrom Q
NAW	NAW Chromosorb P	Anakrom C22U		GasChrom R
AW	AW Chromosorb P	Anakrom C22A		GasChrom RA
AW-S	AW-DMCS Chromosorb P	Anakrom C22AS		GasChrom RZ

[a]   Analabs-Foxboro
[b]   AW, acid-washed; S, silanized; NAW, non-acid-washed.

There are six principal types of Chromosorb supports available, viz.; Chromosorbs A, W, G, P, T, and 750. These are available in various mesh sizes:

Chromosorb A: 20-30 and 60-80 mesh
Chromosorb W: 30-60, 45-60, 60-80, 80-100, 100-120, and 120-140 mesh
Chromosorb G: 45-60, 60-80, 80-100, 100-120, and 120-140 mesh
Chromosorb P: 30-60, 45-60, 80-100, and 100-120 mesh
Chromosorb T: 30-60 and 40-60 mesh
Chromosorb 750: 60-80, 80-100, and 100-120 mesh

The smaller particle size ranges (e.g., 100—120 mesh) offer reduced solute band broadening and are more frequently used in GLC analysis. Some of the characteristics of these Chromosorbs are given below.

Chromosorb A — Chromosorb A is a diatomite support specifically developed for preparative-scale GLC and is superior to other supports in this regard. It exhibits a good capacity for loading liquid phases up to a 25 wt% loading and possesses a structure that will not readily break down with handling and a surface that is not highly adsorptive. It is less sorptive than Chromosorb P, but is more adsorptive than Chromosorb W. The use of relatively large particle sizes (e.g., 20-30 mesh) allows the use of long columns without excessive pressure drops.

Chromosorb W — Chromosorb W is a flux-calcined diatomite support prepared from Johns-Manville Celite diatomaceous silica (e.g., Celite 545) having a medium loading capacity of up to 15 wt%. It possesses a relatively nonadsorptive surface. It is not identical to Celite 545, but is similar to it in performance and properties. It is white in color and more friable than Chromosorb G and must be handled gently to prevent the formation of fine particles than can reduce column efficiency. It is recommended for the preparation of GLC packings used for the analysis of polar solutes. Chromosorb W has several advantages over supports prepared from Celite 545. Supports prepared from Celite 545 have a higher density that can vary from batch to batch; since GLC packings are prepared on the basis of wt% loading, variations in support density will lend to variations in the actual amount of a given liquid phase that is applied to such supports. The special processing used in the preparation of Chromosorb W provides a lighter, more uniform product.

Chromosorb G — Chromosorb G is a harder, less friable, and more dense diatomite support than Chromosorb W supports and possesses a medium capacity for liquid phase loadings: it is 2.4 times heavier than Chromosorb W and has a maximum recommended liquid phase loading of up to 5 wt% (equivalent to a 12 wt% loading on Chromosorb W). Because of its lower surface area and higher density, it is recommended in place of Chromosorb W for low loaded column packings. It is useful for the preparation of high-efficiency, low-loaded column packings, particularly for the analysis of polar solutes.

Chromosorb P — Chromosorb P is a calcined diatomite prepared from Johns-Manville Sil-O-Cel C-22 firebrick and can be used both for analytical and preparative GLC packings. It is more adsorptive than other Chromosorb materials and is used predominantly for the preparation of packings for the analysis of hydrocarbons and solutes of low polarity. The surface energy of this material is generally too high for use in the analysis of polar solutes; however, AW-DMCS treatment and the use of high liquid phase loadings (e.g., 20 to 30 wt%) can sufficiently reduce the surface activity for its use in the analysis of moderately polar solutes.

Chromosorb T — Chromosorb T is a Johns-Manville fluorocarbon resin made from DuPont Teflon® 6, and has been used as a support in several special cases where a highly inert surface is required to prevent peak tailing problems (e.g., water, hydrazine, SO_2, halogens) encountered with conventional diatomite supports. This material, however, produces column packings of relatively poor efficiency.

Chromosorb 750 — Chromosorb 750 is the most inert, least friable diatomite support of the Chromosorbs and is prepared from a specially selected, high-purity diatomaceous earth that has been acid washed and silanized. Owing to its hardness, very little fines are produced during the preparation of GLC packings and packed columns. Its high surface inertness results in minimum adsorption and decomposition effects in the analysis of polar solutes. It is 1.5 times as dense as Chromosorb W type supports and has a maximum liquid phase loading of 7 wt% (corresponding to 11 wt% loading on Chromosorb W). Column efficiencies of greater than 600 theoretical plates per foot have been reported for 5 wt% loaded Chromosorb 750 GLC packings.

V. SOLID SORBENTS USED IN GSC

Despite several unique problems encountered with GSC, it clearly is superior to GLC for the analysis of gases and some low-boiling liquids and for analyses requiring very low and very high column temperatures. Problems unique to GSC include: (1) much greater partition coefficients and retention times due to adsorptive rather than solution retention; (2) nonlinear partition isotherms and peak asymmetry often occur, especially at lower column operating temperatures; (3) solute peak shape and width may be sensitive to sample size and sample composition, necessitating the use of dilute samples; and (4) the sorptive properties of a solid sorbent may vary from batch-to-batch or with column use. On the bright side, experience and the development of a variety of new sorbents over the past decade have markedly advanced the utility of GSC. These include the use of spherical silicas, molecular sieves, graphitized carbons, chemically-bonded adsorbents, and porous polymers.

A. Porous Polymers:
Various types of porous polymers are available for GSC applications, including eight different Chromosorbs (Chromosorb 101 to 108) and eight different Porapaks (P, P-S, Q, Q-S, R, S, N, and T). A brief description of these follows:

Chromosorb 101 — A styrene-divinylbenzene copolymer useful to 275°C for fast, efficient separations of volatile alkanes, other hydrocarbons, and polar solutes (e.g., free fatty acids, alcohols, glycols, aldehydes, ketones, esters, ethers).

Chromosorb 102 — A high surface area styrene-divinylbenzene copolymer useful to 250°C for light and permanent gases and low-molecular-weight solutes, offering little peak tailing from water, alcohols, and many other oxygenated solutes.

Chromosorb 103 — A cross-linked polystyrene useful to 275°C for fast, efficient separations of basic solutes (e.g., amines), as well as amides, hydrazines, and oxygenated solutes (alcohols, aldehydes, ketones). This material, however, does adsorb glycols.

Chromosorb 104 — An acrylonitrile-divinylbenzene polymer useful to 250°C for the analysis of trace water in benzene, for the separation of nitroparaffins, nitriles, xylenols, ammonia, aqueous hydrogen sulfide, and for nitrogen, sulfur, and carbon oxides. This material affords different selectivities than Chromosorbs 101 to 103.

Chromosorb 105 — A polyaromatic polymer useful to 250°C for the separation of acetylene from lower boiling hydrocarbons, of most gases and organics boiling up to 200°C, and of aqueous solutions containing formaldehyde. This material is less polar and has a different selectivity than Chromosorb 104 which can change the retention order of polar solutes.

Chromosorb 106 — A cross-linked polystyrene useful to 250°C which retains benzene and nonpolar organics and has been used for the separation of C-2 to C-5 fatty acids from their corresponding alcohols.

Chromosorb 107 — An intermediate polarity polymerically cross-linked acrylic ester useful to 250°C for efficient separations of various classes of solutes and has been used for formaldehyde analysis.

Chromosorb 108 — A polymerically cross-linked acrylic ester useful to 250°C for the separation of gases and low-molecular-weight polar solutes (e.g.: alcohols, glycols, water, aldehydes, ketones) which offers different retention characteristics than Chromosorbs 101 to 107.

Porapak P — A styrene-divinylbenzene copolymer useful to 250°C for the separation of a wide variety of simple carbonyls, alcohols, and glycols. This material is the least polar of the Porapaks.

Porapak P-S — Surface-silanized Porapak P useful to 250°C that offers reduced solute peak tailing for separations of aldehydes and glycols.

Porapak Q — A divinylbenzene-ethylvinylbenzene copolymer useful to 250°C for separations of volatile hydrocarbons, organic solutes in water, and nitrogen oxides. This material is the most widely used of the Porapaks.

Porapak Q-S — Surface-silanized Porapak Q useful to 250°C that offers reduced solute peak tailing for separations of organic acids and other volatile polar compounds.

Porapak R — A moderate polarity vinyl pyrollidone polymer useful to 250°C, providing long retention and good resolution of ethers, separations of esters, and separations of water from chlorine gas and hydrochloric acid gas.

Porapak S — A vinyl pyridine polymer useful to 250°C for the separation of normal and branched-chain alcohols.

Porapak N — A divinylbenzene-ethylvinylbenzene-ethyleneglycoldimethacrylate polymer useful to 190°C for the analysis of carbon dioxide, ammonia, and water, and for the separation of acetylene from other C-2 hydrocarbons. This material exhibits a high water retention.

Porapak T — An ethyleneglycoldimethacrylate polymer useful to 190°C for the determination of formaldehyde in aqueous solutions. This material has the highest polarity of the Porapaks and exhibits the greatest retention of water.

The relative retention behavior of the Chromosorbs and Porapak porous polymer materials described above is shown in Table 6 using benzene, tertiary-butanol, 2-butanone, and acetonitrile as solute probes.

Table 6
RETENTION INDICES OF TEST SOLUTES FOR
CHROMOSORBS AND PORAPAKS[b]

Porous polymer	Benzene	*t*-Butanol	2-Butanone	Acetonitrile
Chromosorb 101	745	565	645	580
Chromosorb 102	650	525	570	460
Chromosorb 103	720	575	640	565
Chromosorb 104	845	735	860	885
Chromosorb 105	635	545	580	480
Chromosorb 106	605	505	540	405
Chromosorb 107	660	620	650	550
Chromosorb 108	710	645	675	605
Porapak P	765	560	650	590
Porapak Q	630	538	580	450
Porapak Q-S	625	525	565	445
Porapak R	645	545	580	455
Porapak S	645	550	575	465
Porapak N	735	605	705	595
Porapak T	—	675	700	635

From Dave, S., *J. Chromatogr. Sci.*, 7, 389, 1969. With permission.

B. Spherical Silica Adsorbents
1. Porasils

Porasils are spherical siliceous materials that are intermediate to conventional GC materials having low surface areas and large pore diameters and molecular sieves having high surface areas and large pore diameters. To illustrate: Porasil B has a surface area of 185 m^2/g and a mean pore diameter of 150 Å; Porasil C has a surface area of 100 m^2/g and a mean pore diameter of 300 Å. Both of these Porasils have mean total pore volumes of 1 cm^3/g and are available in particle sizes of 80 to 100 and 100 to 150 mesh. Porasils have also been used as supports for preparing stationary phase coated GLC packings at phase loadings from 1 to 40 wt%. Porasil A has been used for heavily loaded packings in preparative GLC owing to the high loading capacity of the Porasils.[317] Porasil C has been used to prepare bonded packings known as Durapaks (see below under chemically bonded adsorbents).

2. Spherosil

Spherosil is a porous, free-flowing, spherical silica that has the absorption properties of silica and is available in various specific surface areas, ranging from 5 to 500 m^2/g. Like the Porasils, Spherosil can be used for the preparation of GLC packings; coating Spherosil with a very thin film of stationary phase provides a column packing that influences its selectivity for polar compounds. Spherosil itself can be used for the GC analysis of low-boiling solutes and gases and has been used for the analysis of high-boiling compounds and unsaturated solutes. As in the case of Porasil C, Spherosil has been used for the preparation of chemically bonded adsorbents (see below). The available grades of Spherosil are as follows:

Grade	Specific surface area (m^2/g)	Pore diameter (Å)
XOA400	300—500	80
XOA200	140—230	150
XOB075	75—125	300
XOB030	37—62	600
XOB015	18—31	1250
XOC005	5—15	3000

C. Molecular Sieves

Molecular sieves are synthetic alkali metal alumino-silicates containing various cations, principally used for the GSC separation of fixed gases and for the drying of carrier gas streams. Separations using molecular sieves are based on adsorption coupled to the molecular diameter of the solutes; solutes are principally adsorbed because their molecular diameter matches the pore diameter of the molecular sieve. The retention properties of molecular sieves can change over prolonged periods of use due to water adsorption; they can be reactivated by heating at various temperatures for various times (e.g.: 350°C for 2 hr; 300°C for 4 hr; 250°C for 12 hr). The types of molecular sieves include 3A, 4A, 5A, 10A, and 13X. They are generally available in mesh ranges of 40 to 60, 60 to 80, 80 to 100, and 100 to 120, as well as in 1/8- and 1/16-in. pellets. Some of the properties of molecular sieves are as follows:

Molecular sieve type	Pore size (Å)	Cation	Water capacity (%)
3A	3	K	20
4A	4	Na	22
5A	5	Ca	21.6
10A	10	Na	28.6

Washed molecular sieves also are available in which the material is carefully cleaned of dust particles by washing with distilled water, followed by heat deactivation. Washed molecular sieves are particularly useful for the analysis of low levels of nitrogen in oxygen.

Spherocarb is a particular type of molecular sieve which is a nonfriable, spherical, carbon molecular sieve. It has been used for GSC analysis of light (C-1 to C-4) hydrocarbons, permanent gases, stack gas effluent components (e.g.: SO_2, H_2S, CO_2), and other pollutant gases. This material has been shown to provide separations of mixtures of saturated and unsaturated hydrocarbons in which the unsaturated hydrocarbons elute prior to the saturated derivatives. It also has been used for separations of mixtures of methane, CO_2 and N_2O, of CO_2, H_2S and SO_2, of krypton and xenon, and for the analysis of commercial grade propane, methane in automotive exhausts, and N_2O in air. Other types of carbon molecular sieves include Carbosphere and Carbosieve. Carbosphere exhibits properties similar to molecular sieves, having a small pore diameter in the angstrom range, a surface area of about 1000 m^2/g and a pore size of about 13 Å. Carbosphere has been found useful for the separation of trace levels of methane and acetylene in ethylene, of methane, acetylene, and ethylene in ethane, and is very nonretentive for water, allowing for rapid separations of low-molecular-weight polar organic solutes (e.g.: formaldehyde and methanol in water). Carbosieves are principally used for the analysis of light gases and have a surface area of 1000 m^2/g. Two types of Carbosieves are available: Carbosieve S (a dense, spherical material for the separation of light hydrocarbons, permanent gases, sulfur gases, and low-molecular-weight organics; and Carbosieve G (which replaces Carbosieve B).

These carbon molecular sieves lack the static electricity problem associated with porous polymers and have a fairly hard density, allowing GC columns to be easily packed. They may, however, be affected by oxygen and contaminants in air and carrier gas streams, and are best stored under nitrogen or argon.

D. Graphitized Carbons

Various graphitized carbon blacks have been utilized as adsorbents for the GSC analysis of light (C-1 to C-10) hydrocarbons. One such series of graphitized carbons is the Carbopacks. These materials can resolve isomeric mixtures (e.g.: *m*- and *p*-cresols; all eight isomers of amyl alcohols) and have been used in separations of free acids, amines, phenols, alcohols, sulfur gases, unsaturated C-4 hydrocarbons, and volatile halogen-containing organics. Carbopack C is a graphitized carbon having a surface area of about 12 m²/g and is similar to Sterling FT graphitized carbon. Carbopack B is a graphitized carbon having a surface area of about 100 m²/g and is similar to Graphon graphitized carbon. The Carbopacks differ markedly from Carbosieve (a carbon molecular sieve; see above), which is not graphitized. These materials have also been used with thin coatings of a stationary phase (generally less than 1% by weight) for the separation of mixtures of volatile solvents. Another type of graphitized carbons are the Graphpacks. These materials are available in surface areas of 10 to 13 m²/g and about 100 to 110 m²/g, offering solute retention almost exclusively by nonspecific adsorption. They typically are used with low stationary phase coatings (less than 0.5 wt%) to modify their surface characteristics for separations of solvent mixtures, phenols, and low-molecular-weight acids and hydrocarbons (C-2 to C-5).

E. Chemically Bonded Adsorbents

The first type of chemically-bonded ("Durapak") adsorbent was a material in which oxypropionitrile was chemically-bonded to Porasil C.[318] This material is of medium polarity, useful in the separation of solutes having widely varying polarities. Two other types of such adsorbents include octane bonded to Porasil C and octadecane bonded to Porasil C. In the octane-Porasil C material, free residual Si-OH groups remain on the silica surface of Porasil C, making this the most polar of the series. Octane-Porasil C has been used for rapid separations of C-1 to C-5 hydrocarbons in which complete resolution of all of the C-4 saturated and unsaturated isomers can be attained. The octadecyl-Porasil C material offers high temperature stability for separations of polycyclic aromatic hydrocarbons and fuel oils. Another type of bonded packing materials is that in which an extensively deactivated diatomaceous earth is coated with a liquid phase which is then heated at high temperatures. The resultant material is solvent extracted, producing an essentially monomolecular film-bonded material having about 0.2% of nonextractable, bonded stationary phase. This type of material has shown a broad utility for the separation of complex mixtures, but may be oxidized at elevated temperatures requiring that traces of oxygen be excluded from the carrier gas stream. The low "bleed" rate and chromatographic efficiency of the chemically bonded phases make them useful for GC-MS analysis. The maximum column operating temperature for these materials under oxygen-free conditions is 260°C. A brief tabulation (courtesy of Alltech Associates) of ΔI values (I on heat-bonded phase at 50°C minus I on 20% squalane column at 120°C) for McReynolds solutes is given below:

Heat-bonded phase	ΔI Values relative to squalane[a]				
	Benz	**1-Butanol**	**2-Pentanone**	**N-P**	**Pyr**
Igepal CO-630	0	222	72	166	89
Triton® X-105	13	253	79	196	106
Ucon 50-HB-2000	5	305	80	234	140
Igepal CO-990	51	270	106	231	145
Triton® X-305	51	296	102	250	165
Igepal CO-990	10	305	120	251	190
Carbowax 20M	121	317	172	280	180
FFAP	151	363	167	341	259
PEGS	241	480	292	494	419

[a] Benz, benzene; N-P, nitropropane; Pyr, pyridine.

F. Other Adsorbents

Other types of adsorbents have been used for GSC analyses. These include purified XAD-2 resin, aluminas, and Tenax. Two Woelm aluminas have been particularly suited for hydrocarbon separations: GSC-120 has been used for the separation for C-1 to C-5 hydrocarbons; GSC-121 (deactivated by alkali treatment) has been used for the separation of longer-chain hydrocarbons in gasoline fractions. Extensive work has been carried out with Tenax, a brief description of which follows.

Tenax is a porous polymer based on 2,6-diphenyl-*p*-phenylene oxide. It has been used for the separation of high-boiling polar solutes (e.g.: aromatic hydrocarbons, ketones, aldehydes, alcohols, amines). Applications in this regard include studies reported by Ponder, J. *Chromatogr.*, 97, 77, 1974, Sakodynski et al., *Chromatographia*, 7, 339, 1974, Daehmen et al., *J. Chromatogr. Sci.*, 13, 79, 1975, and Van Wijk, *J. Chromatogr. Sci.*, 8, 122, 1970, among others. Tenax is also widely used today as an adsorbent for trapping organic air pollutants, followed by thermal desorption in the inlet of a gas chromatography and flame-ionization detection of the trapped organic pollutants. Further applications of the use of adsorbents are given in Table 2.

REFERENCES

1. **Gaspar, G., Olivo, J., and Guiochon, G.,** *Chromatographia*, 11, 321, 1978.
2. **Alekseeva, A. V., Shatilova, L. A., Slepneva, A. T., and Berezovskaya, E. N.,** *Khim. Tekhnol. Topl. Masel*, 6, 31, 1979.
3. **Koci, K.,** *Bul. Shkencave Nat.*, 32, 59, 1978; *Chem. Abstr.*, 90, 171106b, 1979.
4. **Saha, N. C., Jain, S. K., and Dua, R. K.,** *J. Chromatogr. Sci.*, 16, 323, 1978.
5. **Smolkova-Keulemansova, E.,** *Chromatographia*, 11, 70, 1978.
6. **Ravey, M.,** *J. Chromatogr. Sci.*, 16, 79, 1978.
7. **Parcher, J. F., Hansbrough, J. R., and Koury, A. M.,** *J. Chromatogr. Sci.*, 16, 183, 1978.
8. **Nabivach, V. M. and Kirilenko, A. V.,** *Khim. Tverd. Topl. (Moscow)*, 1979 (3), 90; *Chem. Abstr.*, 91, 116878a, 1979.
9. **Kumar, P., Sarowha, S. L. S., and Gupta, P. L.,** *Analyst (London)*, 104, 788, 1979.
10. **Mathews, R. G., Torres, J., and Schwartz, R. D.,** *J. High Resolut. Chromatogr. Chromatogr. Commun.*, 1, 139, 1978.
11. **Kuchhal, R. K. and Mallik, K. L.,** *Anal. Chem.*, 51, 392, 1979.
12. **Lukac, S.,** *Chromatographia*, 12, 17, 1979.
13. **Schneider, W., Frohne, J. C., and Bruderreck, H.,** *J. Chromatogr.*, 155, 311, 1978.
14. **Soroka, J. M.,** *City Univ., N.Y., Diss.*, 1979, 297 pp.; API Abstracts, Petroleum Refining and Petrochemicals, 27, 2019, 1980.
15. **Rooney, T. A.,** *Ind. Res. Dev.*, 20, 143, 1978.
16. **Boksanyi, L. and Kovats, E. S.,** *J. Chromatogr.*, 126, 87, 1976.
17. **Banerjee, D. K., Chawla, S. L., and Malik, K. L.,** *Indian J. Technol.*, 13 (10), 473, 1975.
18. **Ryba, M.,** *J. Chromatogr.*, 123, 327, 1976.
19. **Kulikov, V. I., Volozhin, A. I., et al.,** *Vesti. Akad. Navuk. B.S.S.R., Ser. Khim. Navuk*, 1977 (2), 47.
20. **Parcher, J. F., Weiner, P. H., et al.,** *J. Chem. Eng. Data*, 20, 145, 1975.
21. **Jeknavorian, A. A. and Barry, E. F.,** *J. Chromatogr.*, 101, 299, 1974.
22. **Talalaev, E. I., Sergienko, S. R., and Ovezova, A. A.,** *Izv. Akad. Nauk Turkm. S.S.R., Ser. Fiz.-Tekh., Khim. Geol. Nauk*, 4, 55, 1973; *Chem. Abstr.*, 82, 113891c, 1975.
23. **Hawkes, S., Grossman, D., Hartkopf, A., Eisenhour, T., Leary, J., Parcher, J., Wold, S., and Yancey, J.,** *J. Chromatogr. Sci.*, 13, 115, 1975.
24. **Martire, D. E.,** *Anal. Chem.*, 46, 626, 1974.
25. **Pacakova, V. and Ullmannova, H.,** *Chromatographia*, 7, 75, 1974.
26. **Lekova, K. and Gerasimov, M.,** *Chromatographia*, 7, 69, 1974; 7, 595, 1974.
27. **Bruner, F., Ciccioli, P., Crescentini, G., and Pistolesi, M. T.,** *Anal. Chem.*, 45, 1851, 1973.
28. **Moolchandra, R. and Mallik, K. L.,** *Anal. Chem.*, 44, 2404, 1972.
29. **Belopol'skaya, S. I., Vigdergauz, M. S., and Khodzhaev, G. Kh.,** *Khim. Tekhnol. Topl. Masel*, 17, (4), 56, 1972.
30. **Galtieri, A. and Casalini, A.,** *Riv. Combust.*, 26, 286, 1972.
31. **Saha, N. C. and Nitra, G. D.,** *J. Chromatogr. Sci.*, 11, 419, 1973.

32. **Sojak, L. and Hrivnak, J.,** *J. Chromatogr. Sci.,* 10, 701, 1972.

33. **Nakamura, M.,** *Sekiyu Gakkai Shi,* 16, 51, 1973; *Chem. Abstr.,* 79, 3331t, 1973.

34. **Dupuy, W. E., Hudson, H. R., and Karam, P. A.,** *J. Chromatogr.,* 71, 347, 1972.

35. **Mamedaliev, G. M., Aliev, S. M., et al.,** *Neftekhimiya,* 13, 194, 1973.

36. **Sojak, L., Hrivnak, J., Majer, P., and Janak, J.,** *Anal. Chem.,* 45, 293, 1973.

37. **Mathews, R. G., Schwartz, R. D., Novotny, M., and Zlatkis, A.,** *Anal. Chem.,* 43, 1161, 1971; **Mathews, R. G., Schwartz, R. D., Stouffer, J. E., and Pettitt, B. C.,** *J. Chromatogr. Sci.,* 8, 508, 1970.

38. **Chow, L. E. and Martire, D. E.,** *J. Phys. Chem.,* 75, 2005, 1971.

39. **Paterok, N. and Wandzik, E.,** *Chem. Anal. (Warsaw),* 15, 977, 1970.

40. **Novotny, M., Segura, R., and Zlatkis, A.,** *Anal. Chem.,* 44, 9, 1972.

41. **Robinson, R. E., Coe, R. H., and O'Neal, M. J.,** *Anal. Chem.,* 43, 591, 1971.

42. **Walker, J. Q. and Wolf, C. J.,** *Anal. Chem.,* 42, 1652, 1970.

43. **Robinson, P. G. and Odell, A. L.,** *J. Chromatogr.,* 57, 1, 1971.

44. **Al-Thamir, W. K., Laub, R. J., and Purnell, J. H.,** *J. Chromatogr.,* 142, 3, 1977.

45. **Kuchhal, R. K., Dogra, P. V., and Mallik, K. L.,** *Res. Ind.,* 22, 94, 1977.

46. **Castello, G. and D'Amato, G.,** *J. Chromatogr.,* 175, 27, 1979.

47. **Haken, J. K. and Ho, D. K. M.,** *J. Chromatogr.,* 142, 203, 1977.

48. **Sidorov, R. I., Khvostikova, A. A., and Bakhrusheva, G. I.,** *Zh. Anal. Khim.,* 32, 1650, 1977.

49. **Tarjan, G., Kiss, A., Kocsis, G., Meszaros, S., and Takacs, J. M.,** *J. Chromatogr.,* 119, 327, 1976.

50. **DiCorcia, A., Liberti, A., and Samperi, R.,** *J. Chromatogr.,* 122, 459, 1976.

51. **Laub, R. J. and Pecsok, R. L.,** *Anal. Chem.,* 46, 2251, 1974.

52. **Good, R. J.,** U. S. Patent 3,808,125.

53. **Kuchhal, R. K. and Mallik, K. L.,** *Indian J. Chem.,* 12, 1322, 1974.

54. **Rohrschneider, L.,** *J. Chromatogr. Sci.,* 11, 160, 1973.

55. **Aliev, M. I., Fisher, S. I., Shirinbekova, F. I., and Imanov, Z. T.,** U.S.S.R. Patent 309,294, September, 20, 1971.

56. **Louis, R.,** *Erdoel Kohle,* 24, 88, 1971.

57. **Kovats, E. sz.,** *Column Chromatography,* Swiss Chemists' Association (Distributed by: Sauerlander AG, CH-5001, Aarau, Switzerland), 1970.

58. **Supina, W. R. and Rose, L. P.,** *J. Chromatogr. Sci.,* 8, 214, 1970.

59. **McReynolds, W. O.,** *J. Chromatogr. Sci.,* 8, 685, 1970.

60. **Pichler, H., Ripperger, W., and Schwarz, G.,** *Erdoel Kohle,* 23, 91, 1970.

61. **Sojak, L., Kraus, G., Ostrovsky, I., Kralovicova, E., and Krupcik, J.,** *J. Chromatogr.,* 206, 475, 1981.

62. **Dimov, N. and Papazova, D.,** *Chromatographia,* 12, 720, 1979.

63. **Khitrik, A. A., Tukov, G. V., Terpilovskii, N. N., and Batalov, V. S.,** U.S.S.R. Patent 549736, February 4, 1975; **Tukov, G. V., Nurtdinov, S. Kh., Kalinchuk, L. N., Novikov, V. F., and Zakharov, V. A.,** U.S.S.R. Patent 432384, June 15, 1972.

64. **Boksanyi, L. and Kovats, E. S.,** *J. Chromatogr.,* 126, 87, 1976.

65. **Riedo, F., Fritz, D., Tarjan, G., and Kovats, E. S.,** *J. Chromatogr.,* 126, 63, 1976.

66. **Ryba, M.,** *J. Chromatogr.,* 123, 327, 1976.

67. **Di Corcia, A., Liberti, A., and Samperi, R.,** *J. Chromatogr.,* 122, 459, 1976.

68. **Lekova, K. and Gerasimovf, M.,** *Chromatographia,* 7, 595, 1974; **Badinska, K. and Gerasimovf, M.,** *Khim. Ind. (Sofia),* 44, (6), 250, 1972.

69. **Mitooka, M.,** *Bunseki Kagaku,* 21, (11), 1447, 1972.

70. **Paterok, N. and Kotowski, W.,** *Chem. Anal. (Warsaw),* 16 (4), 801, 1971.

71. **Kuliev, A. M., Mamedov, G. Kh., and Abdullaev, A. I.,** U.S.S.R. Patent 316988, August 5, 1970.

72. **Dimov, N., Shopov, D., and Petkova, T.,** *Proc. 3rd Anal. Chem. Conf.,* 1, 299, 1970.

73. **Beider, T. B. and Bochkareva, T. P.,** *Khim. Prom-st., Ser.: Metody Anal. Kontrolya Kach. Prod. Khim. Prom-sti,* 5, 14, 1979; *Chem. Abstr.,* 91, 186129f, 1979.

74. **Dimov, N. and Papazova, D.,** *Chromatographia,* 12, 443, 1979.

75. **Stancher, B. and Cerma, E.,** *Rass. Chim.,* 27 (3), 117, 1975.

76. **Dimov, M.,** *J. Chromatogr.,* 119, 109, 1976.

77. **Marik, K. and Smolkova, E.,** *J. Chromatogr.,* 91, 303, 1974.

78. **Marik, K. and Smolkova, E.,** *Chromatographia,* 6, 420, 1973.

79. **Guillot, J., Janin, C., and Guyot, A.,** *J. Chromatogr. Sci.,* 11, 375, 1973.

80. **Bruner, F., Ciccioli, P., and Di Corcia, A.,** *Anal. Chem.,* 44, 894, 1972.

81. **Rappoport, S. and Gaeumann, T.,** *Helv. Chm. Acta,* 56, 1145, 1973.

82. **Vollert, U. and Mautsch, M.,** *Chem. Tech. (Leipzig),* 24, 100, 1972.

83. **Petrovic, K. and Vitrovic, D.,** *J. Chromatogr.,* 65, 155, 1972.

84. Takacs, J., Szita, C., and Tarjan, G., *J. Chromatogr.*, 56, 1, 1971.
85. Smolkova, E., Feltl, L., and Vetecka, *Chromatographia*, 12, 5, 1979.
86. Takacs, E. C., Voros, J., and Takacs, J. M., *J. Chromatogr.*, 159, 297, 1978.
87. Castello, G. and D'Amato, G., *J. Chromatogr.*, 107, 1, 1975.
88. Hartkopf, A., *J. Chromatogr. Sci.*, 10, 145, 1972.
89. Eppert, G., Ludwig, H., and Schinke, I., *J. Prakt. Chem.*, 321 (4), 570, 1979.
90. Scheiner, R., *Zh. Anal. Khim.*, 33, 789, 1978.
91. Seroshtan, V. A. and Zakharova, N. V., *Zh. Anal. Khim.*, 34, 1166, 1979.
92. Lukac, S. and Nguyen, T. Q., *Chromatographia*, 12, 231, 1979.
93. Johansen, N. G., *Chromatogr. Newsl.*, 6 (2), 17, 1978.
94. Orav, A., Kuningas, K., Rang, S., and Eisen, O., *Eesti NSV Tead. Akad. Toim.*, *Keem.*, 29, 18, 1980.
95. Chre'tien, J. R. and Dubois, J. E., *Anal. Chem.*, 49, 747, 1977.
96. Rang, S., Kuningas, K., Orav, A., and Eisen, O., *Chromatographia*, 10, 55, 1977.
97. Wampler, F. B., *Anal. Chem.*, 48, 1644, 1976.
98. Hirsch, R. F. and Phillips, C. S. G., *Anal. Chem.*, 49, 1549, 1977.
99. Sojak, L., Hrivnak, J., Ostrovsky, I., and Janak, J., *J. Chromatogr.*, 91, 613, 1974.
100. Guha, O. K. and Janak, J., *J. Chromatog.*, 68, 325, 1972.
101. Kladnig, W., *Chromatographia*, 6, 243, 1973.
102. Sojak, L., Hrivnak, J., Krupcik, J., and Janak, J., *Anal. Chem.*, 44, 1701, 1972.
103. Ryba, M., *Chromatographia*, 5, 23, 1972.
104. Mel'kanovitskaya, S. G., *Gidrokhim. Mater.*, 53, 153, 1972; *Chem. Abstr.*, 77, 22466y, 1972.
105. Lysyj, I. and Newton, P. R., *Anal. Chem.*, 44, 2385, 1972.
106. Rang, S., Eisen, O., and Kuningas, K., *Eesti NSV Tead. Akad. Toim.*, *Keem. Geol.*, 19, 99, 1970.
107. Schmitt, D. L. and Jonassen, H. B., *Anal. Chim. Acta*, 49, 580, 1970.
108. Jaworski, M. and Szewczyk, H., *Zh. Anal. Khim.*, 25, 1215, 1970.
109. Seweryniak, M. and Slonka, T., *Chem. Anal. (Warsaw)*, 23, 529, 1978.
110. Zlatkis, A. and de Andrade, I. M. R., *Chromatographia*, 2, 298, 1969.
111. Sojak, L., Skalak, P., and Janak, J., *J. Chromatogr.*, 65, 137, 1972.
112. Lukac, S. and Nguyen, T. Q., *Chromatographia*, 12, 231, 1979.
113. Welsch, Th., Engewald, W., and Berger, P., *Chromatographia*, 11, 5, 1978.
114. Baiulescu, G. E. and Ilie, V. A., *Anal. Chem.*, 46, 1847, 1974.
115. Ryba, M., *Chromatographia*, 5, 23, 1972.
116. Gil-Av, E. and Schurig, V., *Anal. Chem.*, 43, 2030, 1971.
117. Guha, O. and Janak, J., *J. Chromatogr.*, 68, 325, 1972.
118. Eisen, O. G., Orav, A., and Rang, S., *Chromatographia*, 5, 229, 1972.
119. Otsa, R., *Tezisy Dokl. Resp. 2nd Konf. Molodykh Uch.-Khim.*, 1977, 2.
120. Moiseeva, V. G., Genkin, A. N., and Fursenko, A. V., *Zh. Fiz. Khim.*, 54 (4), 995, 1980.
121. Wasik, S. P. and Tsang, W., *J. Phys. Chem.*, 74 (15), 2970, 1970; Rang, S., Eisen, O., and Kuningas, K., *Eesti NSV Tead. Akad. Tiom. Keem. Geol.*, 19 (2), 99, 1970.
122. Pscheidl, H., Moeller, E., Raedel, C., and Haberland, D., *Z. Chem. (Leipzig)*, 19, 114, 1979.
123. Berezkin, V. G., Fateeva, V. M., Kazakova, Z. A., and Shikalova, I. V., *Zh. Anal. Khim.*, 31, 1753, 1976.
124. Bardyshev, I. I. and Kulikov, V. I., *Gidroliz. Lesokhim. Pron.*, 5, 17, 1972; *Chem. Abstr.*, 77, 172377j, 1972.
125. Vanek, J., Podrouzkova, B., and Landa, S., *J. Chromatogr.*, 52, 77, 1970.
126. Stopp, I., Engewald, W., Kuhn, H., and Welsch, T., *J. Chromatogr.*, 147, 21, 1978.
127. Dimov, N., *J. Chromatogr.*, 148, 11, 1978.
128. Tyson, B. J., *J. Chromatogr.*, 111, 419, 1975.
129. Schomburg, G. and Dielmann, G., *Anal. Chem.*, 45, 1647, 1973.
130. Engewald, W., Epsch, K., Graefe, J., and Weisch, T., *J. Chromatogr.*, 91, 623, 1974.
131. Eisen, O. G., Orav, A., and Rang, S., *Chromatographia*, 5, 229, 1972.
132. Grizzle, P. L. and Coleman, H. J., *Anal. Chem.*, 51, 602, 1979.
133. DiCorcia, A., Samperi, R., and Capponi, G., *J. Chromatogr.*, 160, 147, 1978.
134. Vigdergauz, M. S., Gabitova, R. K., Vigalok, R. V., and Novikova, I. R., *Zavod. Lab.*, 45, 894, 1979.
135. Tejedor, J. N., *J. Chromatogr.*, 177, 279, 1979.
136. Nabivach, V. M., Bur'yan, P., and Macak, J., *Zh. Anal. Khim.*, 33, 1416, 1978; Nabivach, V. M., and Kirilenko, A. V., *Vopr. Khim. Khim. Tekhnol.*, 52, 139, 1978; *Chem. Abstr.*, 91, 109859s, 1979; *Khim. Tverd. Topl. (Moscow)*, 1979 (3), 90; *Chem. Abstr.*, 91, 116878a, 1979.
137. Beernaert, H., *J. Chromatogr.*, 173, 109, 1979.
138. Chmela, Z., Cap, L., and Adamek, J., *Acta Univ. Palacki. Olomuc. Fac. Rerum Nat.*, 1977, 53 (Chem. 16), 163; Chem. Abstr., 90, 33597m, 1979.

139. **Schroeder, H.,** *J. High Resolut. Chromatogr. Chromatogr. Commun.,* 3, 38, 1980.
140. **Nawrocki, J., Szczepaniak, W., and Wasiak, W.,** *J. Chromatogr.,* 178, 91, 1979.
141. **Kuchhal, R. K., Mathur, M. S., and Kumar, P.,** *Chromatographia,* 11, 30, 1978.
142. **Kriz, J., Popl, M., and Mostecky, J.,** *J. Chromatogr.,* 97, 3, 1974.
143. **Nabivach, V. M. and Vargalyuk, V. N.,** *Vopr. Khim. Khim. Tekhnol.,* 34, 119, 1974; *Chem. Abstr.,* 83, 30515c, 1975.
144. **Laub, R. J., Ramamurthy, V., and Pecsok, R.,** *Anal. Chem.,* 46, 1659, 1974.
145. **Langer, S. H.,** *Anal. Chem.,* 44, 1915, 1972.
146. **Sakodinskii, K. I., Klinskaya, N. S., and Panina, L. I.,** *Anal. Chem.,* 45, 1369, 1973.
147. **Sybilska, D., Malinowska, K., Siekierska, M., and Bylina, J.,** *Chem. Anal. (Warsaw),* 17, 1031, 1972.
148. **Il'icheva, L. F. and Il'ichev, G. N.,** *Neftepererab. Neftekhim. (Moscow),* 9, 40, 1973; *Chem. Abstr.,* 80, 50074z, 1974.
149. **Kuchhal, R. K. and Mallik, K. L.,** *Chromatographia,* 5, 278, 1972.
150. **Deur-Siftar, D. and Svob, V.,** *Kem. Ind.,* 20, 549, 1971; *Chem. Abstr.,* 77, 69831d, 1972.
151. **Nikelly, J. G.,** *Anal. Chem.,* 45, 2280, 1973.
152. **Diskina, D. E., Lazareva, I. S., and Zarubina, M. I.,** *Khim. Tekhnol. Topl. Masel,* 17 (6), 61, 1972; **Diskina, D. E., Prokof'ev, K. V., and Kozhaeva, N. G.,** *Neftepererab. Neftekhim. (Moscow),* 10, 34, 1972; *Chem. Abstr.,* 78, 6049j, 1973.
153. **Grushka, E. and Solsky, J. F.,** *Anal. Chem.,* 45, 1836, 1973.
154. **Louis, R.,** *Erdoel Kohle, Erdgas, Petrochem. Brennst.-Chem.,* 25, 582, 1972.
155. **Otvos, I., Iglewski, S., Hunneman, D. H., Bartha, B., Balthazar, Z., and Palvi, G.,** *J. Chromatogr.,* 78, 309, 1973.
156. **Vigalok, R. V. and Vigdergauz, M. S.,** *Izv. Akad. Nauk S.S.R. Ser. Khim.,* 3, 715, 1972.
157. **Vigalok, R. V. and Vigdergauz, M. S.,** *Izv. Akad. Nauk S.S.R., Ser. Khim.,* 3, 718, 1972.
158. **Uno, T. and Okuda, H.,** *Jpn. Anal.,* 19, 1204, 1970.
159. **Richmond, A. B.,** *J. Chromatogr. Sci.,* 9, 571, 1971.
160. **Duerbeck, H. W.,** *Z. Anal. Chem.,* 251, 108, 1970.
161. **Krupcik, J., Liska, O., and Sojak, L.,** *J. Chromatogr.,* 51, 119, 1970.
162. **Stuckey, C. L.,** *J. Chromatogr. Sci.,* 9, 575, 1971.
163. **Cook, L. E. and Raushel, F. M.,** *J. Chromatogr.,* 65, 556, 1972.
164. **Vigalok, R. V., Gabitova, R. K., Anoshina, N. P., Palikhov, N. A., Maidachenko, G. G., and Vigdergauz, M. S.,** *Zh. Anal. Khim.,* 31, 644, 1976.
165. **Witkiewicz, Z. and Popiel, S.,** *J. Chromatogr.,* 154, 60, 1978.
166. **Witkiewicz, Z., Suprynowicz, Z., and Dawbrowski, R.,** *J. Chromatogr.,* 175, 37, 1979.
167. **Haak, K. E., Oelert, H. H., Siegert, H., and Zajontz, J.,** *Erdoel Kohle, Erdgas, Petrochem. Brennst.-Chem.,* 30, 567, 1977.
168. **Svob, V. and Deur-Siftar, D.,** *J. Chromatogr.,* 91, 677, 1974.
169. **Sybilska, D., Malinowska, K., Siekierska, M., and Bylina, J.,** *Chem. Anal. (Warsaw),* 17, 1031, 1972.
170. **Vidal-Madjar, C. and Guiochon, G.,** *J. Chromatogr. Sci.,* 9, 664, 1971.
171. **Grushka, E. and Solsky, J. F.,** *Anal. Chem.,* 45, 1836, 1973.
172. **Taramasso, M. and Fuchs, P.,** *J. Chromatogr.,* 49, 70, 1970.
173. **Porcaro, P. J. and Shubiak, P.,** *J. Chromatogr. Sci.,* 9, 1971.
174. **Castells, R. C.,** *Chromatographia,* 6, 57, 1973.
175. **Lyuter, A. V., Rekhviashvili, A. N., and Gol'dshtein, Yu. A.,** U.S.S.R. Patent 309293, July 9, 1965 (7/9/71).
176. **Strand, J. W. and Andren, A. W.,** *Anal. Chem.,* 50, 1508, 1978.
177. **Frycka, J.,** *Chromatographia,* 11, 413, 1978.
178. **Tesarik, K., Fryska, J., and Ghyczy, S.,** *J. Chromatogr.,* 148, 223, 1978.
179. **Mutton, I. M.,** *J. Chromatogr.,* 172, 438, 1979.
180. **Zielinski, W. L., Jr. and Janini, G. M.,** *Proc. ORNL Workshop, Exposure Polynucl. Aromat. Hydrocarbons Coal Convers. Processes,* 2, 83, 1977.
181. **Janini, G. M., Muschik, G. M., Schroer, J. A., and Zielinski, W. L., Jr.,** *Anal. Chem.,* 48, 1879, 1976.
182. **Janini, G. M., Johnston, K., and Zielinski, W. L., Jr.,** *Anal. Chem.,* 47, 670, 1975; **Janini, G. M., Muschik, G. M., and Zielinski, W. L., Jr.,** *Anal. Chem.,* 48, 809, 1976.
183. **Radecki, A., Lamparczyk, H., and Kaliszan, R.,** *Chromatographia,* 12, 595, 1979.
184. **Zielinski, W. L., Jr. and Janini, G. M.,** *Proc. ORNL Workshop, Exposure Polynucl. Aromat. Hydrocarbons Coal Convers. Processes,* 2, 83, 1977.
185. **Saluste, S., Klesment, I., and Kivirahk, S.,** *Eesti NSV Tead. Akad. Toim., Keem.,* 28, 7, 1979; **Shlyakhov, A. F., Novikova, N. V., and Koreshkova, R. I.,** *Zavod. Lab.,* 45, 103, 1979.
186. **Lee, M., Vassilaros, D. L., White, C. M., and Novotny, M.,** *Anal. Chem.,* 51, 768, 1979.

187. **Blomberg, L., Buijten, J., Gawdzik, J., and Wannman, T.,** *Chromatographia,* 11, 521, 1978.
188. **Lysyuk, L. S. and Korol, A. N.,** *Chromatographia,* 10, 712, 1977.
189. **Grimmer, G., Boehnke, H.,** *Fresenius' Z. Anal. Chem.,* 261, 310, 1972; **Grimmer, G., Hildebrandt, A., and Boehnke, H.,** *Erdoel Kohle, Erdgas, Petrochem. Brennst.-Chem.,* 25, 442, 1972; 25, 531, 1972.
190. **Vigdergauz, M. S. and Vigalok, R. V.,** *Neftekhimiya,* 11, 141, 1971.
191. **Bhatia, K.,** *Anal. Chem.,* 43, 609, 1971.
192. **Janini, G. M., Shaikh, B., and Zielinski, W. L., Jr.,** *J. Chromatogr.,* 132, 136, 1977; **Strand, J. W. and Andren, A. W.,** *Anal. Chem.,* 50, 1508, 1978.
193. **Frycka, J.,** *J. Chromatogr.,* 170, 459, 1979.
194. **Kaliszan, R. and Lamparczyk, H.,** *J. Chromatogr. Sci.,* 16, 246, 1978.
195. **Lee, M. L., Vassilaros, D. L., White, C. M., and Novotny, M.,** *Anal. Chem.,* 51, 768, 1979.
196. **Deaconeasa, V., Constantinesau, T., and Trestianu, S.,** *Rev. Chim.,* 28, 777, 1977.
197. **Wasik, S. P. and Chesler, S.,** *J. Chromatogr.,* 122, 451, 1976.
198. **Chiavari, G. and Pastorelli, L.,** *Chromatographia,* 7, 30, 1974; **Chitour, S. E. and Vergnaud, J. M.,** *J. Chromatogr.,* 89, 295, 1974.
199. **Lane, D. A., Moe, H. K., and Morris, K.,** *Anal. Chem.,* 45, 1776, 1973.
200. **Janini, G. M., Muschik, G. M., Schroer, J. A., and Zielinski, W. L., Jr.,** *Anal. Chem.,* 48, 1879, 1976.
201. **Willis, D. E.,** *Anal. Chem.,* 50, 827, 1978.
202. **Krawiec, Z., Gonnord, M. -F., Guiochon, G., and Chretien, J. R.,** *Anal. Chem.,* 51, 1655, 1979.
203. **Feltl, L., Smolkova, E., and Skurovcova, M.,** *Collect. Czech. Chem. Commun.,* 44, 1116, 1979.
204. **DiCorcia, A. and Giabbai, M.,** *Anal. Chem.,* 50, 1000, 1978.
205. **Ciccioli, P., Hayes, J. M., Rinaldi, G., Denson, K. B., and Meinschein, W. G.,** *Anal. Chem.,* 51, 400, 1979.
206. **Jaworski, M. and Herzog, G.,** *Chem. Anal. (Warsaw),* 24, 475, 1979.
207. **Khots, M. S., Yarmakov, M. R., and Polyakova, A. A.,** *Zh. Khim.,* 34, 2022, 1979.
208. **Soroka, J. M.,** *City Univ., N.Y., Diss.,* 1979, 297 pp.; API Abstracts, Petroleum Refining and Petro-chemicals, 27, 2019, 1980; **Hirsch, R. F., Gaydosh, R. J., and Chretein, J. R.,** *Anal. Chem.,* 52, 723, 1980.
209. **Snowdon, L. R. and Peake, E.,** *Anal. Chem.,* 50, 379, 1978.
210. **Sawatzky, H., George, A. E., et al.,** *Fuel,* 55, 16, 1976; 55, 329, 1976.
211. **Stepukhovich, A. D. and Bolotina, N. E., et al.,** *Neftekhimiya,* 16 (5), 789, 1976.
212. **Wolf, F., Heyer, W., Jamowski, F., and Finger, E.,** *Z. Chem.,* 17 (11), 423, 1977.
213. **Smolkova-Keulemansova, E.,** *Chromatographia,* 11, 70, 1978.
214. **Di Corcia, A. and Samperi, R.,** *J. Chromatogr.,* 117, 199, 1976.
215. **Tanaka, K., Ishizuka, T., and Sunahara, H.,** *Bunseki Kagaku,* 25 (3), 183, 1976.
216. **Di Corcia, A., Liberti, A., and Samperi, R.,** *J. Chromatogr.,* 122, 459, 1976.
217. **Long, M., Raverdino, V., et al.,** *J. Chromatogr.,* 117, 305, 1976.
218. **Allulli, S., Tomassini, N., Bertoni, G., and Bruner, F.,** *Anal. Chem.,* 48, 1259, 1976.
219. **Bruening, W., Bruening, I. M. R. de A., and Scofield, A. L.,** *Anal. Chem.,* 46, 1908, 1974.
220. **Datar, A. G. and Ramanathan, P. S.,** *J. Chromatogr.,* 114, 29, 1975.
221. **Sarkar, M. K. and Haselden, G. G.,** *J. Chromatogr.,* 104, 425, 1975.
222. **Kumar, A., Dua, R. D., and Sarkar, M. K.,** *J. Chromatogr.,* 107, 190, 1975.
223. **Noles, G. T. and Lieberman, M. L.,** *J. Chromatogr.,* 114, 211, 1975; **Nand, S. and Sarkar, M. K.,** *J. Chromatogr.,* 89, 73, 1974.
224. **Westberg, H. H., Rasmussen, R. A., and Holdres, M.,** *Anal. Chem.,* 46, 1852, 1974.
225. **Kuchhal, R. K. and Mallik, K. L.,** *Chromatographia,* 8, 640, 1975.
226. **Baiulescu, G. E. and Llie, V. A.,** *Anal. Chem.,* 46, 1847, 1974.
227. **Fuller, E. N.,** *Anal. Chem.,* 44, 1747, 1972.
228. **Belyakova, L. D., Kiselev, A. V., and Soloyan, G. A.,** *Zh. Anal. Khim.,* 27, 1182, 1972.
229. **Mathus, D. S. and Saha, N. C.,** *Technology,* 9, 23, 1972.
230. **Bruening, W. and Bruening, I. M. R. de A.,** *Anal. Chem.,* 45, 1169, 1973.
231. **Ottenstein, D. M. and Supina, W. R.,** *Am. Chem. Soc. Div. Fuel Chem. Prepr.,* 16, 29, 1972.
232. **Raulin, F. and Toupance, G.,** *J. Chromatogr.,* 90, 218, 1974.
233. **Garilli, F., Fabiani, L., Filia, U., and Cusi, V.,** *J. Chromatogr.,* 77, 3, 1973.
234. **Hirsch, R. F., Stober, H. C., et al.,** *Anal. Chem.,* 45, 2100, 1973.
235. **Mathur, D. S. and Saha, N. C.,** *Technology,* 9, 23, 1972.
236. **Little, J. N. and Dark, W. A.,** *J. Chromatogr. Sci.,* 8, 647, 1970.
237. **Kirkland, J. J. and DeStefano, J. J.,** *J. Chromatogr. Sci.,* 8, 309, 1970.
238. **Pflaum, R. T. and Cook, L. E.,** *J. Chromatogr.,* 50, 120, 1970.
239. **Neumann, M. G. and Hertl, W.,** *J. Chromatogr.,* 65, 467, 1972.

240. Seide, H., Serfas, O., and Assmann, K., *Plaste Kaut.*, 16, 440, 1969.
241. Duerbeck, H. W., *Z. Anal. Chem.*, 250, 377, 1970.
242. Sokolowski, S., Leboda, R., Slonka, T., and Swewryniak, M., *J. Chromatogr.*, 139, 237, 1977.
243. Anon., *Res. Ind.*, 19 (3), 111, 1974.
244. Berezkin, V. G. and Gavrichev, V. S., *J. Chromatogr.*, 116, 9, 1976.
245. Moriguchi, S., Naito, K., and Takei, S., *J. Chromatogr.*, 131, 19, 1977.
246. DiCorcia, A., Samperi, R., and Capponi, G., *J. Chromatogr.*, 121, 370, 1976.
247. Carson, J. W., Lege, G., and Irizarry, J., *J. Chromatogr. Sci.*, 13, 168, 1975.
248. DiCorcia, A. and Samperi, R., *J. Chromatogr.*, 107, 99, 1975.
249. Tsitsishvili, G. V., Andronikashvili, T. G., and Banakh, O. S., *Soobshch. Akad. Nauk Gruz. S.S.R.*, 74, 345, 1974.
250. Nabiev, N. I. and Baghirov, R. A., *Tr. Azerb. Fil. Vses. Nauchono-Issled. Inst. Prirod. Gazov*, 2, 139, 1973.
251. Kuman, A., Dua, R. E., and Sarkar, M. K., *J. Chromatogr.*, 107, 190, 1975.
252. Grober, A., Kada, T., and Sarosi, G., *Proc. 3rd Anal. Chem. Conf.*, (Budapest), Vol. 1, 1970, 223.
253. Nonaka, A., *Bunseki Kagaku*, 20, 416, 1971.
254. Han, R. U. and Li, U. M., *Punsok Hwahak*, 1980 (4), 12.
255. Castello, G. and D'Amato, G., *J. Chromatogr.*, 196, 245, 1980.
256. Castello, G. and D'Amato, G., *Ann. Chim. (Rome)*, 69 (9—10), 541, 1979.
257. Castello, G., D'Amato, G., and Canciani, G., *Ann. Chim. (Rome)*, 68 (3—4), 255, 1978.
258. Cicioli, P., Hayes, J. M., Rinaldi, G., Denson, K. B., and Meinschein, W. G., *Anal. Chem.*, 51, 400, 1979.
259. Long, M., Raverdino, V., Di Tullio, G., and Tomarchio, L., *J. Chromatogr.*, 117, 305, 1976.
260. Berezkin, V. G. and Gavrichev, V. S., *J. Chromatogr.*, 116, 9, 1976.
261. Soulages, N. L. and Brieva, A. M., *J. Chromatogr.*, 101, 365, 1974.
262. Gavrichev, V. S., Berezkin, V. G., and Shkolina, L. A., *Neftepererab. Neftekhim. (Moscow)*, 6, 29, 1974.
263. Garilli, F., Fabiani, L., Filia, U., and Cusi, V., *J. Chromatogr.*, 77, 3, 1973.
264. Okamura, J. P. and Sawyer, D. T., *Anal. Chem.*, 45, 80, 1973.
265. Kratzsch, T. and Fischer, W., *Z. Chem.*, 12 (5), 182, 1972.
266. Waksmundzki, A., Suprynowicz, Z., Leboda, R., Rayss, J., and Kusak, R., *Chem. Anal. (Warsaw)*, 16 (1), 43, 1971.
267. Rang, S., Pilt, A., and Eisen, O., *Eesti, NSV Tead. Akad. Tiom. Keem. Geol.*, 19 (3), 211, 1970.
268. Koci, K., *Bul. Shkencave Nat.*, 32, 59, 1978; *Chem. Abstr.*, 90, 171106b, 1979.
269. Lukac, S., *J. Chromatogr.*, 166, 287, 1978.
270. Snowdon, L. R. and Peake, E., *Anal. Chem.*, 50, 379, 1978.
271. Aue, W. A., Kapila, S., and Gerhardt, K. O., *J. Chromatogr.*, 78, 228, 1973.
272. Jamieson, I. L., *J. Appl. Chem. Biotechnol.*, 22, 1157, 1972.
273. Eisen, O., Kiselev, A. V., Pilt, A., Rang, S., and Shcherbakova, K. D., *Chromatographia*, 4, 448, 1971.
274. Manara, G. and Taramasso, M., *J. Chromatogr.*, 65, 349, 1972.
275. Aue, W. A., Hastings, C. R., and Kapila, S., *J. Chromatogr.*, 77, 299, 1973.
276. Ross, W. D. and Jefferson, R. T., *J. Chromatogr. Sci.*, 8, 386, 1970.
277. Peterson, R. M. and Rodgers, J., *Chromatographia*, 5, 13, 1972.
278. Kuliev, A. M., Nabiev, N. I., Baghirov, R. A., and Farkhadov, T. S., *Neftepererab. Neftekhim. (Kiev)*, 6, 27, 1974.
279. Wasik, S. P. and Brown, R. L., *Anal. Chem.*, 48, 2218, 1976.
280. Magidman, P., Barford, R. A., Saunders, D. H., and Rothbar, H. L., *Anal. Chem.*, 48, 44, 1976.
281. Betti, A., Lodi, G., Bighi, C., and Dondi, F., *J. Chromatogr.*, 106, 291, 1975.
282. Carson, J. W., Lege, G., and Irizarry, J., *J. Chromatogr. Sci.*, 13, 168, 1975.
283. Kraitr, M. and Komers, R., *Anal. Chem.*, 46, 974, 1974.
284. Sawatzky, H., George, A. E., and Smiley, G. T., *Am. Chem. Soc. Div. Petrol. Chem. Prepr.*, 18, 99, 1973.
285. Schurig, V., Chang, R. C., Zlatkis, A., Gil-Av, E., and Mikes, F., *Chromatographia*, 6, 223, 1973; Gil-Av, E. and Schurig, V., *Anal. Chem.*, 43, 2030, 1971.
286. Pilt, A., Rang, S., and Eisen, O., *Izv. Akad. Nauk S.S.S.R. Ser. Khim.*, 21, 108, 1972.
287. Grob, R. L. and McGonigle, E. J., *J. Chromatogr.*, 59, 13, 1971.
288. Gil-Av, E. and Schurig, V., *Anal. Chem.*, 43, 2030, 1971.
289. Kugucheva, E. E. and Alekseeva, A. V., *Khim. Tekhnol. Topl. Masel*, 15 (8), 48, 1970.
290. Carson, J. W., Lege, G., and Irizarry, J., *J. Chromatogr. Sci.*, 13, 168, 1975.
291. Rang, S. A., Eisen, O. G., Kiselev, A. V., Meister, A. E., and Shcherbakova, K. D., *Chromatographia*, 8, 327, 1975.

292. **Zershchikova, V. P., Dorogochinskii, A. Z., Pecheritsa, S. A., and Balashova, V. V.,** *Tr. Groznensk. Neft. Inst.,* 33, 121, 1971.
293. **Andronikashvili, T. G., Tsitsishvili, G. V., and Sabelashvili, Sh. D.,** *J. Chromatogr.,* 58, 47, 1971.
294. **Kalashnikova, E. V. and Scherbakova, K. D.,** *J. Chromatogr.,* 91, 695, 1974.
295. **Engewald, W., Epsch, K., Graefe, J., and Weisch, T.,** *J. Chromatogr.,* 91, 623, 1974.
296. **Garilli, A., Fabiana, L., Filia, U., and Cusi, V.,** *J. Chromatogr.,* 77, 3, 1973.
297. **Peterson, R. M. and Rodgers, J.,** *Chromatographia,* 5, 13, 1972.
298. **Bertoni, G., Ciccioli, P., Severini, C., and Bruner, F.,** *Chromatographia,* 11, 55, 1978.
299. **Engewald, W., Wennrich, L., and Poerschmann, J.,** *Chromatographia,* 11, 434, 1978.
300. **Ohzeki, K. and Kambara, T.,** *J. Chromatogr.,* 174, 204, 1979.
301. **Gonnord, M. F., Vidal-Madjar, C., and Guiochon, G.,** *J. Chromatogr. Sci.,* 12, 839, 1974.
302. **Vernon, F.,** *J. Chromatogr.,* 60, 406, 1971.
303. **Engewald, W., Wennrich, L., and Porschmann, J.,** *Chromatographia,* 11, 434, 1978.
304. **Matsumoto, H., Futami, H., Morita, F., and Morita, Y.,** *Bull. Chem. Soc. Jpn.,* 44, 3170, 1971.
305. **Narmetova, G. and Ryabova, N. D.,** *Uzbek. Khim. Zh.,* 1971, 40.
306. **Vernon, F.,** *J. Chromatogr.,* 60, 406, 1971.
307. **Matsumoto, H., Futami, H., Kato, F., and Morita, Y.,** *Bull. Chem. Soc. Jpn.,* 44 (11), 3170, 1971.
308. **Frycka, J.,** *Chromatographia,* 8, 413, 1975.
309. **Sawatzky, H., George, A. E., and Smiley, G. T.,** *Am. Chem. Soc. Div. Petrol. Chem. Prepr.,* 18, 99, 1973.
310. **Vernon, F.,** *J. Chromatogr.,* 60, 406, 1971.
311. **Lane, D. A., Moe, H. K., and Katz, M.,** *Anal. Chem.,* 45, 1776, 1973; **Frycka, J.,** *J. Chromatogr.,* 65, 432, 1972.
312. **Nonaka, A.,** *Jpn. Anal.,* 20, 416, 1971; 20, 422, 1971.
313. **Popl, M., Dolansky, V., and Mostecky, J.,** *J. Chromatogr.,* 117, 117, 1976.
314. **Frycka, J.,** *Chromatogrphia,* 8, 413, 1975.
315. **Zoccolillo, L. and Salomoni, F.,** *J. Chromatogr.,* 106, 103, 1975.
316. **Frycka, J.,** *J. Chromatogr.,* 65, 341, 1972.
317. **Asshauer, J. and Halasz, I.,** *Anal. Chem.,* 45, 1142, 1973.
318. **Halasz, I. and Sebastian, I.,** *Agnew. Chem. (Int. Ed. Engl.),* 8, 453, 1969.

Chapter 8

RETENTION DATA

Selected studies reporting GC retention data on hydrocarbons are presented in Section I of this chapter. In addition, 45 tables of GC retention data on hydrocarbons are presented in Section II arranged according to the major hydrocarbon classes.

I. STUDIES REPORTING GC RETENTION DATA ON HYDROCARBONS

Studies reporting GC retention data on hydrocarbons are cited in Table 1. The literature citations for these studies are categorized under: (1) General Hydrocarbon Analyses; (2) Studies on Alkanes; (3) Studies on Alkenes and Alkynes; (4) Studies on Cyclic Hydrocarbons; (5) Studies on Aromatic Hydrocarbons; and (6) Studies on Polycyclic Aromatic Hydrocarbons.

II. HYDROCARBON RETENTION DATA TABLES

Selected GC retention data for the major hydrocarbon classes are presented in Tables 2 through 46. A special effort was made to standardize these data as much as possible by the use of retention indices. Several data sets were converted to retention indices from other retention units (e.g., relative retention, log relative retention, retention in minutes). Hence, these retention data tables should represent a consistent data set for the GC analysis of hydrocarbons, based principally on retention index. These data cover numerous hydrocarbons, representing several thousand data entries using both packed and capillary columns.

Specifically, the tables provide retention data for various GC columns and conditions for: C4 to C13 alkanes (Tables 2 to 7); C-2 to C-16 alkenes (Tables 8 to 12); C-6 to C-14 alkynes (Tables 13 to 17); C5 to C16 cyclic hydrocarbons (Tables 18 to 27); C-6 to C-16 alkyl benzenes (Tables 28 to 33, and Tables 35 and 36); C-8 to C-11 aryl alkenes (Table 34); biphenyls and diphenylalkanes (Table 37); 2 to 7 ring polycyclic aromatic hydrocarbons (Tables 38 to 44); alkylated naphthalenes (Table 45); and terpene hydrocarbons (Table 46).

Table 1
SELECTED GC RETENTION DATA ON HYDROCARBONS

Comments	Ref.
General Hydrocarbon Analyses	
26 hydrocarbons	1
Various hydrocarbons	2—11
Factor analysis	12—13
C-11—C-36 hydrocarbons	14
C-6—C-8 hydrocarbons	15,16
70 hydrocarbons	17
Retention as a function of stationary phase	18
Retention index standards	19
21 hydrocarbons in alkane stationary phases	20
63 hydrocarbons: squalane capillary column	21
25 hydrocarbons	22
Retention index reproducibility	23
Computer identification from retention data	24
Hydrocarbon isomers	25
Retention data from solute structures	26
Calculated vs. experimental retention data	27
11 C-4 hydrocarbons	28
Retention data vs. boiling point data	29
Prediction of retention data: 39 hydrocarbons	30
Isomeric hydrocarbons	31
Saturated hydrocarbons	32
900 retention volumes	33
Retention volume treatment for hydrocarbons	34
Stationary phase classification by retention data	35
Retention vs. solute structure	36,37
Determination of precise retention data	38
Carbon number vs. retention of hydrocarbon homologs: 38 alkanes, 18 cycloalkanes, 58 alkenes, 5 cycloalkenes, 8 alkadienes	39
Carbon number vs. retention indices of hydrocarbons: alkanes, alkenes, alkadienes (gasoline), cyclics	40
Retention data of 20 C-7—C-8 alkanes and 9 C8 olefins	41
Retention data of alkenes and C-6—C-11 alkyl benzenes	42
Retention data of alkanes, olefins, cyclics, and aromatics on 37 stationary phases	43
Effective use of retention indices	44—46
Development of highly precise retention data	47,48
Identification of solute structure from retention data	49
GC stability in the identification of solutes from retention data	50
Solute identification from solute retention	51
Retention of 195 nonolefinic hydrocarbons on Lukopren G	52

Table 1 (continued)
SELECTED GC RETENTION DATA ON HYDROCARBONS

Comments	Ref.
Literature compilation of hydrocarbon retention data	53
Retention data errors in GC temperature programing	54
GC retention control	55
Errors in retention index measurements	56—58
Reproducibility of retention indices	59
Retention vs. solute structure	60,61
Hydrocarbon retention in various stationary phases	62
Retention calculation for saturated hydrocarbons	63
Hydrocarbon retention vs. stationary phase structure	64,65
Identification of hydrocarbons from retention indices	66
Review of identification of hydrocarbons from retention index data	67
C-5—C-8 saturated hydrocarbons on mixed stationary phases	68
Identification of isomeric hydrocarbons from retention index data	69
Temperature dependence of hydrocarbon retention indices on squalane and SE-30	70
Retention index tables for the GC analysis of hydrocarbons	71
Prediction of hydrocarbon retention data	72
Selection of optimum column temperature for separation of hydrocarbon mixtures	73
Calculation of retention volumes for unsaturated and aromatic hydrocarbons by a molecular-statistical method	74

Studies on Alkanes

Retention data on alkanes	75—87
Iso-alkanes	88,89
65 alkanes and alkenes	90
C-6—C-9 alkanes	91
Nonanes	92
n-Hexane	93
Retention data on 300 C-7—C-24 alkanes	94
C-10 alkanes	95
C-23 and C-24 *n*-alkanes	96
C-5—C-18 alkanes	97
C-9—C-11 alkanes	98
Normal and iso-alkanes	99
Homologous alkane series	100
16 C-8—C-10 isomeric alkanes	101
Trimethyl and tetramethyl C-13, C-14, and C-16 alkanes: retention prediction	102
Alkanes: retention anomolies	103
Identification of alkane isomers by retention indices	104
GC retention vs. alkane structure	105

Table 1 (continued)
SELECTED GC RETENTION DATA ON HYDROCARBONS

Comments	Ref.
GC retention vs. iso-alkane structure	106,107
Iso-decanes	108
C-6—C-8 branched alkanes on SF-96	109
Retention volumes of iso-octane at pressures up to 100 atmospheres	110
Prediction of specific retention volumes for *n*-octane on 65 stationary phases	111
Partition coefficients of octane on 81 stationary phases	112
Retention vs. structure of branched paraffins	113

Studies on Alkenes and Alkynes

65 alkenes and alkanes	90
Retention data on alkenes	114—126
C-10-C-12 cis-trans alkenes	127
Retention of all 85 *n*-alkenes up to C-14	128
Unsaturated octanes	129
Retention data on alkadienes on two columns	130
Retention vs. alkene structure	131—134
n-Alkenes on 3 stationary phases	135
Retention indices vs. structure for 9 *n*-undecenes	136
Separation of 2,4-*cis/trans*-hexadienes	137
Retention data of alkynes	138,139
Retention vs. structure of *n*-alkynes	140
Temperature dependence of retention indices for *n*-alkynes	141

Studies on Cyclic Hydrocarbons

Retention data on cycloalkanes	142—146
Saturated and unsaturated alkylcyclopentanes	147
Alkyl-cyclohexane isomers	148
Cyclic alkanes and alkenes	149
Cyclic *cis/trans*-alkanes and -alkenes	150
C-9—C-10 alkylcyclohexenes	151
Tricyclic saturated hydrocarbons	152
31 adamantanoid hydrocarbons	153
Dicyclopentadienes	154
30 C-5—C-9 cyclakenes	155
Retention vs. structure of monosubstituted cyclopentenes and cyclohexenes	156
Retention vs. structure of cycloalkanes	157,158
Retention vs. structure of bicycloalkanes	159
Retention of 156 saturated and unsaturated cyclopropanes on squalane	160
C-6—C-13 cycloalkanes, cycloalkenes, and cycloalkadienes on graphatized carbon black and squalane	161
Retention vs. structure for saturated and unsaturated alkyl-cyclopentanes and methylcyclopentanes	162

Table 1 (continued)
SELECTED GC RETENTION DATA ON HYDROCARBONS

Comments	Ref.
Retention of cyclopentenes and cyclohexenes on 3 stationary phases	163
Identification of naphthene hydrocarbons from retention index data	164
Retention vs. structure for tricyclanes	165
Retention vs. structure for mono- and sequiterpenic hydrocarbons	166

Studies on Aromatic Hydrocarbons

Comments	Ref.
Retention data on miscellaneous aromatic hydrocarbons	167— 180
C-9—C-14 alkylbenzenes	181
meta/para-Xylenes	182,183
Alkylaromatics	184
Alkylbenzenes	185,186
Biphenyls and diphenylalkanes	187
Methylbenzenes	188
Benzene	189,190
Benzene, tolune, xylenes	191
C-6—C-9 aromatic hydrocarbons and styrenes	192
Alkyl benzenes and alkenyl benzenes	193
Methyl biphenyls	194
Retention data on 23 isomeric C-6—C-10 aromatics	195
C-6—C-10 and C-11 alkylbenzenes	196
Di- and trisubstituted benzenes	197
Up to C-8 alkylbenzenes	198
Retention vs. structure for alkylbenzenes	199
Retention prediction on any stationary phase for aromatic hydrocarbons	200
Retention vs. structure for alkylbenzenes	201— 203
Retention vs. structure for benzene homologs	204
GC temperature and pressure dependence of retention index increments for alkyl benzenes	205
Quasi-retention indices for aromatic hydrocarbons	206
Biphenyls and diphenylalkanes on 3 stationary phases	207
Retention data of alkylbenzenes on 4 stationary phases	208
Predicted retention index values for benzene derivatives	209
Retention volumes of toluene at pressures up to 100 atmospheres	210
Partition coefficients of toluene on 81 stationary phases	211
Retention volume changes of hydrocarbons with changes in sodium content in zeolites	212
Retention vs. structure for benzene and benzene derivatives	213
Aromatic hydrocarbons as quantitative analytical standards on 5 columns at column temperatures between 125—175°C	214

Table 1 (continued)
SELECTED GC RETENTION DATA ON
HYDROCARBONS

Comments	Ref.
Studies on Polycyclic Aromatic Hydrocarbons (PAH)	
Retention data of miscellaneous PAH	215—218
Retention data of 209 PAH	219
Naphthalenes	220
Eighteen C-12 alkyl naphthalenes	221
Retention data of 51 PAH	222
Retention data of 10 PAH	223
Retention data of PAH on a liquid crystal stationary phase	224
Analysis of PAH in air	225
New retention index scale for 1—4 ring PAH	226
Factors affecting GC retention of PAH	227
Retention indices of PAH: capillary GC	228

Table 2
ALKANES (C-4—C-9)

Column packing	P1	P1	P2	P2	P3	P3
Temperature (°C)	50	70	70	80	70	80
Reference	1	1	2	2	2	2

Compound	I					
n-Butane	400	400				
2,2-Dimethylpropane	412	413				
2-Methylbutane	475	475				
n-Pentane	500	500				
2,2-Dimethylbutane	537	538				
2-Methylpentane	570	570	570	571	569	570
3-Methylpentane	584	585	585	586	585	585
2,2-Dimethylbutane			538	540	537	538
2,3-Dimethylbutane			569	570	568	568
n-Hexane	600	600				
2,2-Dimethylpentane	626	627	627	628	625	626
2,3-Dimethylpentane			674	674	673	673
2,4-Dimethylpentane			632	631	629	629
3,3-Dimethylpentane	659	662	662	663	661	662
2,2,3-Trimethylbutane	640	643	643	645	641	642
2-Methylhexane	667	667				
2,2,4-Trimethylpentane				694		691
n-Heptane	700	700				
2,2-Dimethylhexane	719	720		721		718
2,5-Dimethylhexane				730		728
2,4-Dimethylhexane				733		732
2,2,3-Trimethylpentane	737	740		742		741
3,3-Dimethylhexane				746		746
2,3,4-Trimethylpentane				756		755
2,3-Dimethylhexane				762		761
2-Methyl-3-ethylpentane				764		763
2,2,3-Trimethylpentane				765		765
2-Methylheptane	765	765		766		764
4-Methylheptane				767		768
3-Methylheptane	772	773		773		772
2,2,3,3-Tetramethylbutane				773		771
3-Ethylhexane				774		773
3,4-Dimethylhexane				774		773
3-Methyl-3-ethylpentane				780		780
n-Octane	800	800				
2,2,3-Trimethylhexane	822	824				
2,2,3,3-Tetramethylpentane	852	858				
n-Nonane	900	900				

Column Packing: P1 = 100-m by 0.25-mm I.D. stainless steel capillary column coated with squalane. P2 = 3-m by 2.2-mm I.D. column packed with 20% loading of squalane. P3 = 3-m by 2.2-mm I.D. column packed with 20% loading of *n*-triacontane (*n*-C$_{30}$).

REFERENCES

1. **Cramers, C. A., Rijks, J. A., Pacakova, V., and Ribeiro de Andrade, I.,** *J. Chromatogr.*, 51, 13, 1970.
2. **Castello, G. and D'Amato, G.,** *J. Chromatogr.*, 175, 27, 1979.

Table 3
ALKANES (C-5—C-9)

Column packing	P1	P1	P2	P3	P4
Temperature (°C)	50	100	50	25	27
Reference	1	1	2	3	3

Compound	I				
2,2-Dimethylpropane			412	397	411
2-Methylbutane	475	476	475	472	474
n-Pentane				500	500
2,2-Dimethylbutane	540	544	537	530	535
2,3-Dimethylbutane			567	566	566
2-Methylpentane	570	571		566	569
3-Methylpentane	589	591	584	585	583
n-Hexane				600	600
2,2-Dimethylpentane	625	629	626	619	624
2,4-Dimethylpentane	629	632	630	622	629
2,2,3-Trimethylbutane			640	639	637
3,3-Dimethylpentane			659	659	656
2-Methylhexane	666	667	667	663	666
2,3-Dimethylpentane			672	675	670
3-Methylhexane	679	681	676	676	676
3-Ethylpentane	692	695	686	691	685
n-Heptane				700	700
2,2,4-Trimethylpentane	693	700	689	678	688
2,2-Dimethylhexane	716	720	719	711	718
2,5-Dimethylhexane	727	729	728	720	727
2,4-Dimethylhexane	733	736	732	726	731
2,2,3-Trimethylpentane			737	738	733
3,3-Dimethylhexane				741	740
2,3,4-Trimethylpentane			752	755	749
2,3,3-Trimethylpentane				765	754
2,3-Dimethylhexane			760	761	759
2-Methyl-3-ethylpentane			761	764	758
2-Methylheptane	764	757	765	759	764
4-Methylheptane			767	765	766
3,4-Dimethylhexane			771	775	768
3-Methylheptane	774	776	772	770	772
2,2,4,4-Tetramethylpentane			773		
3-Methyl-3-ethylpentane			774	779	769
3-Ethylhexane				774	770
n-Octane				800	800
2,2,4-Trimethylhexane			789		
2,4,4-Trimethylhexane			808		
2,3,5-Trimethylhexane	814	820	812		
2,2-Dimethylheptane			815		
2,4-Dimethylheptane	819	821			
2,2,3,4-Tetramethylpentane			820		
2,2,3-Trimethylhexane			813		
2,2-Dimethyl-3-ethylpentane			822		
3,3-Dimethylheptane			836		
2,4-Dimethyl-3-ethylpentane			837		
2,3,4-Trimethylhexane			849		
2,2,3,3-Tetramethylpentane			852		
2,3,3,4-Tetramethylpentane			858		
4-Ethylheptane	859	862			
2-Methyloctane	864	865			
3-Methyloctane	872	873	870		
3,3-Diethylpentane			877		

Table 3 (continued)
ALKANES (C-5—C-9)

Column Packing: P1 = 50-m by 0.25-mm I.D. stainless steel capillary column coated with di-*n*-butyl-tetrachlorophthalate. P2 = Capillary column coated with squalane. P3 = Dimethylsulfolane. P4 = Squalane.

REFERENCES

1. **Ryba, M.,** *J. Chromatogr.,* 123, 327, 1976.
2. **Vanheertum, R.,** *J. Chromatogr. Sci.,* 13, 150, 1975.
3. **Ladon, A. W.,** *J. Chromatogr.,* 99, 203, 1974.

Table 4
ALKANES (C-6—C-8)

Column packing	P1	P1	P2	P2	P2
Temperature (°C)	70	100	80	100	120
Reference	1	1	2	2	2
Compound	I[a]				
n-Hexane			600	600	600
2-Methylpentane	570	571	569	569	568
3-Methylpentane	585	586	585	586	587
2,2-Dimethylbutane	538	541	540	542	543
2,3-Dimethylbutane	569	571	568	570	571
n-Heptane			700	700	700
2-Methylhexane	667	668	668	669	670
3-Methylhexane	677	678	677	678	680
3-Ethylpentane	687	689	685	686	687
2,2-Dimethylpentane	627	629	632	634	635
2,3-Dimethylpentane	674	676	672	675	677
2,4-Dimethylpentane	630	632	632	633	633
3,3-Dimethylpentane	662	666	661	664	668
2,2,3-Trimethylbutane	643	647	642	646	651
n-Octane			800	800	800
2-Methylheptane	765	766	766	767	767
3-Methylheptane	773	774	774	775	776
4-Methylheptane	768	768	768	769	770
3-Ethylhexane	773	775	775	776	777
2,2-Dimethylhexane	720	722	724	726	728
2,3-Dimethylhexane	762	762	761	764	767
2,4-Dimethylhexane	733	734	736	737	739
2,5-Dimethylhexane	729	730	733	735	736
3,3-Dimethylhexane	746	750	749	753	755
3,4-Dimethylhexane	773	776	775	777	782
2-Methyl-3-ethylpentane	764	767	773	775	778
3-Methyl-3-ethylpentane	778	783	775	779	782
2,2,3-Trimethylpentane	740	745	743	747	754
2,2,4-Trimethylpentane	692	695	696	699	702
2,3,3-Trimethylpentane	763	770	760	765	770
2,3,4-Trimethylpentane	755	759	745	749	753
2,2,3,3-Tetramethylbutane			727	733	739

Column Packing: P1 = Capillary column coated with squalane. P2 = 4-m by 1/4-in. O.D. column packed with 20% SF-96 methylsilicone on 60—80 mesh DMCS Chromosorb P.

[a]   Retention indices for all temperatures for column packing P2 were calculated from specific retention volume data in Reference 2.

REFERENCES

1. **Dimov, N.,** *J. Chromatogr.,* 119, 109, 1976.
2. **Castello, G. and D'Amato, G.,** *J. Chromatogr.,* 107, 1, 1975.

Table 5
ALKANES (C-6—C-8)

Column packing	P1	P1	P1	P1	P1	P2	P3
Temperature (°C)	80	90	100	110	120	80	80
Reference	1	1	1	1	1	2	2
Compound				I			
2-Methylpentane	570	570	571	571	571	569	571
3-Methylpentane	586	586	586	587	588	585	585
2,2-Dimethylbutane	542	543	545	546	547	539	539
2,3-Dimethylbutane	569	570	572	573	574	568	568
2-Methylhexane	668	669	669	670	670		
3-Methylhexane	677	678	679	679	680		
3-Ethylpentane							688
2,2-Dimethylpentane	628	629	630	630	632	626	627
2,3-Dimethylpentane	672	674	675	676	677		670
2,4-Dimethylpentane	632	633	633	634	634	632	632
3,3-Dimethylpentane	661	662	664	666	668		657
2,2,3-Trimethylbutane	642	644	646	649	651		639
2-Methylheptane							766
3-Methylheptane	774	775	775	776	776	775	774
2,2-Dimethylhexane	724	725	726	726	728		724
2,3-Dimethylhexane	761	762	764	765	767		763
2,4-Dimethylhexane	736	737	737	738	739	736	
3,3-Dimethylhexane	744	746	748	750	751		
2,2,3-Trimethylpentane						738	
2,2,4-Trimethylpentane						693	
2,3,3-Trimethylpentane	760	762	765	767	770		
2,3,4-Trimethylpentane						754	
2,2,3,3-Tetramethylbutane	727	730	733	736	739		

Column Packing: P1 = 4-m by 1/4-in. column packed with 20% SF-96 methyl-silicone on 60—80 mesh DMCS Chromosorb P. P2 = 4-m by 4-mm I.D. stainless steel column packed with 15% SE-30 on 80—100 mesh AW-DMCS Chromosorb W. P3 = 2-m by 4-mm I.D. stainless steel column packed with 15% OV-1 on 80—100 mesh AW-DMCS Chromosorb W.

REFERENCES

1. **Castello, G., Berg, M., and Lunardelli, M.,** *J. Chromatogr.,* 79, 23, 1973.
2. **Dimov, N. and Papazova, D.,** *Chromatographia,* 12, 443, 1979.

Table 6
ALKANES (C-7—C-10)

Column packing	P1	P1[a]	P2	P3
Temperature (°C)	80	80	50	95
Reference	1	2	3	4
Compound		I		
2,2-Dimethylpentane	630			
2,4-Dimethylpentane	633			
2,2,3-Trimethylbutane	643			
3,3-Dimethylpentane	662			
2-Methylhexane	669			
2,3-Dimethylpentane	673			

Table 6 (continued)
ALKANES (C-7—C-10)

Compound	I			
3-Methylhexane	680			
3-Ethylpentane	685			
n-Heptane	700			
2,2,4-Trimethylpentane	696			
2,2-Dimethylhexane	725			
2,2,3,3-Tetramethylbutane	731			
2,5-Dimethylhexane	733			
2,4-Dimethylhexane	737			
2,2,3-Trimethylpentane	743			
2,3,4-Trimethylpentane	745			
3,3-Dimethylhexane	747			
2,3,3-Trimethylpentane	760			
2,3-Dimethylhexane	764	760		
2-Methylheptane	766	765		
4-Methylheptane	768	768		
3,4-Dimethylhexane	775			
3-Methylheptane	774	773		
3-Ethylhexane	775			
3-Methyl-3-ethylpentane	775			
n-Octane	800			
2,2,4,4-Trimethylpentane	785	785		
2,2,5-Trimethylhexane	786	786		797
2,2,4-Trimethylhexane	790	790		803
2,4,4-Trimethylhexane	814	814		
2,3,5-Trimethylhexane	816	816		821
2,2,3,4-Trimethylpentane	816	816		
2,2-Dimethylheptane	822	822		826
2,4-Dimethylheptane	825	825	822	827
2,2,3-Trimethylhexane	826	826		831
2-Methyl-4-ethylhexane	830	830		
2,2-Dimethyl-3-ethylpentane		830		
2,6-Dimethylheptane	835	835	827	836
4,4-Dimethylheptane	835	835	828	
2,5-Dimethylheptane	841	841	833	840
3,5-Dimethylheptane	841	841	834	840
3,3-Dimethylheptane	845	845		844
2,4-Dimethyl-3-ethylpentane	845	845		847
2,3,3-Trimethylhexane	849	849		
2-Methyl-3-ethylhexane	854	854		
2,2,3,3-Tetramethylpentane	858	858		
2,3,4-Trimethylhexane	858	858		
3,3,4-Trimethylhexane	861	861		857
3-Methyl-4-ethylhexane	862	862		
4-Ethylheptane	862	862	858	861
2,3-Dimethylheptane		862		859
2,3,3,4-Tetramethylpentane	864	864		862
3-Methyl-3-ethylhexane				858
3,4-Dimethylheptane	865	865		858
4-Methyloctane	871	871		865
2-Methyloctane	872	872		870
3-Ethylheptane	875	875		868
3-Methyloctane	877	877		872
2,3-Dimethyl-3-ethylpentane	878	878		
3,3-Diethylpentane	885	885	878	880
2,2,4,5-Tetramethylpentane				887
n-Nonane	900	900		

Table 6 (continued)
ALKANES (C-7—C-10)

Compound	I	
2,2-Dimethyl-3-ethylhexane	830	
2,3-Dimethyloctane	862	958
2,4,6-Trimethylheptane		886
2,2,4-Trimethylheptane		890
2,2,5-Trimethylheptane		901
2,4,4-Trimethylheptane		902
2,2,4-Trimethyl-3-ethylpentane	901	
2,2,4,4-Tetramethylhexane		910
3,3,5-Trimethylheptane		918
2,2,3-Trimethylheptane	913	
2,4-Dimethyloctane		921
2,6-Dimethyloctane		930
2,2,3,4,4-Pentamethylpentane	918	
2-Methyl-4-ethylheptane		924
4-Propylheptane		929
2,2,3,4-Tetramethylhexane		932
4-Isopropylheptane		934
2,7-Dimethyloctane		936
2,3,4-Trimethylheptane		938
2,3,3-Trimethylheptane		939
2,3,5-Trimethylheptane		941
3,6-Dimethyloctane		944
2,3,4,4-Tetramethylhexane		946
3,3,4-Trimethylheptane		946
3,4-Dimethyl-3-ethylhexane		948
4,5-Dimethyloctane		949
3,4,5-Trimethylheptane		950
3,4-Dimethyloctane		951
4-Ethyloctane		954
3-Methyl-3-ethylheptane		954
2,3-Dimethyl-3-ethylhexane		956
5-Methylnonane		960
4-Methylnonane		964
2,2,3,3,4-Pentamethylpentane	949	964
3,3-Diethylhexane		964
3-Ethyloctane		968
2-Methylnonane	964	968
3-Methylnonane		972
2,2,3-Trimethyl-3-ethylpentane		978
n-Decane		1000

Column Packing: P1 = 4-m by 1/4-in. O.D. column packed with 20% SF-96 on 60—80 mesh Chromosorb P. P2 = 3.0 m by 2.0 mm I.D. aluminum column packed with 10.0 wt% squalane on 60—80 mesh Chromosorb W. P3 = 150 m by 0.25 mm I.D. copper capillary column coated with BM-4 vacuum oil.

[a] Retention indices in this data column were calculated from specific retention volume data in Reference 2.

REFERENCES

1. **Castello, G., Lunardelli, M., and Berg, M.,** *J. Chromatogr.,* 76, 31, 1973.
2. **Castello, G. and D'Amato, G.,** *J. Chromatogr.,* 116, 249, 1976.
3. **Takacs, J., Szita, C., and Tarjan, G.,** *J. Chromatogr.,* 56, 1, 1971.
4. **Sultanov, N. T. and Arustamova, L. G.,** *J. Chromatogr.,* 115, 553, 1975.

Table 7
ALKANES (C-12—C-13)

Column packing	P1
Temperature (°C)	100
Reference	1

Compound	I	Compound	I
2-Methylundecane	1164	3,4-Dimethylundecane	1247
3-Methylundecane	1170		1248
4-Methylundecane	1159	3,5-Dimethylundecane	1207
5-Methylundecane	1154		1211
6-Methylundecane	1152	4,5-Dimethylundecane	1230
2,3-Dimethylundecane	1251		1234
2,4-Dimethylundecane	1208	4,6-Dimethylundecane	1193
2,5-Dimethylundecane	1210		1196
2,6-Dimethylundecane	1210	4,7-Dimethylundecane	1207[a]
2,7-Dimethylundecane	1216	5,6-Dimethylundecane	1223
2,8-Dimethylundecane	1221		1227
2,9-Dimethylundecane	1233	5,7-Dimethylundecane	1190
2,10-Dimethylundecane	1227		1198
		6,6-Dimethylundecane	1200[a]

Column Packing: P1 = 100-m by 0.25-mm I.D. capillary column coated with squalane.

[a]   Taken from: Schomburg, G. and Henneberg, D., *Z. Anal. Chem.*, 236, 279, 1968 (referenced in Lombosi, T. S., *J. Chromatogr.*, 119, 307, 1976.

REFERENCE

1. **Schomburg, G. and Dielmann, G.**, *J. Chromatogr. Sci.*, 11, 151, 1973.

Table 8
ALKENES (C-2—C-6)

Column packing	P1	P2	P3	P4	P4	P4
Temperature (°C)	27	25	40	22.5	56	100
Reference	1	2	3	4	4	4

Compound	I					
Ethylene			200			
Propylene			333			
Isobutylene			430			
1-Butene			432	384		
Propadiene (allene)			443			
Acetylene			447			
trans-2-Butene			449	415		
cis-2-Butene			471	402		
1,3-Butadiene			507	410		
3-Methyl-1-butene	449	493	551			
1,4-Pentadiene	462		570		462	
1-Pentene	481	533			479	
1-Pentyne						463
2-Methyl-1-butene	488	546				
2-Methyl-1,3-butadiene	496	613				
trans-2-Pentene	500	555			493	
cis-2-Pentene	505	564			481	
2-Methyl-2-butene	514	574				

Table 8 (continued)
ALKENES (C-2—C-6)

Compound	I			
trans-1,3-Pentadiene	514	644	639	539
1,3-Cyclopentadiene	517		680	
cis-1,3-Pentadiene	523	659	651	526
Cyclopentene			629	
4-Methyl-1-pentene	548		597	
3-Methyl-1-pentene	549			
4-Methyl-*cis*-2-pentene	555			
2,3-Dimethyl-1-butene	557			
4-Methyl-*trans*-2-pentene	562			
2,3-Dimethylbutene	566			
2-Methyl-1-pentene	579			
Cyclohexene				535
Cyclohexadiene				539
1-Hexyne				560
1,5-Hexadiene				560
1-Hexene	581		630	577
1,4-*cis*-Hexadiene	586			
cis-3-Hexene	592			
trans-3-Hexene	592			
trans-2-Hexene	597			
2-Methyl-2-pentene	598			
3-Methylcyclopentene	600			
3-Methyl-*cis*-2-pentene	602			
cis-2-Hexene	603			

Column Packing: P1 = 100-m by 0.25-mm I.D. capillary column coated with squalane. P2 = 4.5-m by 3-mm I.D. glass column packed with 25% phenyl-glycid ether. P3 = 10-m by $^1/_8$-in. column (two 5-m columns) packed with 10% mixed bis-lactams on 80—100 mesh AW Chromosorb P, plus a 0.5-m by $^1/_8$-in. column packed with 3% OV-101 on 80—100 mesh AW-DMCS Chromosorb G (to improve the separation of butadiene from *n*-pentane). P4 = Column packed with graphitized thermal carbon black.

REFERENCES

1. **Schomburg, G. and Dielmann, G.,** *J. Chromatogr. Sci.,* 11, 151, 1973.
2. **Lukac, S. and Nguyen, T. Q.,** *Chromatographia,* 12, 231, 1979.
3. **Ravey, M.,** *J. Chromatogr. Sci.,* 16, 79, 1978.
4. **Kalashnikova, E. V., Kiselev, A. V., Poshkus, D. P., and Shcherbakova, K. D.,** *J. Chromatogr.,* 119, 233, 1976.

Table 9
ALKENES (C-3—C-8)

Column packing	P1[a]	P2[a]	P3	P4	P5	P6
Temperature (°C)	25	27	50	50	50	50
Reference	1	1	2	3	4	4
Compound			**I**			
Propene	354	287				
2-Methylpropene	457	383				
1-Butene	449	385				
trans-2-Butene	476	407				
cis-2-Butene	493	417				
3-Methyl-1-butene	506	449				
1,3-Butadiene	534	386				
1-Pentene	546	481				
2-Methyl-1-butene	563	488				
trans-2-Pentene	569	501				
cis-2-Pentene	578	505				
2-Methyl-2-butene	594	514		514		
3,3-Dimethyl-1-butene	555	505				
4-Methyl-1-pentene	606	549				
2,3-Dimethyl-1-butene	626	558				
3-Methyl-1-pentene				551		
4-Methyl-*cis*-2-pentene	615	556		556		
4-Methyl-*trans*-2-pentene	618	562				
2-Methyl-1,3-butadiene	647	496				
3-Methyl-1,2-butadiene	651	512				
2-Methyl-1-pentene	652	580				
1,5-Hexadiene	693	563		563		
1-Hexene	648	582		582		
Cyclopentene	652	547				
2-Ethyl-1-butene	666	592		592		
cis-3-Hexene	659	593				
trans-3-Hexene	654	593				
1,2-Pentadiene	672	526				
2-Methyl-2-pentene	672	598		598		
3-Methyl-*cis*-2-pentene				603		
2,3-Pentadiene	674	532				
trans-1,3-pentadiene	679	515				
cis-1,3-pentadiene	691	524				
3-Methyl-*trans*-2-pentene	691	613				
trans-2-Hexene	661	597	597	597		
cis-2-Hexene	675	604	604	604		
2-Methyl-*cis*-2-pentene	681	602				
4,4-Dimethyl-1-pentene	652	603		605		
2,3-Dimethyl-2-butene	712	625				
4,4-Dimethyl-*trans*-2-pentene	662	616		615		
3,3-Dimethyl-1-pentene	677	624		626		
2,3,3-Trimethyl-1-butene	691	626				
4,4-Dimethyl-*cis*-2-pentene	692	633		636		
3,4-Dimethyl-1-pentene	689	635				
2,4-Dimethyl-1-pentene	699	636				
2,4-Dimethyl-2-pentene	699	641				
3-Methyl-1-hexene	698	643		645		
3-Ethyl-1-pentene	696	645				
2,3-Dimethyl-1-pentene	712	649				
5-Methyl-1-hexene	713	649				
2-Methyl-*trans*-3-hexene	694	648		647		
3-Methyl-2-ethyl-1-butene	722	659				
4-Methyl-*cis*-2-hexene	709	654		655		

Table 9 (continued)
ALKENES (C-3—C-8)

Compound	I				
4-Methyl-1-hexene	719	657		658	
4-Methyl-*trans*-2-hexene	707	656		657	
5-Methyl-*trans*-2-hexene	713	660		660	
1-Methylcyclopentene	745	642			
3,4-Dimethyl-*trans*-2-pentene	744	678			
2-Methyl-1-hexene	750	678		678	
3-Methyl-*trans*-3-hexene	756	685		691	
1-Heptene	747	682			
3-Methyl-*cis*-2-hexene	770	701		693	
2,2-Dimethyl-*trans*-3-hexene					693
2,5-Dimethyl-*trans*-3-hexene					695
2-Ethyl-1-pentene	750	682			
3-Methyl-*trans*-2-hexene	766	693			
3-Methyl-*cis*-3-hexene	759	692			
2-Methyl-2-hexene	760	692		691	
cis-3-Heptene	754	690	690	690	
trans-3-Heptene	744	688	688	688	
trans-2-Heptene			698	698	
2,4,4-Trimethyl-1-pentene				704	
2,4,4-Trimethyl-2-pentene				715	
3-Ethyl-2-pentene	771	697			
Cyclohexene	779	667			
2,3-Dimethyl-2-pentene	783	702			
3-Ethylcyclopentene	798				
1-Methylcyclohexene	866				
2-Methyl-3-ethyl-1-pentene				735	
2,3-Dimethyl-1-hexene				739	
2-Methyl-*trans*-3-heptene			741	741	
2,5-Dimethyl-2-hexene				750	
2,3,4-Trimethyl-2-pentene				766	
2-Methyl-3-ethyl-2-pentene				778	
1,7-Octadiene				763	816
1-Octene			781	782	808
trans-4-Octene			784	784	
trans-3-Octene				789	814
trans-2-Octene			798	798	
1,3(*trans*)-Octadiene				809	868

Column Packing: P1 = Dimethylsulfolane. P2 = Squalane. P3 = Squalane. P4 = Capillary column coated with squalane. P5 = 80-m by 0.23-mm glass capillary column coated with squalane. P6 = 80-m by 0.23-mm glass capillary column coated with Ucon LB-550-X.

[a]   Retention indices in this data column were calculated from relative retention data in Reference 1.

REFERENCES

1. **Ladon, A. W.,** *J. Chromatogr.,* 99, 203, 1974.
2. **Tarjan, G.,** *J. Chromatogr. Sci.,* 14, 309, 1976.
3. **Vanheertum, R.,** *J. Chromatogr. Sci.,* 13, 150, 1975.
4. **Welsch, Th., Engewald, W., and Berger, P.,** *Chromatographia,* 11, 5, 1978.

Table 10
ALKENES (C-6—C-14)

Column packing	P1	P1	P1	P1	P1	P1	P1	P1
Temperature (°C)	20	30	40	60	80	100	120	140
Reference	1	1	1	1	1	1	1	1
Compound					I			
1-Hexene	619	620	621		624	626		
cis-3-Hexene	629	630	631		634	636		
trans-3-Hexene	632	632	632		630	629		
trans-2-Hexene	635	636	635		636	637		
cis-2-Hexene	643	644	645		648	649		
1-Heptene	718	718	719	719	720	720		
trans-3-Heptene	724	724	724	724	723	723		
cis-3-Heptene	727	726	728	729	730	731		
trans-2-Heptene	736	736	736	737	737	738		
cis-2-Heptene	742	743	744	744	745	746		
1-Octene			818	819	820	821		
trans-4-Octene			818	819	819	820		
cis-4-Octene			822	824	826	827		
cis-3-Octene			824	826	826	828		
trans-3-Octene			825	826	826	826		
trans-2-Octene			836	837	838	839		
cis-2-Octene			842	843	845	847		
1-Nonene			918	919	920	920		
trans-4-Nonene			918	919	920	920		
cis-4-Nonene			919	920	922	923		
cis-3-Nonene			923	924	925	926		
trans-3-Nonene			924	924	924	924		
trans-2-Nonene			935	935	936	936		
cis-2-Nonene			941	942	944	944		
cis-5-Decene				1015	1017	1018		
trans-4-Decene				1017	1018	1018		
cis-4-Decene				1017	1019	1020		
1-Decene				1018	1019	1020		
trans-5-Decene				1018	1019	1020		
cis-3-Decene				1021	1023	1024		
trans-3-Decene				1023	1023	1023		
trans-2-Decene				1036	1036	1036		
cis-2-Decene				1041	1043	1044		
cis-5-Undecene				1112	1113	1115	1116	
cis-4-Undecene				1114	1116	1117	1119	
trans-4-Undecene				1115	1116	1117	1118	
trans-5-Undecene				1116	1117	1118	1118	
1-Undecene				1118	1119	1120	1121	
cis-3-Undecene				1120	1121	1123	1124	
trans-3-Undecene				1123	1123	1123	1123	
trans-2-Undecene				1136	1135	1136	1137	
cis-2-Undecene				1141	1142	1144	1145	
cis-6-Dodecene					1210	1212	1214	
cis-5-Dodecene					1210	1212	1214	
trans-6-Dodecene					1215	1216	1218	
cis-4-Dodecene					1215	1216	1219	
trans-5-Dodecene					1215	1217	1218	
trans-4-Dodecene					1216	1216	1217	
1-Dodecene					1219	1220	1221	
cis-3-Dodecene					1221	1222	1224	
trans-3-Dodecene					1222	1223	1223	
trans-2-Dodecene					1236	1236	1237	

Table 10 (continued)
ALKENES (C-6—C-14)

Compound	I		
cis-2-Dodecene	1242	1244	1245
cis-6-Tridecene	1308	1310	1312
cis-5-Tridecene	1309	1312	1314
trans-6-Tridecene	1314	1316	1316
cis-4-Tridecene	1315	1316	1319
trans-5-Tridecene	1315	1317	1318
trans-4-Tridecene	1315	1317	1317
1-Tridecene	1319	1321	1322
cis-3-Tridecene	1322	1323	1325
trans-3-Tridecene	1322	1323	1323
trans-2-Tridecene	1335	1337	1337
cis-2-Tridecene	1342	1344	1346
cis-7-Tetradecene	1404	1406	1409
cis-6-Tetradecene	1405	1408	1410
cis-5-Tetradecene	1408	1411	1412
trans-7-Tetradecene	1411	1413	1414
trans-6-Tetradecene	1412	1414	1415
trans-5-Tetradecene	1414	1416	1416
cis-4-Tetradecene	1414	1417	1418
trans-4-Tetradecene	1415	1416	1416
1-Tetradecene	1420	1422	1423
cis-3-Tetradecene	1421	1422	
trans-3-Tetradecene	1422	1422	1422
trans-2-Tetradecene	1436	1436	1437
cis-2-Tetradecene	1441	1444	

Column Packing: P1 = 45-m by 0.25-mm I.D. stainless-steel capillary column coated with polyphenyl ether.

REFERENCES

1. **Rang, S., Kuningas, K., Orav, A., and Eisen, O.,** *Chromatographia,* 10, 55, 1977.

Table 11
ALKENES (C-10—C-11)

Column packing	P1	P1	P2	P2	P2	P2
Temperature (°C)	50	100	86	100	115	130
Reference	1	1	2	2	2	2
Compound			I			
2-Methyl-*cis*-3-nonene	947	950				
2-Methyl-*trans*-3-nonene	948	951				
2-Methyl-*trans*-4-nonene	959	961				
2-Methyl-*cis*-4-nonene	967	970				
2,2-Dimethyl-*trans*-3-nonene	984	986				
2,2-Dimethyl-*trans*-4-nonene	1002	1005				
2-Methyl-2-nonene	1010	1010				
2,2-Dimethyl-*cis*-4-nonene	1017	1023				
2,2-Dimethyl-*cis*-3-nonene			1026	1031		
cis-5-Undecene			1078	1078	1079	1080
cis-4-Undecene			1080	1080	1081	1082
trans-5-Undecene			1081	1081	1081	1082
trans-4-Undecene			1082	1082	1082	1084
1-Undecene			1082	1082	1083	1084
cis-3-Undecene			1085	1085	1086	1087
trans-3-Undecene			1086	1085	1085	1086
trans-2-Undecene			1097	1097	1097	1097
cis-2-Undecene			1101	1102	1102	1103

Column Packing: P1 = 200-m by 0.2-mm I.D. stainless-steel capillary column coated with squalane. P2 = 50-m by 0.25-mm I.D. stainless-steel capillary column coated with di-*n*-butyl tetrachlorophthalate.

REFERENCES

1. **Sojak, L., Majer, P., Skalak, P., and Janak, J.,** *J. Chromatogr.*, 65, 137, 1972.
2. **Ryba, M.,** *J. Chromatogr.*, 123, 327, 1976.

Table 12
ALKENES (C-12—C-16)

Column packing	P1	P1	P1	P1	P2
Temperature (°C)	100	120	140	160	130
Reference	1	1	1	1	2
Compound			I		
trans-6-Dodecene	1236	1237	1238	1240	
trans-5-Dodecene	1237	1238	1239	1240	
trans-4-Dodecene	1237	1238	1239	1239	
cis-6-Dodecene	1240	1242	1244	1246	
cis-5-Dodecene	1240	1242	1244	1246	
cis-4-Dodecene	1245	1247	1248	1250	
trans-3-Dodecene	1246	1246	1246	1246	
1-Dodecene	1251	1252	1253	1254	
cis-3-Dodecene	1252	1254	1256	1257	
trans-2-Dodecene	1265	1266	1266	1267	
cis-2-Dodecene	1279	1281	1283	1285	
trans-6-Tridecene	1333	1335	1336	1337	
trans-5-Tridecene	1336	1337	1338	1338	
trans-4-Tridecene	1337	1337	1338	1339	
cis-6-Tridecene	1335	1337	1340	1342	
cis-5-Tridecene	1338	1340	1342	1343	
cis-4-Tridecene	1344	1346	1347	1349	
trans-3-Tridecene	1345	1346	1346	1346	
1-Tridecene	1351	1352	1353	1354	
cis-3-Tridecene	1352	1353	1355	1356	
trans-2-Tridecene	1365	1365	1365	1366	
cis-2-Tridecene	1379	1381	1382	1385	
trans-7-Tetradecene	1431	1432	1434	1435	
cis-7-Tetradecene	1431	1433	1435	1437	
trans-6-Tetradecene	1432	1433	1434	1435	
cis-6-Tetradecene	1433	1435	1437	1439	
trans-5-Tetradecene	1435	1436	1437	1438	
cis-5-Tetradecene	1436	1438	1441	1443	
trans-4-Tetradecene	1436	1437	1438	1438	
cis-4-Tetradecene	1443	1445	1447	1449	
trans-3-Tetradecene	1445	1445	1446	1446	
cis-3-Tetradecene	1450	1452			
1-Tetradecene	1451	1452	1453	1454	
trans-2-Tetradecene	1465	1466	1466	1466	
cis-2-Tetradecene	1479	1480			
cis-7-Pentadecene					1468
cis-6-Pentadecene					1470
cis-5-Pentadecene					1474
trans-7-Pentadecene					1475
trans-6-Pentadecene					1477
cis-4-Pentadecene					1480
trans-5-Pentadecene					1480
trans-4-Pentadecene					1480
1-Pentadecene					1485
trans-3-Pentadecene					1485
cis-3-Pentadecene					1486
trans-2-Pentadecene					1498
cis-2-Pentadecene					1504
cis-8-Hexadecene					1564
cis-7-Hexadecene					1565
cis-6-Hexadecene					1568
trans-8-Hexadecene					1572

Table 12 (continued)
ALKENES (C-12—C-16)

Compound	I
cis-5-Hexadecene	1573
trans-7-Hexadecene	1574
trans-6-Hexadecene	1575
cis-4-Hexadecene	1579
trans-5-Hexadecene	1579
trans-4-Hexadecene	1580
1-Hexadecene	1584
trans-3-Hexadecene	1585
cis-3-Hexadecene	1586
trans-2-Hexadecene	1597
cis-2-Hexadecene	1604

Column Packing: P1 = 100-m by 0.25-mm I.D. capillary column coated with polyethylene glycol 4000. P2 = 200-m by 0.2-mm I.D. stainless-steel capillary column coated with squalane.

REFERENCES

1. **Rang, S., Kuningas, K., Orav, A., and Eisen, O.,** *Chromatographia,* 10, 55, 1977.
2. **Sojak, L., Hrivnak, J., Ostrovsky, I., and Janak, J.,** *J. Chromatogr.,* 91, 613, 1974.

Table 13
ALKYNES (C-7—C-8)

Column packing	P1	P2	P3	P4	P5
Temperature (°C)	50	50	50	50	140
Reference	1	1	1	1	1
Compound	**I**				
1-Heptyne	685	794	838	916	662
1-Octyne	784	894	938	1012	759
2-Octyne	845	936	971	1034	769
3-Octyne	821	902	934	988	746
4-Octyne	814	892	923	977	743
1,3-Octadiyne	848	1052	1136	1273	779
1,4-Octadiyne	812	1028	1113	1262	720
1,5-Octadiyne	801	994	1071	1204	718
1,6-Octadiyne	825	1022	1105	1248	717
1,7-Octadiyne	769	993	1086	1248	713
1-Octen-7-yne	766	901	962	1058	737

Column Packing: P1 = 80-m by 0.23-mm I.D. glass capillary column coated with squalane. P2 = 80-m by 0.23-mm I.D. glass capillary column coated with Ucon LB-550-X. P3 = 80-m by 0.23-mm I.D. glass capillary column coated with Ucon 50-HB-280-X polar. P4 = 50-m by 0.23-mm I.D. glass capillary column coated with Carbowax 1000. P5 = 1.5-m by 1.6-mm I.D. glass column packed with 0.16—0.20 mm GTR Sterling MT.

REFERENCE

1. **Welsch, Th., Engewald, W., and Berger, P.,** *Chromatographia,* 11, 5, 1978.

Table 14
n-ALKYNES (C-6—C-12)

Column packing	P1	P1	P1	P1	P1	P1
Temperature (°C)	86	90	100	110	120	130
Reference	1	1	1	1	1	1
Compound	**I**					
1-Hexyne	587	584	584	584	584	584
2-Hexyne	642	640	639	638	638	638
3-Hexyne	624	623	622	620	620	619
1-Heptyne	686	684	684	684	684	684
2-Heptyne	744	743	743	742	742	741
3-Heptyne	718	717	716	716	715	714
1-Octyne		784	784	784	784	784
2-Octyne		842	842	841	841	840
3-Octyne		818	817	816	816	815
4-Octyne		811	811	810	810	809
1-Nonyne		884	884	884	884	884
2-Nonyne		941	941	941	940	940
3-Nonyne		916	915	915	914	913
4-Nonyne		910	910	910	909	909
1-Decyne		984	984	984	984	984
2-Decyne		1041	1041	1041	1040	1040
3-Decyne		1014	1014	1013	1013	1012
4-Decyne		1008	1007	1007	1006	1006
5-Decyne		1008	1008	1008	1008	1007
1-Undecyne				1084	1084	1084
2-Undecyne				1140	1140	1140
3-Undecyne				1112	1112	1111
4-Undecyne				1105	1104	1104
5-Undecyne				1104	1104	1104
1-Dodecyne				1184	1185	1184
2-Dodecyne				1240	1240	1239
3-Dodecyne				1211	1211	1210
4-Dodecyne				1203	1203	1202
5-Dodecyne				1202	1201	1201
6-Dodecyne				1200	1200	1200

Column Packing: P1 = 100-m by 0.25-mm I.D. stainless-steel capillary column coated with squalane.

REFERENCES

1. **Rang, S., Kuningas, K., Orav, A., and Eisen, O.,** *J. Chromatogr.,* 119, 451, 1976.

Table 15
n-ALKYNES (C-10—C-14)

Column packing	P1	P1	P1	P1	P1
Temperature (°C)	110	130	150	170	190
Reference	1	1	1	1	1
Compound	**I**				
1-Decyne	995	996	997		
2-Decyne	1052	1052	1051		
3-Decyne	1020	1019	1017		
4-Decyne	1012	1012	1011		
5-Decyne	1013	1013	1013		
1-Undecyne	1095	1096	1095	1096	
2-Undecyne	1151	1152	1151	1150	
3-Undecyne	1119	1119	1117	1116	
4-Undecyne	1110	1111	1110	1110	
5-Undecyne	1110	1110	1109	1109	
1-Dodecyne		1195	1195	1197	
2-Dodecyne		1251	1250	1250	
3-Dodecyne		1218	1216	1216	
4-Dodecyne		1208	1209	1208	
5-Dodecyne		1208	1207	1208	
6-Dodecyne		1207	1206	1206	
1-Tridecyne			1296	1297	1298
2-Tridecyne			1351	1351	1351
3-Tridecyne			1316	1315	1315
4-Tridecyne			1307	1307	1307
5-Tridecyne			1306	1306	1306
6-Tridecyne			1303	1304	1304
1-Tetradecyne			1396	1397	1398
2-Tetradecyne			1451	1451	1451
3-Tetradecyne			1415	1414	1414
4-Tetradecyne			1407	1406	1407
5-Tetradecyne			1404	1404	1405
6-Tetradecyne			1401	1401	1402
7-Tetradecyne			1400	1400	1401

Column Packing: P1 = 50-m × 0.25-mm I.D. stainless-steel capillary column coated with Apiezon L.

REFERENCE

1. **Rang, S., Kuningas, K., Orav, A., and Eisen, O.,** *J. Chromatogr.*, 119, 451, 1976.

Table 16
n-ALKYNES (C-9—C-14)

Column packing	P1	P1	P1	P1	P1
Temperature (°C)	90	110	130	150	170
Reference	1	1	1	1	1
Compound	**I**				
1-Nonyne	1003	1003	1004		
2-Nonyne	1066	1068	1068		
3-Nonyne	1033	1032	1032		
4-Nonyne	1022	1022	1023		
1-Decyne	1103	1104	1104		
2-Decyne	1168	1169	1171		
3-Decyne	1132	1132	1132		
4-Decyne	1120	1121	1121		
5-Decyne	1121	1121	1122		
1-Undecyne	1202	1203	1204	1205	
2-Undecyne	1267	1268	1269	1270	
3-Undecyne	1230	1231	1231	1231	
4-Undecyne	1218	1219	1220	1221	
5-Undecyne	1218	1218	1219	1220	
1-Dodecyne		1303	1304	1305	
2-Dodecyne		1368	1368	1370	
3-Dodecyne		1329	1330	1330	
4-Dodecyne		1317	1318	1319	
5-Dodecyne		1316	1317	1318	
6-Dodecyne		1316	1317	1317	
1-Tridecyne		1403	1403	1405	1405
2-Tridecyne		1468	1468	1470	1471
3-Tridecyne		1429	1429	1429	1430
4-Tridecyne		1415	1416	1417	1417
5-Tridecyne		1414	1415	1416	1416
6-Tridecyne		1412	1414	1415	1416
1-Tetradecyne		1502	1503	1505	1506
2-Tetradecyne		1568	1569	1570	1571
3-Tetradecyne		1529	1529	1529	1529
4-Tetradecyne		1515	1516	1516	1517
5-Tetradecyne		1513	1513	1515	1515
6-Tetradecyne		1509	1511	1512	1514
7-Tetradecyne		1508	1509	1511	1512

Column Packing: P1 = 45-m by 0.25-mm I.D. stainless-steel capillary column coated with polyphenyl ether.

REFERENCE

1. **Rang, S., Kuningas, K., Orav, A., and Eisen, O.,** *J. Chromatogr.*, 119, 451, 1976.

Table 17
n-ALKYNES (C-6—C-14)

Column packing	P1	P1	P1	P1	P1	P1	P1	P1
Temperature (°C)	60	80	100	110	120	140	150	160
Reference	1	1	1	1	1	1	1	1
Compound				I				
1-Hexyne	837	833		827				
2-Hexyne		862		860				
3-Hexyne	832	829		825				
1-Heptyne	938	934		929				
2-Heptyne	964	964		965				
3-Heptyne	916	913		910				
1-Octyne	1036	1034		1031				
2-Octyne	1063	1064		1065				
3-Octyne	1013	1012		1012				
4-Octyne	1000	999		998				
1-Nonyne		1135	1134	1133	1132			
2-Nonyne		1162	1163	1163	1163			
3-Nonyne		1109	1110	1109	1108			
4-Nonyne		1095	1096	1095	1094			
1-Decyne		1235	1234	1233	1233	1232		
2-Decyne		1262	1263	1263	1264	1265		
3-Decyne		1208	1207	1207	1207	1207		
4-Decyne		1192	1192	1192	1192	1192		
5-Decyne		1190	1191	1191	1191	1191		
1-Undecyne			1334	1333	1333	1332	1332	1331
2-Undecyne			1362	1363	1364	1365	1365	1366
3-Undecyne			1306	1306	1306	1306	1305	1305
4-Undecyne			1289	1289	1289	1290	1289	1289
5-Undecyne			1286	1286	1286	1287	1286	1286
1-Dodecyne			1434	1434	1434	1433	1432	1433
2-Dodecyne			1463	1464	1464	1465	1466	1467
3-Dodecyne			1406	1406	1406	1406	1405	1405
4-Dodecyne			1387	1387	1387	1388	1388	1388
5-Dodecyne			1382	1383	1383	1384	1384	1384
6-Dodecyne			1381	1382	1383	1382	1383	1383
1-Tridecyne				1533	1533	1533	1532	1533
2-Tridecyne				1562	1563	1565	1565	1567
3-Tridecyne				1505	1505	1505	1505	1505
4-Tridecyne				1486	1486	1487	1487	1487
5-Tridecyne				1480	1480	1481	1481	1482
6-Tridecyne				1477	1478	1479	1479	1480
1-Tetradecyne				1633	1633	1633	1632	1633
2-Tetradecyne				1664	1664	1666	1665	1667
3-Tetradecyne				1604	1604	1605	1604	1605
4-Tetradecyne				1584	1585	1586	1586	1586
5-Tetradecyne				1578	1578	1580	1580	1581
6-Tetradecyne				1574	1574	1576	1576	1577
7-Tetradecyne				1572	1572	1574	1574	1575

Column Packing: P1 = 100-m by 0.25-mm I.D. stainless-steel capillary column coated with polyethylene glycol 4000.

REFERENCE

1. **Rang, S., Kuningas, K., Orav, A., and Eisen, O.,** *J. Chromatogr.,* 119, 451, 1976.

Table 18
CYCLIC HYDROCARBONS (C-5—C-8)

Column packing	P1	P2
Temperature (°C)	25	27
Reference	1	1

Compound	I[a]	
Cyclopentane	612	563
Cyclopentene	652	547
Methylcyclopentane	662	624
1-Methylcyclopentene	745	642
Cyclohexane	700	658
Cyclohexene	779	667
1,1-Dimethylcyclopentane	699	669
1-*cis*-3-Dimethylcyclopentane	701	679
1-*trans*-3-Dimethylcyclopentane	707	683
1-*trans*-2-Dimethylcyclopentane	712	686
1-*cis*-2-Dimethylcyclopentane	754	716
Methylcyclohexane	751	721
Ethylcyclopentane	766	729
1,1,3-Trimethylcyclopentane	731	720
1-*trans*-2-*cis*-4-Trimethylcyclopentane	747	738
1-*trans*-2-*cis*-3-Trimethylcyclopentane	759	744
1-*cis*-2-*trans*-4-Trimethylcyclopentane	788	769
Cycloheptane	841	786
Isopropylcyclopentane	839	807
Propylcyclopentane	861	826
Ethylcyclohexane	861	828

Column Packing: P1 = Dimethylsulfolane. P2 = Squalane.

[a]   Retention indices were calculated from log relative retention data in Reference 1.

REFERENCE

1. **Ladon, A. W.,** *J. Chromatogr.,* 99, 203, 1974.

Table 19
CYCLIC HYDROCARBONS (C-5—C-9)

Column packing	P1	P1
Temperature (°C)	50	70
Reference	1	1

Compound	I	
1-*trans*-2-Dimethylcyclopropane	479	480
Ethylcyclopropane	510	511
1-*cis*-2-Dimethylcyclopropane	515	516
1,1,2-Trimethylcyclopropane	549	550
Cyclopentane	566	568
1,1,2,2-Tetramethylcyclopropane	620	622
Ethylcyclobutane	621	623
Methylcyclopentane	628	631
Cyclohexane	663	667
1-*cis*-3-Dimethylcyclopentane	683	686
1-*trans*-3-Dimethylcyclopentane	687	690
1-*trans*-2-Dimethylcyclopentane	689	692
1,1,3-Trimethylcyclopentane	724	727
Methylcyclohexane	726	731
Ethylcyclopentane	734	737
1-*trans*-2-*cis*-3-Trimethylcyclopentane	748	751
1,1,2-Trimethylcyclopentane	763	768
1-*cis*-2-*trans*-4-Trimethylcyclopentane	773	777
1-*cis*-3-Dimethylcyclohexane	785	790
1,1-Dimethylcyclohexane	787	793
1-*trans*-2-Dimethylcyclohexane	802	808
1-*cis*-2-*cis*-3-Trimethylcyclopentane	802	808
1-*trans*-3-Dimethylcyclohexane	806	811
1-*cis*-2-Dimethylcyclohexane	829	835
Ethylcyclohexane	834	840
1,1,3-Trimethylcyclohexane	840	846

Column Packing: P1 = 100-m by 0.25-mm I.D. stainless-steel capillary column coated with squalane.

REFERENCE

1. **Cramers, C. A., Rijks, J. A., Pacakova, V., and Ribeiro de Andrade, I.,** *J. Chromatogr.,*, 51, 13, 1970.

Table 20
CYCLIC HYDROCARBONS (C-5—C-10)

Column packing	P1	P2	P3	P1	P2	P3
Temperature (°C)	60	60	60	40	40	40
Reference	1	1	1	1	1	1
Compound			I			
Cyclopentane				563	583	586
Cyclopentene				548	589	608
1,3-Cyclopentadiene				521	602	646
Norbornane				745	774	790
Norbornene				707	758	781
Norbornadiene				681	762	794
exo-Dicyclopentadiene	984	1068	1109			
endo-Dicyclopentadiene	987	1077	1119			
exo-1,2-Dihydrodicyclopentadiene	1007	1071	1101			
endo-1,2-Dihydrodicyclopentadiene	1017	1083	1113			
exo-9,10-Dihydrodicyclopentadiene	1024	1087	1119			
endo-9,10-Dihydrodicyclopentadiene	1038	1104	1136			
exo-Tetrahydrodicyclopentadiene	1044	1086	1108			
endo-Tetrahydrodicyclopentadiene	1071	1116	1141			

Column Packing: P1 = 70-m by 0.23-mm I.D. glass capillary column coated with squalane. P2 = 70-m by 0.23-mm I.D. glass capillary column coated with Ucon LB-550-X. P3 = 70-m by 0.23-mm I.D. glass capillary column coated with Ucon 50-HB-280-X polar.

REFERENCE

1. **Stopp, I., Engewald, W., Kuhn, H., and Welsch, Th.**, *J. Chromatogr.*, 147, 21, 1978.

Table 21
CYCLIC HYDROCARBONS (C-5—C-12)

Column packing	P1	P1	P2[a]	P3
Temperature (°C)	50	100	130	120
Reference	1	1	2	2
Compound		I		
Cyclopentane	574	584	587	587
Methylcyclopentane	640	649		
Cyclohexane	672	687	700	692
Ethylcyclopentane	749	760		
1,1-Dimethylcyclopentane	685	698		
1-*cis*-3-Dimethylcyclopentane	694	703		
1-*trans*-2-Dimethylcyclopentane	707	716		
n-Propylcyclopentane	845	856		
Methylcyclohexane	743	759		
Cycloheptane			846	832
Ethylcyclohexane	852	869		
Cyclooctane			979	959
n-Propylcyclohexane	944	960		
Cyclononane			1093	
n-Butylcyclohexane	1039	1055		
tert-Butylcyclohexane	996	1019		
Cyclodecane	1198	1170		
trans-Decalin	1089	1117		
cis-Decalin	1125	1155		
trans-(1,2)-Octalin	1126	1152		
trans-(2,3)-Octalin	1145	1171		
(1,9)-Octalin	1131	1158		
Tetralin	1274	1296		
cis-Hydrindan	1011	1036		
Indan	1153	1168		
Cycloundecane			1292	
Cyclododecane			1384	1354

Column Packing: P1 = 50-m by 0.25-mm I.D. stainless-steel capillary column coated with di-*n*-butyltetrachlorophthalate. P2 = Squalane. P3 = 1.5—4-m by ¹/₈-in. O.D. column packed with 10% diisodecyl phthalate on 60—80 mesh Chromosorb W.

[a] Retention indices for Column P2 taken from Kovats, E. sz., *Helv. Chim. Acta,* 41, 1915, 1958.

REFERENCES

1. **Ryba, M.,** *J. Chromatogr.,* 123, 327, 1976.
2. **Grenier-Loustalot, M. F., Bonastre, J., Potin, M., and Grenier, P.,** *J. Chromatogr.,* 138, 63, 1977.

Table 22
CYCLIC HYDROCARBONS (C-5—C-15)

Column packing	P1	P2	P3	P4
Temperature (°C)	100	100	100	100
Reference	1	1	1	1

Compound	I			
Methylcyclopentane	636		678	
Ethylcyclopentane	743		789	817
n-Propylcyclopentane	839	852	883	912
n-Butylcyclopentane	937	950	980	1009
n-Pentylcyclopentane	1036	1050	1078	1107
n-Hexylcyclopentane	1135	1150	1178	1207
n-Heptylcyclopentane	1234	1249	1275	1305
Cyclopentene	560			705
3-Methyl-1-cyclopentene	610		680	736
1-Methyl-1-cyclopentene	651		730	792
3-Ethyl-1-cyclopentene	722		789	849
1-Ethyl-1-cyclopentene	753		832	891
3-*n*-Propyl-1-cyclopentene	818		888	944
1-*n*-Propyl-1-cyclopentene	840		916	972
3-Isopropyl-1-cyclopentene	803		870	928
1-Isopropyl-1-cyclopentene	816		884	938
3-Allyl-1-cyclopentene	797		901	985
1-Allyl-1-cyclopentene	822		932	1020
3-Isobutyl-1-cyclopentene	881		944	999
1-Isobutyl-1-cyclopentene	892		955	1003
3-*n*-Butyl-1-cyclopentene	916		986	1042
1-*n*-Butyl-1-cyclopentene	940		1015	1070
3-Isopentyl-1-cyclopentene	978		1041	1094
1-Isopentyl-1-cyclopentene	984			1108
3-*n*-Pentyl-1-cyclopentene	1015		1085	1140
1-*n*-Pentyl-1-cyclopentene	1036		1112	1166
3-*n*-Hexyl-1-cyclopentene	1116		1185	1240
1-*n*-Hexyl-1-cyclopentene	1135		1210	1263
3-*n*-Heptyl-1-cyclopentene	1214		1284	1338
1-*n*-Heptyl-1-cyclopentene	1234		1308	1362
Cyclohexane	674	692	728	766
Methylcyclohexane	738	754	783	814
Ethylcyclohexane	848	865	892	920
n-Propylcyclohexane	940	956	984	1019
Isopropylcyclohexane	933	951	981	
Isobutylcyclohexane	992	1004	1028	1058
sec-Butylcyclohexane	1032	1046	1083	1120
n-Butylcyclohexane	1037	1053	1082	1116
n-Pentylcyclohexane	1137	1151	1181	1214
n-Hexylcyclohexane	1236	1249	1280	1312
Cyclohexene	683		780	850
3-Methyl-1-cyclohexene	743		825	888
4-Methyl-1-cyclohexene	746		830	893
1-Methyl-1-cyclohexene	773		864	932
3-Ethyl-1-cyclohexene	853		940	1004
4-Ethyl-1-cyclohexene	856		943	1009
1-Ethyl-1-cyclohexene	867		958	1023
1-Isopropyl-1-cyclohexene	924		1008	1064
3-Isopropyl-1-cyclohexene	933		1018	1081
3-*n*-Propyl-1-cyclohexene	948		1032	1097
4-*n*-Propyl-1-cyclohexene	949		1034	1099
1-*n*-Propyl-1-cyclohexene	954		1039	1100

Table 22 (continued)
CYCLIC HYDROCARBONS (C-5—C-15)

Compound	I		
1-Isobutyl-1-cyclohexene	1003	1075	1129
3-Isobutyl-1-cyclohexene	1005	1081	1142
3-*sec*-Butyl-1-cyclohexene	1033	1120	1185
3-*n*-Butyl-1-cyclohexene	1045	1129	1193
4-*n*-Butyl-1-cyclohexene	1046	1132	1196
1-*n*-Butyl-1-cyclohexene	1050	1136	1198
3-*n*-Pentyl-1-cyclohexene	1144	1228	1291
1-*n*-Pentyl-1-cyclohexene	1145	1232	1292
1-*n*-Hexyl-1-cyclohexene	1243	1329	1388
3-*n*-Hexyl-1-cyclohexene	1244	1327	1390
1-*n*-Heptyl-1-cyclohexene	1342	1428	1486
3-*n*-Heptyl-1-cyclohexene	1344	1426	1489
1-*n*-Octyl-1-cyclohexene		1526	1583
3-*n*-Octyl-1-cyclohexene		1526	1587
1-*n*-Nonyl-1-cyclohexene		1624	1680
3-*n*-Nonyl-1-cyclohexene		1626	1685

Column Packing: P1 = 100-m by 0.25-mm I.D. capillary column coated with squalane. P2 = 50-m by 0.25-mm I.D. capillary column coated with Apiezon L. P3 = 45-m by 0.25-mm I.D. capillary column coated with polyphenylether. P4 = 100-m by 0.25-mm I.D. capillary column coated with polyethylene glycol 4000.

REFERENCE

1. **Rang, S., Orav, A., Kuningas, K., and Eisen, O.,** *Chromatographia,* 10, 115, 1977.

Table 23
CYCLIC HYDROCARBONS (C-6—C-13)

Column packing	P1	P2	P3
Temperature (°C)	90	90	220
Reference	1	1	1

Compound	I		
cis-Bicyclo[3.1.0]hexane	678	741	528
cis-Bicyclo[4.1.0]heptane	797	870	577
cis-Bicyclo[3.2.0]heptane	787	838	565
cis-Bicyclo[5.1.0]octane	888	929	673
cis-Bicyclo[4.2.0]octane	880	931	655
cis-Bicyclo[3.3.0]octane	870	922	658
cis-Bicyclo[6.1.0]nonane	996	1060	768
cis-Bicyclo[5.2.0]nonane	990	1043	769
cis-Bicyclo[4.3.0]nonane[a]	991	1046	750
trans-Bicyclo[4.3.0]nonane[b]	961	1007	800
cis-Bicyclo[7.1.0]decane	1110	1187	861
trans-Bicyclo[7.1.0]decane	1090	1162	872
cis-Bicyclo[6.2.0]decane	1097	1154	866
cis-Bicyclo[5.3.0]decane	1095	1150	867
cis-Bicyclo[4.4.0]decane[c]	1100	1159	842
trans-Bicyclo[4.4.0]decane[d]	1064	1109	883
cis-Bicyclo[8.1.0]undecane	1216	1292	957
trans-Bicyclo[8.1.0]undecane	1194	1269	968
cis-Bicyclo[5.4.0]undecane	1226	1278	942
trans-Bicyclo[5.4.0]undecane	1206	1253	968
cis-Bicyclo[9.1.0]dodecane	1310	1386	1082
trans-Bicyclo[9.1.0]dodecane	1288	1355	1100
cis-Bicyclo[6.4.0]dodecane	1299	1350	1057
cis-Bicyclo[5.5.0]dodecane	1312	1369	1077
trans-Bicyclo[5.5.0]dodecane	1299	1350	1093
cis-Bicyclo[10.1.0]tridecane	1404	1473	1197
trans-Bicyclo[10.1.0]tridecane	1382	1450	1207

Column Packing: P1 = 100-m by 0.23-mm I.D. glass capillary column coated with squalane. P2 = 100-m by 0.23-mm I.D. glass capillary column coated with Ucon 50-HB-280X polar. P3 = 2-m by 1.9-mm I.D. stainless-steel column packed with 0.2—0.315 mm graphitized thermal carbon black T-168.

[a]   *cis*-hydrindan.
[b]   *trans*-hydrindan.
[c]   *cis*-decalin.
[d]   *trans*-decalin.

REFERENCE

1. **Engewald, W., Epsch, K., Welsch, Th., and Graefe, J.,** *J. Chromatogr.,* 119, 119, 1976.

Table 24
CYCLIC HYDROCARBONS (C-6—C-13)

Column packing	P1	P1	P1	P1	P2	P2	P2	P2
Temperature (°C)	42.5	70	80	120	110	160	200	220
Reference	1	1	1	1	1	1	1	1
Compound				**I**				
Cyclohexane	660	668			509			
Cyclohexene	669	677			532			
1,3-Cyclohexadiene	652	655			537			
1,4-Cyclohexadiene	687	694			568			
Cycloheptane	791		804			606		
Cycloheptene	774		785			607		
1,3-Cycloheptadiene	796					642		
1,4-Cycloheptadiene	784					623		
Cyclooctane	908		925			688		
cis-Cyclooctene	883		895			672		
trans-Cyclooctene			899			686		
1-*cis*-3-*cis*-Cyclooctadiene	874		889			674		
1-*cis*-4-*cis*-Cyclooctadiene	882		895			674		
1-*cis*-5-*cis*-Cyclooctadiene	907		915			686		
Cyclononane				1049			792	
cis-Cyclononene				1029			767	
trans-Cyclononene				1024			779	
Cyclodecane				1147			908	
cis-Cyclodecene				1130			870	
trans-Cyclodecene				1120			879	
Cycloundecane				1235				1019
cis-Cycloundecene				1223				991
trans-Cycloundecene				1214				991
Cyclododecane				1326				1149[a]
cis-Cyclododecene				1315				1111[a]
trans-Cyclododecene				1306				1099[a]
Cyclotridecane								1268[b]
cis-Cyclotridecene								1230[b]
trans-Cyclotridecene								1221[b]

Column Packing: P1 = 100-m by 0.23-mm I.D. glass capillary column coated with squalane. P2 = 2-m by 1.9-mm I.D. stainless-steel column packed with 0.2—0.315 mm graphatized thermal carbon black T-168.

[a]   Column temperature, 240°C.
[b]   Column temperature, 280°C.

REFERENCE

1. **Engewald, W., Epsch, K., Graefe, J., and Welsch, Th.**, *J. Chromatogr.*, 91, 623, 1974.

Table 25
CYCLIC HYDROCARBONS (C-5—C-10)

Column packing	P1	P2	P3
Temperature (°C)	80	80	100
Reference	1	1	2

Compound	I		
Hexylcyclopropane	913	944	
5-Hexenylcyclopropane	895	955	
5,6-Methylene-hexylcyclopropane	1027	1088	
Heptylcyclopropane	1013	1044	
5-Methyl-hexylcyclopropane	977	1005	
4-Methyl-hexylcyclopropane	983	1012	
3-Methyl-hexylcyclopropane	975	1003	
Cyclopentane			426
Cyclohexane			508
Methylcyclopentane			524
Cyclohexene			535
Cyclohexadiene			539
Bicyclo[2.2.1]heptane[a]			534
Bicyclo[2.2.1]-2-heptene[b]			517
endo-5-Methyl-bicyclo[2.2.1]-2-heptene			578
exo-5-Methyl-bicyclo[2.2.1]-2-heptene			597
1-Methyl-bicyclo[2.2.1]heptane			586
endo-2-Methyl-bicyclo[2.2.1]heptane			592
exo-2-Methyl-bicyclo[2.2.1]heptane			606
2-Methylene-bicyclo[2.2.1]heptane			594
7,7-Dimethyl-bicyclo[2.2.1]heptane			643
endo-5-Vinyl-bicyclo[2.2.1]-2-heptene			648
exo-5-Vinyl-bicyclo[2.2.1]-2-heptene			669
endo-1,2-Dimethyl-bicyclo[2.2.1]heptane			649
exo-1,2-Dimethyl-bicyclo[2.2.1]heptane			668
trans-5-Ethylidene-bicyclo[2.2.1]-2-heptene			666
cis-5-Ethylidene-bicyclo[2.2.1]-2-heptene			689
1,4-Dimethyl-bicyclo[2.2.1]heptane			666
endo-endo-2,3-Dimethyl-bicyclo[2.2.1]heptane			667
trans-2,3-Dimethyl-bicyclo[2.2.1]heptane			675
exo-exo-2,3-Dimethyl-bicyclo[2.2.1]heptane			694
exo-2,2,3-Trimethyl-bicyclo[2.2.1]heptane			678
endo-2,2,3-Trimethyl-bicyclo[2.2.1]heptane			697
2,2-Dimethyl-3-methyl-bicyclo[2.2.1]heptane			694
endo-2-Ethyl-bicyclo[2.2.1]heptane			689
exo-2-Ethyl-bicyclo[2.2.1]heptane			703
endo-Trimethylene-bicyclo[2.2.1]heptane			717

Column Packing: P1 = 100-m by 0.25-mm I.D. capillary column coated with squalane. P2 = 50-m by 0.25-mm I.D. glass capillary column coated with polypropyleneglycol. P3 = Column packed with graphitized thermal carbon black.

[a] Norbornane.
[b] Norbornene.

REFERENCES

1. **Schomburg, G., and Dielmann, G.,** *J. Chromatogr. Sci.,* 11, 151, 1973.
2. **Kalashnikova, E. V., Kiselev, A. V., Poshkus, D. P., and Shcherbakova, K. D.,** *J. Chromatogr.,* 119, 233, 1976.

Table 26
CYCLIC HYDROCARBONS (C-9—C-14)

The retention index **I** for each compound is given under the following column packing / temperature (°C) conditions, all with Reference 1:

Compound	Stereoisomer[a]	P1 / 100	P2 / 100	P1 / 125	P2 / 125	P1 / 150	P2 / 150	P1 / 175	P2 / 175	P1 / 200
Bicyclo[4.3.0]nonane	1	963	1086	975	1102	987	1125			
	2	996	1138	1010	1158	1023	1183			
Bicyclo[4.4.0]decane	1	1061	1195	1075	1218	1089	1244			
	2	1101	1260	1116	1285	1132	1314			
Tricyclo[7.2.1.0^{5,12}]dodecane	1				1430	1274	1462	1291	1494	1310
	2				1466	1295	1499	1314	1533	1334
	3				1504	1317	1539	1338	1574	1359
	4				1504	1323	1539	1344	1574	1365
	5				1554	1352	1591	1374	1630	1396
Tricyclo[7.3.0.0^{2,6}]dodecane	1				1428		1458		1484	
	2				1452		1480		1505	
	3				1477	1307	1508	1326	1535	1343
	4				1503	1323	1537	1343	1566	1362
	5				1514	1328	1547	1348	1578	1365
	6				1550	1352	1585	1373	1619	1391
Tricyclo[7.3.1.0^{5,13}]tridecane	1				1553	1377	1590	1400	1628	1424
	2				1614	1416	1652	1442	1691	1468
	3				1641	1429	1681	1456	1721	1482
Tricyclo[8.3.0.0^{4,9}]tridecane	1				1592	1409	1628	1433	1660	1456
	2				1613	1422	1652	1446	1687	1469
	3				1636	1437	1674	1461	1709	1484
	4				1647	1444	1685	1468	1720	1491
	5				1676	1461	1718	1486	1755	1511
Tricyclo[8.4.0.0^{3,8}]tetradecane	1				1658	1481	1696	1506	1734	1531
	2				1680	1494	1718	1519	1755	1545
	3				1705	1513	1744	1537	1782	1562
	4				1761	1549	1804	1576	1844	1602
	5				1777	1559	1819	1587	1860	1616
Tricyclo[8.4.0.0^{2,7}]tetradecane	1				1672	1493	1712	1515	1751	1540
	2				1705	1513	1747	1538	1786	1564
	3				1741	1537	1786	1563	1826	1589
	4				1765	1552	1810	1575	1850	1602
	5				1796	1568	1845	1596	1886	1624

Column Packing: P1 = 50-m by 0.25—0.30-mm I.D. stainless-steel capillary coated with SE-30. P2 = 50-m by 0.25—0.30-mm I.D. stainless-steel capillary coated with carbowax 20M.

[a] Of the stereoisomers analyzed, the only absolute configurations known are for three of the tricyclo[8.4.0.0^{3,8}]tetradecane (perhydroanthracene) stereoisomers — for these, the structural designations are: number 1, trans-syn-trans; number 3, cis-trans; and number 4, cis-syn-cis.

REFERENCE

1. **Vanek, J., Podrouzkova, B., and Landa, S.,** *J. Chromatogr.,* 52, 77, 1970.

Table 27
CYCLIC HYDROCARBONS (C-10—C-16)

Column packing	P1	P1	P1	P2	P2	P2	P3	P3
Temperature (°C)	200	170	140	140	125	110	125	110
Reference	1	1	1	1	1	1	1	1
Compound				I				
Adamantane	1216	1188	1162	1114	1104	1095	1405	1363
1-Methyladamantane	1232	1205	1179	1135	1125	1116	1374	1339
1,3-Dimethyladamantane	1245	1218	1194	1154	1144	1136	1344	1312
1,3,5-Trimethyladamantane	1254	1230	1204	1170	1161	1153	1311	1282
1,3,5,7-Tetramethyladamantane	1255	1234	1214	1185	1176	1169	1274	1251
2-Methyladamantane	1300	1270	1242	1192	1182	1175	1475	1432
1-Ethyladamantane	1367	1337	1309	1259	1249	1240	1515	1471
1-Ethyl-3,5,7-trimethyladamantane	1368	1348		1294	1285		1400	1375
1-Ethyl-3-methyladamantane	1371	1343		1273	1263		1478	1438
1-Ethyl-3,5-dimethyladamantane	1372	1348	1321	1285	1276	1267	1439	1407
Homoadamantane	1379	1345	1313	1255	1244	1234	1577	1527
2-Ethyladamantane	1384	1356	1329	1280	1270	1262	1549	1508
2-Isopropyladamantane	1449	1419		1342	1333		1600	1556
1-Propyladamantane	1450	1422		1346	1335	1326	1584	1544
Tetracyclododecane	1456	1420	1386	1322	1311	1300	1666	1616
2-Propyladamantane	1463	1436		1363	1353	1346	1618	1575
1-Isopropyladamantane	1472	1441		1358	1347	1337	1623	1578
1,3-Diethyl-5,7-dimethyladamantane	1481	1456			1390		1520	
1,3-Diethyladamantane	1491	1461		1387	1377		1607	1568
1,3-Diethyl-5-methyl-adamantane	1492	1463			1384		1565	
2-Isobutyladamantane	1500	1473						
1-Butyladamantane	1542	1514						
2-Butyladamantane	1556	1529						
1-*tert*-Butyladamantane	1559	1525						
1-*sec*-Butyladamantane	1570	1538						
2-(1-Adamantyl)-pentane	1646	1616						
3-(1-Adamantyl)-pentane	1653	1620						
Diamantane	1676	1631						
4-Methyldiamantane	1691	1647						
1-Methyldiamantane	1731	1683						
3-Methyldiamantane	1750	1704						

Column Packing: P1 = 50-m by 0.2-mm I.D. stainless-steel capillary column coated with apiezon L. P2 = 50-m by 0.2-mm I.D. stainless-steel capillary column coated with squalane. P3 = 50-m by 0.2-mm I.D. stainless-steel capillary column coated with tetrakis-*O*-(2-cyanthyl)-pentaerythritol.

REFERENCE

1. **Hala, S., Eyem, J., Burkhard, J., and Landa, S.** *I. Chromatogr. Sci.*, 8, 203, 1970.

Table 28
ALKYL BENZENES (C-6—C-10)

Column packing	P1	P1	P1	P2	P2	P2	P3	P3	P4
Temperature (°C)	160	130	100	160	130	100	130	100	65
Reference	1	1	1	1	1	1	1	1	2
Compound					**I**				
Benzene	700	690	680	675	669	662	656	649	655
Toluene	812	800	790	779	772	766	764	757	758
Styrene									875
Ethylbenzene	903	899	880	878	868	862	856	848	849
1,4-Dimethylbenzene									858
1,3-Dimethylbenzene									858
1,2-Dimethylbenzene									879
Allylbenzene									929
n-Propylbenzene	992	980	968	971	960	953	946	937	938
i-Propylbenzene	961	949	936	941	931	923	916	909	910
1-Methyl-3-ethylbenzene									948
1-Methyl-4-ethylbenzene									948
1,3,5-Trimethylbenzene									952
α-Methylstyrene									963
1-Methyl-2-ethylbenzene									964
3-Methylstyrene									973
1,2,4-Trimethylbenzene									975
4-Methylstyrene									978
2-Methylstyrene									991
i-Allylbenzene									1004
1,2,3-Trimethylbenzene									1005
n-Butylbenzene	1091	1078		1069	1059	1050	1046	1038	
i-Butylbenzene	1043	1029	1018	1025	1015	1005	1000	991	993
sec-Butylbenzene									993
tert-Butylbenzene	1029	1013	1002	1012	1000	990	984	975	
β-Methylallylbenzene									1022
1-Methyl-3-allylbenzene									1028
1,3-Diethylbenzene									1032
1-Methyl-4-allylbenzene									1033
1,4-Diethylbenzene									1036
α-Ethylstyrene									1058
β-Ethylstyrene									1059
1-Methyl-3-α-methylstyrene									1062
3-Ethylstyrene									1064
1-Methyl-3-tert-butylbenzene									1068
1-Methyl-4-α-methylstyrene									1069
4-Ethylstyrene									1072
1,3-Dimethyl-5-vinylbenzene									1073
1,4-Dimethyl-2-vinylbenzene									1076
1,3-Divinylbenzene									1085
α,β-Dimethylstyrene									1091
1,4-Divinylbenzene									1096
1,2-Dimethyl-4-vinylbenzene									1100
1-Methyl-3-isoallylbenzene									1102
1-Methyl-4-isoallylbenzene									1104
1-Methyl-2-isoallylbenzene									1116
3-Isopropylstyrene									1117
4-Isopropylstyrene									1136
4-Vinyl-α-methylstyrene									1218

Table 28 (continued)
ALKYL BENZENES (C-6—C-10)

Column packing: P1 = 6-ft by ¹/₄-in. O.D. copper column packed with 20% Apiezon L on 80—100 mesh AW-DMCS Chromosorb W. P2 = 6-ft by ¹/₄-in. O.D. copper column packed with 20% SE-30 on 80—100 mesh AW-DMCS Chromosorb W. P3 = 6 ft by ¹/₄-in. O.D. copper column packed with 20% Squalane on 80—100 mesh AW-DMCS Chromosorb W. P4 = 25.5-m by 0.50-mm I.D. glass capillary column coated with SE-30.

REFERENCES

1. **Cook, L. E. and Raushel, F. M.**, *J. Chromatogr.*, 65, 556, 1972.
2. **Svob, V., Deur-Siftar, D., and Cramers, C. A.**, *J. Chromatogr.*, 91, 659, 1974.

Table 29
ALKYL BENZENES (C-6—C-11)

Column packing	P1	P1
Temperature (°C)	50	100
Reference	1	1
Compound	I	
Benzene	774	783
Toluene	886	893
Ethylbenzene	962	972
1,2-Dimethylbenzene	1027	1033
1,3-Dimethylbenzene	990	995
1,4-Dimethylbenzene	995	1000
n-Propylbenzene	1043	1055
i-Propylbenzene	1008	1020
1-Ethyl-2-methylbenzene	1092	1101
1-Ethyl-3-methylbenzene	1058	1067
1-Ethyl-4-methylbenzene	1064	1074
1,2,3-Trimethylbenzene	1168	1175
1,2,4-Trimethylbenzene	1129	1132
1,3,5-Trimethylbenzene	1085	1088
n-Butylbenzene	1143	1155
tert-Butylbenzene	1069	1084
1-*n*-Propyl-4-methylbenzene	1145	1155
1-*i*-Propyl-4-methylbenzene	1104	1116
1,2-Diethylbenzene	1154	1163
1,3-Diethylbenzene	1126	1136
1,4-Diethylbenzene	1139	1148
1-Ethyl-3,4-dimethylbenzene	1190	1197
1-Ethyl-2,4-dimethylbenzene	1187	1194
1-Ethyl-3,5-dimethylbenzene	1147	1152
1-Ethyl-2,5-dimethylbenzene	1187	1192
1,2,3,4-Tetramethylbenzene	1308	1312
1,2,3,5-Tetramethylbenzene	1259	1263
1,2,4,5-Tetramethylbenzene	1258	1260
n-Pentylbenzene	1241	1252

Column Packing: P1 = 50-m by 0.25-mm I.D. stainless-steel capillary column coated with di-*n*-butyl tetrachlorophthalate.

REFERENCE

1. **Ryba, M.**, *J. Chromatogr.*, 123, 327, 1976.

Table 30
ALKYL BENZENES (C-6—C-16)

Column packing	P1	P2	P3	P3	P4	P5	P6	P7
Temperature (°C)	100	65	70	90	90	80	80	80
References	1	1	1	1	1	2	2	2
Compound				I				
Benzene	650	655	780			644	781	1128
Toluene	758	758	881			752	883	1219
Ethylbenzene	848	849	971			842	974	1290
p-Xylene	862	858	978			856	980	1302
m-Xylene	863	858	981			858	984	1306
o-Xylene	884	879	1008			877	1011	1358
i-Propylbenzene	908	910	1029			901	1032	1320
n-Propylbenzene	936	938	1056			930	1060	1350
1-Methyl-3-ethylbenzene	949	948	1069			943	1071	1370
1-Methyl-4-ethylbenzene	951	948	1069			945	1071	1372
1-Methyl-2-ethylbenzene	965	963	1090			958	1092	1415
1,3,5-Trimethylbenzene	968	952	1079			963	1084	1388
tert-Butylbenzene	973	975	1096			967	1099	1373
1,2,4-Trimethylbenzene	988	975	1095			980	1107	1436
i-Butylbenzene	990	993	1105			982	1109	1363
sec-Butylbenzene	990	993	1108			983	1112	1372
1-Methyl-3-isopropylbenzene	1002	998	1119			997	1125	1396
1-Methyl-4-isopropylbenzene	1011	1003	1124			1004	1130	1403
1,2,3-Trimethylbenzene	1012	1005	1135			1004	1141	1493
1-Methyl-2-isopropylbenzene	1016	1017	1140			1010	1145	1439
1,3-Diethylbenzene	1029	1032	1149			1023	1151	1428
1-Methyl-3-*n*-propylbenzene	1033	1032	1150			1028	1153	1425
n-Butylbenzene	1036	1037	1157			1029	1160	1431
1,4-Diethylbenzene	1039	1037	1158			1034	1160	1439
1-Methyl-4-*n*-propylbenzene	1039	1035	1153			1033	1157	1430
1,2-Diethylbenzene	1039	1043	1165			1033	1167	1465
1-Methyl-2-*n*-propylbenzene	1046	1046	1168			1039	1172	1468
1,3-Dimethyl-5-ethylbenzene	1048	1043	1165			1043	1167	1448
1,4-Dimethyl-2-ethylbenzene	1060	1060	1181			1054	1183	1487
1-Methyl-3-*tert*-butylbenzene	1062	1068	1184			1051	1186	1444
1,3-Dimethyl-4-ethylbenzene	1067	1060	1186			1060	1189	1493
tert-Pentylbenzene	1070	1070	1189			1062	1193	1447
1,2-Dimethyl-4-ethylbenzene	1072	1066	1192			1065	1194	1499
1,3-Dimethyl-2-ethylbenzene	1072	1070	1200			1065	1202	1527
1-Methyl-4-*tert*-butylbenzene	1076	1072	1194		1355	1069	1195	1455
sec-Pentylbenzene	1078	1082	1200			1072	1201	1437
1,2-Dimethyl-3-ethylbenzene	1088	1082		1220	1431	1081	1218	1544
1,2,4,5-Tetramethylbenzene		1095		1232	1446	1100	1229	1558
1,2,3,5-Tetramethylbenzene	1098		1239	1456	1105	1236	1565	
1,3-Diisopropylbenzene	1135		1243	1376	1115	1241	1459	
n-Pentylbenzene	1135		1261	1433	1129	1258	1506	
1,2,3,4-Tetramethylbenzene	1126		1271		1126	1266	1619	
1,4-Diisopropylbenzene	1152		1271	1418	1149	1274	1494	
1,3-Dimethyl-5-*tert*-butylbenzene	1152		1275	1431	1147	1272	1513	
1,3,5-Triethylbenzene	1158		1315	1476	1188	1314	1547	
Pentamethylbenzene			1391		1251	1390	1738	
1,3,5-Triisopropylbenzene				1495	1283	1406	1556	

Table 30 (continued)
ALKYL BENZENES (C-6—C-16)

Column Packing: P1 = 50-m by 0.2-mm I.D. stainless-steel capillary column coated with squalane. P2 = 25.5-m by 0.50-mm I.D. glass capillary column coated with SE-30. P3 = 10.5-m by 0.25-mm I.D. stainless-steel capillary column coated with Citroflex A-4. P4 = 10-m by 0.25-mm I.D. stainless-steel capillary column coated with Carbowax 6000. P5 = 100-m by 0.25-mm I.D. stainless-steel capillary column coated with squalane. P6 = 50-m by 0.25-mm I.D. stainless-steel capillary column coated with acetyltributyl citrate. P7 = 50-m by 0.25-mm I.D. stainless-steel capillary column coated with 1,2,3-triscyanoethoxypropane.

REFERENCES

1. **Svob, V. and Deur-Siftar, D.,** *J. Chromatogr.,* 91, 677, 1974.
2. **Sojak, L. and Rijks, J. A.,** *J. Chromatogr.,* 119, 505, 1976.

Table 31
ALKYL BENZENES (C-6—C-14)

Column packing	P1	P1	P1	P1	P2
Temperature (°C)	250	270	300	320	100
References	1	1	1	1	2
Compound			I		
Benzene	561				572
Toluene	699				704
Ethylbenzene	762				761
m-Xylene	824				828
o-Xylene	836				838
p-Xylene	840				838
i-Propylbenzene	798				793
n-Propylbenzene	852				849
1-Methyl-4-ethylbenzene	882				
1-Methyl-2-ethylbenzene	888				
1-Methyl-3-ethylbenzene	894				
1,3,5-Trimethylbenzene	945				949
1,2,4-Trimethylbenzene	970				966
1,2,3-Trimethylbenzene	979				973
tert-Butylbenzene		845			
sec-Butylbenzene		862			
i-Butylbenzene		910			
1-Methyl-4-isopropylbenzene		918			
1,3-Diethylbenzene		941			
n-Butylbenzene		945			936
1-Methyl-4-*n*-propylbenzene		988			
1,4-Diethylbenzene		954			
1,2-Diethylbenzene		941			
1-Methyl-2-*n*-propylbenzene		980			
1,4-Dimethyl-2-ethylbenzene		1019			
1,3-Dimethyl-4-ethylbenzene		1015			
1,2-Dimethyl-4-ethylbenzene		1021			
1,2,3,5-Tetramethylbenzene		1107			1101
1,2,4,5-Tetramethylbenzene		1111			1097
1,2,3,4-Tetramethylbenzene		1131			
sec-Pentylbenzene		933			
n-Pentylbenzene		1045			1018
1-Ethyl-4-*n*-propylbenzene		1047			
1-Methyl-4-*n*-butylbenzene		1078			

Table 31 (continued)
ALKYL BENZENES (C-6—C-14)

Compound	I	
Pentamethylbenzene	1271	
1,3-Diisopropylbenzene	998	
1,4-Diisopropylbenzene	1015	
1,3-Dimethyl-5-*tert*-butylbenzene	1021	
1-*i*-Propyl-4-*n*-propylbenzene	1079	
1,3-Di-*n*-propylbenzene		1121
n-Hexylbenzene		1140
1-Ethyl-4-*n*-butylbenzene		1142
1,4-Di-*n*-propylbenzene		1149
1-Methyl-4-*n*-pentylbenzene		1178
1-*n*-Propyl-4-*n*-butylbenzene		1240
1-Ethyl-4-*n*-pentylbenzene		1244
1-Methyl-4-*n*-hexylbenzene		1273
Hexamethylbenzene		1414
1,4-Di-*tert*-butylbenzene		1091
1-*tert*-Butyl-4-isobutylbenzene		1205
1-*tert*-Butyl-4-*sec*-butylbenzene		1138
1,4-Di-*sec*-butylbenzene		1182
1-*sec*-Butyl-4-isobutylbenzene		1235
1,4-Diisobutylbenzene		1281
1-*tert*-Butyl-4-*n*-butylbenzene		1216
1-*sec*-Butyl-4-*n*-butylbenzene		1245
1-Isobutyl-4-*n*-butylbenzene		1309
1,4-Di-*n*-butylbenzene		1334
1-*n*-Propyl-4-n-pentylbenzene		1347

Column Packing: P1 = Column packed with graphitized thermal carbon black. P2 = Column packed with graphitized thermal carbon black.

REFERENCES

1. **Engewald, W., Wennrich, L., and Porschmann, J.,** *Chromatographia,* 11, 434, 1978.
2. **Kalashnikova, E. V., Kiselev, A. V., Poshkus, D. P., and Shcherbakova, K. D.,** *J. Chromatogr.,* 119, 233, 1976.

Table 32
ALKYL BENZENES (C-7—C-10)

Column packing	P1	P1	P1	P2	P2
Temperature (°C)	60	72	82	92	115
Reference	1	1	1	1	1
Compound			I		
Toluene					761
1,2-Dimethylbenzene					888
1,3-Dimethylbenzene	1153	1164	1172	863	867
1,4-Dimethylbenzene	1147	1158	1167	861	866
1-Methyl-4-ethylbenzene	1228	1229	1249	950	956
1-Methyl-3-ethylbenzene	1230	1240	1249	948	953
1-Methyl-2-ethylbenzene	1263	1267	1285	964	969
1-Methyl-3-*n*-propylbenzene	1299	1311	1320	1033	1037
1-Methyl-4-*n*-propylbenzene	1300	1312	1323	1039	1044
1-Methyl-2-*n*-propylbenzene	1329	1342	1353	1045	1051
1-Methyl-3-isopropylbenzene	1268	1278	1289	1002	1006
1-Methyl-4-isopropylbenzene	1270	1280	1290	1010	1015
1-Methyl-2-isopropylbenzene	1299	1313	1320	1015	1025

Column Packing: P1 = 45-m by 0.2-mm I.D. capillary column coated with polyethylene glycol 400. P2 = 45-m by 0.2-mm I.D. or 200-m by 0.2-mm I.D. capillary column coated with squalane.

REFERENCE

1. **Krupcik, J., Liska, O., and Sojak, L.,** *J. Chromatogr.,* 51, 119, 1970.

Table 33
ALKYL BENZENES (C-7—C-14)

Column packing	P1	P2	P3	P1	P1	P4	P1	P5
Temperature (°C)	100	100	100	86	115	86	80	80
Reference	1	1	1	1	1	1	1	1
Compound				I				
Toluene	760	862	920	754	761	837	752	883
Ethylbenzene	850	950	1007	844	851	923	842	974
1,4-Dimethylbenzene	865	960	1017	858	865	936	856	980
1,3-Dimethylbenzene	865	964	1022	860	866	940	858	984
1,2-Dimethylbenzene	886	992	1053	880	888	965	877	1010
i-Propylbenzene	908	1004	1060	904	910		901	1032
n-Propylbenzene	938	1037	1092	933	940	1012	930	1060
1-Methyl-3-ethylbenzene	950	1048	1105	946	952	1025	943	1071
1-Methyl-4-ethylbenzene	954	1048	1105	948	955	1027	945	1071
1-Methyl-2-ethylbenzene	966	1069	1131	961	969	1047	958	1092
tert-Butylbenzene	972	1068	1123				967	1099
sec-Butylbenzene	990	1085	1138				983	1112
i-Butylbenzene	991	1084	1134				982	1109
1,3-Diethylbenzene	1027	1126	1182	1025				
n-Butylbenzene	1037	1134	1192	1033	1040		1029	1160
1-Methyl-1-isopropylbenzene							1004	1130
1-Methyl-3-*n*-propylbenzene							1028	1153
1-Methyl-4-*n*-propylbenzene	1040	1135	1188				1033	1157
1,4-Diethylbenzene	1040	1135	1190				1034	1160

Table 33 (continued)
ALKYL BENZENES (C-7—C-14)

Compound	I						
1-Methyl-2-*n*-propylbenzene	1046	1146	1207	1042	1050	1039	1172
1-Methyl-4-*tert*-butylbenzene						1069	1195
sec-Pentylbenzene	1078	1172	1224			1072	1201
1-Ethyl-4-*n*-propylbenzene	1126	1218	1272				
n-Pentylbenzene	1133	1233	1289		1138	1129	1258
1-Methyl-4-*n*-butylbenzene	1139	1234	1288				
1,4-Diisopropylbenzene	1154	1239	1289			1149	1274
1-Isopropyl-4-*n*-propylbenzene	1182	1270	1320				
1,3-Di-*n*-propylbenzene	1196	1268	1337				
1,4-Di-*n*-propylbenzene	1212	1302	1353				
1-Ethyl-4-*n*-butylbenzene	1227	1319	1372				
n-Hexylbenzene	1231	1332	1387				
1-Methyl-4-*n*-pentylbenzene	1236	1331	1385				
1,4-Di-*tert*-butylbenzene	1281	1360	1408				
1-*tert*-Butyl-4-isobutylbenzene	1291	1368	1412				
1-*tert*-Butyl-4-*sec*-butylbenzene	1291	1371	1418				
1,4-Di-*sec*-butylbenzene	1304	1384	1428				
1-*sec*-Butyl-4-isobutylbenzene	1306	1384	1426				
1,4-Diisobutylbenzene	1308	1384	1425				
1-*n*-Propyl-4-*n*-butylbenzene	1311	1402	1452				
1-Ethyl-4-*n*-pentylbenzene	1322	1415	1468				
1-Methyl-4-*n*-hexylbenzene	1333	1429	1482				
1-*tert*-Butyl-4-*n*-butylbenzene	1345	1429	1479				
1-*sec*-Butyl-4-*n*-butylbenzene	1358	1440	1492				
1-Isobutyl-4-*n*-butylbenzene	1360	1442	1488				
1,4-Di-*n*-butylbenzene	1411	1502	1552				
1-*n*-Propyl-4-*n*-pentylbenzene	1419	1498	1548				
1-Ethyl-4-*n*-hexylbenzene	1421	1512	1567				

Column Packing: P1 = squalane. P2 = Ucon LB-550X. P3 = Ucon 50-HB-280X. P4 = dibutylphthalate. P5 = acetyltributyl citrate.

REFERENCE

1. **Tejedor, J. N.,** *J. Chromatogr.,* 177, 279, 1979.

<table>
<tr><td colspan="2" align="center">**Table 34**
ARYL ALKENES (C-8—C-11)</td></tr>
</table>

Column packing	P1
Temperature (°C)	65
Reference	1

Compound	I
Styrene	875
Allylbenzene	929
α-Methylstyrene	963
3-Methylstyrene	973
4-Methylstyrene	978
2-Methylstyrene	991
Isoallylbenzene	1004
Indene	1018
β-Methylallylbenzene	1022
1-Methyl-3-allylbenzene	1028
1-Methyl-4-allylbenzene	1033
α-Ethylstyrene	1058
1-Methyl-3-α-methylstyrene	1061
3-Ethylstyrene	1064
1-Methyl-4-α-methylstyrene	1066
4-Ethylstyrene	1072
3,5-Dimethylstyrene	1073
2,5-Dimethylstyrene	1076
2,4-Dimethylstyrene	1080
1,3-Divinylbenzene	1085
2-Ethylstyrene	1086
α,β-Dimethylstyrene	1091
1,4-Divinylbenzene	1096
2,3-Dimethylstyrene	1097
β-Ethylstyrene	1098
3,4-Dimethylstyrene	1100
1-Methyl-3-isoallylbenzene	1102
1-Methyl-4-isoallylbenzene	1104
3-Isopropylstyrene	1117
2,5-Dimethylstyrene	1118
4-Isopropylstyrene	1136
1-Vinyl-3-α-methylstyrene	1196
1-Vinyl-4-α-methylstyrene	1218

Column Packing: P1 = 25.5-m by 0.50-mm I.D. glass capillary column coated with SE-30.

REFERENCE

1. **Svob, V. and Deur-Siftar, D.**, *J. Chromatogr.*, 91, 677, 1974.

<table>
<tr><td colspan="2" align="center">**Table 35**
ALKYL BENZENES (C-9—C-12)</td></tr>
</table>

Column packing	P1
Temperature (°C)	120
Reference	1

Compound	I[a]
1-Methyl-4-butylbenzene	1310
1-Methyl-4-*tert*-butylbenzene	1313
tert-Pentylbenzene	1321
1,4-Dimethyl-2-ethylbenzene	1341
1,2,3-Trimethylbenzene	1342
1,3-Dimethyl-4-ethylbenzene	1348
1-Methyl-3-*tert*-butylbenzene	1348
1,2-Dimethyl-4-ethylbenzene	1354
1-Methyl-3,5-diethylbenzene	1364
1,4-Diisopropylbenzene	1368
1,3-Dimethyl-2-ethylbenzene	1383
n-Pentylbenzene	1383
3,4-Dimethyl-isopropylbenzene	1392
1,2-Dimethyl-3-ethylbenzene	1406
1,2,4,5-Tetramethylbenzene	1418
1,2,3,5-Tetramethylbenzene	1430
1,2,3,4-Tetramethylbenzene	1505
1,2,3,4-Tetrahydronaphthalene	1549

Column Packing: P1 = 300-ft by 0.01-in. I.D. stainless-steel column coated with 1,2,3-tris-2-cyanoethoxypropane, followed by a 150-ft by 0.01-in. I.D. stainless-steel column coated with silicone DC-550.

[a] Retention indices calculated from retention data given in Reference 1.

REFERENCE

1. **Stuckey, C. L.**, *J. Chromatogr. Sci.*, 9, 575, 1971.

Table 36
ALKYL BENZENES (C-11—C-16)

Column packing	P1	P2
Temperature (°C)	95.4	95.2
Reference	1	1

Compound	I	
1-Ethyl-2-propylbenzene	1115	1247
1-Methyl-2-butylbenzene	1140	1272
1,2-Di-*n*-propylbenzene	1188	1318
1-Ethyl-2-butylbenzene	1208	1340
n-Hexylbenzene	1228	1359
1-Methyl-2-pentylbenzene	1235	1368
1-Propyl-2-butylbenzene	1279	1409
1-Ethyl-2-pentylbenzene	1302	1435
4-Phenyloctane	1311	
3-Phenyloctane	1322	
n-Heptylbenzene	1328	1460
1-Methyl-2-hexylbenzene	1332	1465
2-Phenyloctane	1358	
1,2-Di-*n*-butylbenzene	1370	1502
1-Propyl-2-pentylbenzene	1372	1502
1-Ethyl-2-hexylbenzene	1398	1530
n-Octylbenzene	1427	1560
1-Methyl-2-heptylbenzene	1431	1565
1-Butyl-2-pentylbenzene	1461	1592
1-Propyl-2-hexylbenzene	1467	1598
1-Ethyl-2-heptylbenzene	1496	1630
n-Nonylbenzene	1527	1660
1-Methyl-2-octylbenzene	1530	1665
1,2-Di-*n*-pentylbenzene	1552	1684
1-Butyl-2-hexylbenzene	1555	1686
1-Propyl-2-heptylbenzene	1564	1696
1-Ethyl-2-octylbenzene	1595	1729
n-Decylbenzene	1627	1760
1-Methyl-2-nonylbenzene	1630	1764

Column Packing: P1 = 20-m capillary column coated with squalane. P2 = 20-m capillary column coated with acetyl tri-*n*-butyl citrate.

REFERENCE

1. **Sojak, L., Janak, J., and Rijks, J. A.,** *J. Chromatogr.*, 135, 71, 1977.

Table 37
BIPHENYLS AND DIPHENYLALKANES (C-12—C-15)

Column packing	P1	P1	P1	P2	P2	P2	P3	P3
Temperature (°C)	200	185	170	200	185	170	185	170
Reference	1	1	1	1	1	1	1	1
Compound					I			
Biphenyl	1475	1459	1446	1737	1719	1705	2078	2029
2-Methylbiphenyl	1464	1449	1438	1709	1694	1683	2000	1960
3-Methylbiphenyl	1571	1556	1545	1833	1812	1801	2167	2120
4-Methylbiphenyl	1587	1570	1558	1845	1824	1811	2182	2134
2,3-Dimethylbiphenyl	1588	1574	1562	1842	1827	1811	2137	2100
2,4-Dimethylbiphenyl	1569	1556	1545	1813	1800	1784	2100	2063
2,5-Dimethylbiphenyl	1557	1544	1535	1800	1788	1774	2086	2050
2,6-Dimethylbiphenyl	1489	1477	1465	1721	1705	1693	1983	1948
3,4-Dimethylbiphenyl	1713	1697	1683	1981	1962	1946	2324	2279
3,5-Dimethylbiphenyl	1667	1652	1641	1928	1912	1898	2252	2211
2,2'-Dimethylbiphenyl	1471	1458	1446	1700	1685	1671	1950	1916
2,3'-Dimethylbiphenyl	1548	1536	1526	1797	1780	1768	2074	2037
2,4'-Dimethylbiphenyl	1570	1556	1546	1818	1800	1787	2100	2061
3,3'-Dimethylbiphenyl	1669	1655	1642	1928	1910	1895	2254	2212
3,4'-Dimethylbiphenyl	1683	1668	1656	1941	1922	1906	2267	2224
4,4'-Dimethylbiphenyl	1697	1681	1669	1952	1933	1916	2281	2236
2-Ethylbiphenyl	1513	1500	1488	1757	1743	1729	2033	1994
3-Ethylbiphenyl	1647	1633	1621	1909	1892	1877	2234	2190
4-Ethylbiphenyl	1678	1661	1647	1938	1920	1903	2268	2220
2-Isopropylbiphenyl	1524	1512	1500	1763	1750	1738	2015	1980
3-Isopropylbiphenyl	1687	1671	1660	1939	1923	1909	2244	2204
4-Isopropylbiphenyl	1730	1713	1700	1986	1968	1953	2300	2255
3,5-Diisopropylbiphenyl				2077	2066			
3,3'-Diisopropylbiphenyl				2132	2119			
3,4'-Diisopropylbiphenyl				2191	2175			
4,4'-Diisopropylbiphenyl				2239	2223			
Diphenylmethane	1503	1488	1477	1774	1756	1745	2092	2047
2-Methyldiphenylmethane	1601	1583	1570	1868	1850	1835	2179	2135
3-Methyldiphenylmethane	1591	1576	1565	1862	1847	1834	2168	2128
4-Methyldiphenylmethane	1611	1594	1582	1876	1859	1844	2185	2144
1,1-Diphenylethane	1548	1532	1522	1818	1800	1790	2116	2074
1,2-Diphenylethane	1593	1576	1563	1862	1841	1827	2171	2126
1,4-Diphenylbutane	1810	1793	1780	2085	2066	2052	2398	2355
trans-Stilbene	1803	1784		2109	2088	2071	2547	
cis-Stilbene	1585	1570		1866	1847	1832	2205	
α-Methylstilbene	1776	1760		2071	2053	2038	2446	

Column Packing: P1 = 50-m by 0.25-mm I.D. stainless-steel capillary column coated with Apiezon L. P2 = 50-m by 0.25-mm I.D. stainless-steel capillary column coated with polyphenyl ether. P3 = 50-m by 0.25-mm I.D. stainless-steel capillary column coated with SP-1000.

REFERENCE

1. **Kriz, J., Popl, M., and Mostecky, J.,** *J. Chromatogr.*, 97, 3, 1974.

Table 38
POLYCYCLIC AROMATIC HYDROCARBONS
(2—7 RINGS)

Column packing	P1
Temperature	(linear temperature program; see footnote)
Reference	1

Compound	I[a]
Naphthalene	1172
Tetrahydroacenaphthene	1357
Biphenyl	1364
o,o'-Bitolyl	1387
Acenaphthylene	1426
1,8-Dimethylnaphthalene	1450
Acenaphthene	1461
Perhydrophenanthrene	1503
Fluorene	1555
m,m'-Bitolyl	1574
9-Methylfluorene	1579
p,p'-Bitolyl	1590
Octahydroanthracene	1667
2-Methylfluorene	1673
1-Methylfluorene	1679
Octahydrophenanthrene	1693
Phenanthrene	1742
Anthracene	1752
2-Methylphenanthrene	1860
2-Methylanthracene	1872
1-Methylphenanthrene	1883
9-Methylanthracene	1901
9-Butylanthracene	1934
Fluoranthene	2011
Pyrene	2057
9,10-Dimethylanthracene	2084
p-Terphenyl	2154
1,2-Benzofluorene	2161
Retene	2176
2,3-Benzofluorene	2178
Triptycene	2179
1-Methylpyrene	2212
Benzo[*c*]phenanthrene	2336
o-Quaterphenyl	2365
9-Phenylanthracene	2374
9,10-Benzophenanthrene	2392
1,2-Benzanthracene	2400
Chrysene	2400
2,3-Benzanthracene	2426
3,4-Benzofluoranthene	2694
11,12-Benzofluoranthene	2702
7,12-Dimethylbenz[*a*]anthracene	2711
Benzo[*a*]pyrene	2778
Perylene	2800
20(3)-Methylcholanthrene	2900
m-Quaterphenyl	2923
7-Methylbenz[*ac*banthracene	2939
9,10-Diphenylanthracene	3000
o-Phenylenepyrene	3081
1,2,5,6-Dibenzanthracene	3095
1,2,3,4-Dibenzanthracene	3103

Table 38 (continued)
POLYCYCLIC AROMATIC HYDROCARBONS
(2—7 RINGS)

Compound	I[a]
1,2,7,8-Dibenzphenanthrene	3140
Benzo[*ghi*]perylene	3146
Ananthrene	3183
1,2,3,4-Dibenzopyrene	3468
Coronene	3523
1,2,4,5-Dibenzopyrene	3540
3,4,9,10-Dibenzopyrene	3570
3,4,8,9-Dibenzopyrene	3588

Column Packing: P1 = 33.3- or 16.6-m by 0.5- or 0.25-mm I.D. capillary column coated with SE-52, run either with a 5-min hold at 50°C, followed by a 6°/min temperature program from 50—320°C, or with a 5-min hold at 70°C, followed by a 4°/min temperature program from 70—320°C.

[a]   Retention indices (I) are based on a *n*-alkane scale using C-10—C-36 even numbered *n*-alkanes.

REFERENCE

1. **Beernaert, H.,** *J. Chromatogr.,* 173, 109, 1979.

Table 39
POLYCYCLIC AROMATIC HYDROCARBONS
(3—5 RINGS)

Column packing	P1	P1	P1	P1	P1	P1	P1	P1	P1	P1
Temperature (°C)	190	200	210	220	235	240	250	260	265	270
Reference	1	1	1	1	1	1	1	1	1	1

Compound	Retention time (min)									
Phenanthrene	12.9	9.6	7.5	5.6		3.2				
Anthracene	17.6	12.9	10.0	7.4		4.1				
Triphenylene					14.9		10.2		9.0	
Benz[*a*]anthracene					22.3		14.6		12.3	
Chrysene					29.1		18.7		15.6	
Benzo[*e*]pyrene								22.9		18.3
Benzo[*a*]pyrene								36.6		28.6

Column Packing: P1 = Phenanthrene and anthracene were run on a 6-ft by 0.25-in. O.D. glass column packed with 5% *N,N′*-bis[*p*-methoxybenzylidene]-α,α′-bi-*p*-toluidine (BMBT) on 100—120 mesh HP Chromosorb W at a He carrier flow rate of 21 mℓ/min; all other compounds were run on a 6-ft by 0.25-in. O.D. glass column packed with 2.5% of the same phase on the same support (He carrier flow rate of 20 mℓ/min for the middle three compounds; 57 mℓ/min for the last two compounds).

REFERENCE

1. **Janini, G. M., Johnston, K., and Zielinski, W. L., Jr.,** *Anal. Chem.,* 47, 670, 1975.

Table 40
POLYCYCLIC AROMATIC HYDROCARBONS
(5 RINGS)

Column packing	P1	P1	P1	P1	P1	P2	P2	P2	P2	P2	P2
Temperature (°C)	255	260	265	270	275	250	255	260	265	270	275
Reference	1	1	1	1	1	1	1	1	1	1	1
Compound	**Retention time (min)**										
Benzo[*k*]-fluoranthene	1.59	1.86	2.65	3.11	2.78	3.87	4.67	4.59	4.06	3.52	3.01
Benzo[*e*]pyrene	1.94	2.19	2.96	3.37	3.05	4.55	5.33	5.23	4.59	3.96	3.41
Perylene	2.36	2.67	3.67	4.22	3.78	5.47	6.49	6.35	5.57	4.79	4.11
Benzo[*a*]pyrene	2.65	3.23	4.71	5.46	4.87	6.38	7.90	7.75	6.74	5.79	4.92

Column Packing: P1 = 4-ft by 2-mm I.D. glass column packed with 2.5% *N,N′*-bis[p-phenylbenzylidene]-α,α′-bi-*p*-toluidine (BPhBT) on 100—120 mesh HP Chromosorb W at a He carrier flow rate of 30 mℓ/min. P2 = 4-ft by 2-mm I.D. glass column packed with 2.5% of a 1:1 mixture of BPhBT and *N,N′*-bis[*p*-hexyloxybenzylidene]-α,α′-bi-*p*-toluidine (BHxBT) on the same packing at the same carrier flow rate.

REFERENCE

1. **Janini, G. M., Muschick, G. M., Schroer, J. A., and Zielinski, W. L., Jr.,** *Anal. Chem.*, 48, 1879, 1976.

Table 41
POLYCYCLIC AROMATIC
HYDROCARCONS (2—7 RINGS)

Column packing	P1	P1	P2	P2
Temperature (°C)	230	270	230	270
Reference	1	1	2	2
Compound	**I[a]**			
Naphthalene	1784		1255	
Acenaphthene	2161		1550	
Fluorene	2456		1645	
Phenanthrene	2800		1836	
Anthracene	2874		1846	
Fluoranthrene	3204		2091	
Pyrene	3301		2139	
Triphenylene		4017		2525
Benz[a]anthracene		4169		2516
Chrysene		4198		2526
Benzo[k]fluoranthene		4629		2802
Benzo[e]pyrene		4650		2858
Perylene		4739		2888
Benzo[a]pyrene		4834		2870
3-Methylcholanthrene		4932		2959
Dibenz[ac]anthracene		5099		3142
Benzo[ghi]perylene		5262		3185
Dibenz[ah]anthracene		5325		3137
Anthanthrene		5439		3215
Benzo[b]chrysene		5488		3159
Coronene		5821		3498

Column Packing: P1 = 2-m by 2-mm I.D. glass column packed with N,N'-bis[p-methoxybenzylidene]-α,α'-bi-p-toluidine (BMBT) on 100—120 mesh HP Chromosorb W. P2 = 50-m by 0.3—0.5-mm I.D. glass capillary column coated with OV-101.

[a]   Retention indices (I) are based on an *n*-alkane scale.

REFERENCE

1. **Radecki, A., Lamparczyk, H., and Kaliszan, R.,** *Chromatographia,* 12, 595, 1979.
2. **Grimmer, G. and Bohnke, H.,** *J. Assoc. Off. Anal. Chem.,* 58, 725, 1975.

Table 42
POLYCYCLIC AROMATIC
HYDROCARBONS
(2—7 RINGS)

Column packing	P1	P2
Temperature (°C)	230	230
Reference	1	1
Compound	**I[a]**	
Naphthalene	1255	1499
2-Methylnaphthalene	1354	1595
1-Methylnaphthalene	1380	1628
Acenaphthylene	1519	1808
Acenaphthene	1550	1833
Fluorene	1645	1935
Phenanthrene	1836	2171
Anthracene	1846	2179
3-Methylphenanthrene	1938	2265
2-Methylphenanthrene	1945	2274
2-Methylanthracene	1955	2281
4-Methylphenanthrene	1959	2308
9-Methylphenanthrene	1959	2302
4,5-Dimethylphenanthrene	1963	2314
1-Methylanthracene	1966	2297
9-Methylanthracene	1999	2354
3,6-Dimethylphenanthrene	2037	2356
Fluoranthrene	2091	2462
Pyrene	2139	2529
Benzo[a]fluorene	2221	2607
Benzo[b]fluorene	2236	2621
Benzo[c]fluorene	2236	2628
4-Methylpyrene	2247	2629
1-Methylpyrene	2273	2265
Benzo[c]phenanthrene	2464	2881
Benzo[ghi]fluoranthene	2473	2881
Benzo[a]anthracene	2516	2935
Triphenylene	2525	2953
Chrysene	2526	2953
12-Methylbenz[a]anthracene	2631	3092
7-Methylbenz[a]anthracene	2669	3116
Benzo[b]fluoranthene	2795	3249
Benzo[k]fluoranthene	2802	3256
Benzo[j]fluoranthene	2796	3258
Benzo[e]pyrene	2858	3339
Benzo[a]pyrene	2870	3353
Perylene	2888	3383
3-Methylcholanthrene	2959	3421
Dibenz[aj]anthracene	3112	3573
Indeno[1,2,3-cd]pyrene	3131	3610
Dibenz[ah]anthracene	3137	3597
Dibenz[ac]anthracene	3142	3601
Benzo[b]chrysene	3159	3632
Picene	3159	3640
Benzo[ghi]perylene	3185	3656
Anthanthrene	3215	3695
Coronene	3498	3968
Naphtho[1,2,3,4-def]chrysene	3507	3971
Dibenzo[fg,op]tetracene	3508	3983
Benzo[rst]pentaphene	3526	4002
Dibenzo[b,def]chrysene	3537	4017

Table 42 (continued)
POLYCYCLIC AROMATIC
HYDROCARBONS
(2—7 RINGS)

Column Packing: P1 = 50-m by 0.3—0.5-mm glass capillary column coated with OV-101. P2 = 50-m by 0.3—0.5-mm glass capillary column coated with OV-17.

a   Retention indices (I) are based on an *n*-alkane scale.

REFERENCE

1.   **Grimmer, G. and Bohnke, H.**, *J. Assoc. Off. Anal. Chem.*, 58, 725, 1975.

Table 43
POLYCYCLIC AROMATIC HYDROCARBONS
(2—6 RINGS)

Column packing	P1
Temperature	(linear temperature program; see footnote)
Reference	1

Compound	I[a]
1,2-Dihydronaphthalene	197
1,4-Dihydronaphthalene	197
Tetralin	197
Naphthalene	*200*
2-Methylnaphthalene	218
1-Methylnaphthalene	221
1,2,2a,3,4,5-Hexahydroacenaphthylene	233
Biphenyl	234
2-Ethylnaphthalene	236
1-Ethylnaphthalene	237
2,6-Dimethylnaphthalene	238
2,7-Dimethylnaphthalene	238
2-Methylbiphenyl	239
1,3-Dimethylnaphthalene	240
1,7-Dimethylnaphthalene	241
1,6-Dimethylnaphthalene	241
2,2′-Dimethylbiphenyl	242
Diphenylmethane	243
2,3-Dimethylnaphthalene	244
1,4-Dimethylnaphthalene	244
1,5-Dimethylnaphthalene	245
Acenaphthylene	245
1,2-Dimethylnaphthalene	246
1,8-Dimethylnaphthalene	250
2-Ethylbiphenyl	251
Acenaphthene	251
4-Methylbiphenyl	255
3-Methylbiphenyl	255
2,3,6-Trimethylnaphthalene	263
1-Methylacenaphthylene	265
2,3,5-Trimethylnaphthalene	266
Fluorene	268
3,3′-Dimethylbiphenyl	272
9-Methylfluorene	272

Table 43 (continued)
POLYCYCLIC AROMATIC HYDROCARBONS
(2—6 RINGS)

Compound	I[a]
4,4'-Dimethylbiphenyl	275
9,10-Dihydroanthracene	285
9-Ethylfluorene	285
9,10-Dihydrophenanthrene	287
1,2,3,4,5,6,7,8-Octahydroanthracene	288
2-Methylfluorene	288
1-Methylfluorene	289
1,2,3,4,5,6,7,8-Octahydrophenanthrene	292
1,2,3,4-Tetrahydrophenanthrene	297
Phenanthrene	*300*
Anthracene	302
1-Phenylnaphthalene	315
1,2,3,10b-Tetrahydrofluoranthene	316
9-*n*-Propylfluorene	318
3-Methylphenanthrene	319
2-Methylphenanthrene	320
2-Methylanthracene	322
o-Terphenyl	322
4H-Cyclopenta[*def*]phenanthrene	322
9-Methylphenanthrene	323
4-Methylphenanthrene	323
1-Methylanthracene	323
1-Methylphenanthrene	324
9-*n*-Butylfluorene	329
9-Methylanthracene	329
4,5,9,10-Tetrahydropyrene	330
4,5-Dihydropyrene	330
2-Phenylnaphthalene	333
9-Ethylphenanthrene	337
2-Ethylphenanthrene	337
3,6-Dimethylphenanthrene	338
2,7-Dimethylphenanthrene	339
1,2,3,6,7,8-Hexahydropyrene	339
Fluoranthene	344
9-Isopropylphenanthrene	346
1,8-Dimethylphenanthrene	346
9-*n*-Hexylfluorene	349
9-*n*-Propylphenanthrene	350
Pyrene	351
9,10-Dimethylanthracene	355
9-Methyl-10-ethylphenanthrene	360
m-Terphenyl	361
p-Terphenyl	366
Benzo[*a*]fluorene	367
11-Methylbenzo[*a*]fluorene	367
9,10-Diethylphenanthrene	368
1-Methyl-7-isopropylphenanthrene	369
Benzo[b]fluorene	369
4-Methylpyrene	370
2-Methylpyrene	370
4,5,6-Trihydrobenz[*de*]anthracene	371
1-Methylpyrene	374
5,12-Dihydronaphthacene	382
9,10-Dimethyl-3-ethylphenanthrene	382
1-Ethylpyrene	385

Table 43 (continued)
POLYCYCLIC AROMATIC HYDROCARBONS
(2—6 RINGS)

Compound	I[a]
2,7-Dimethylpyrene	386
1,2,3,4,5,6,7,8,9,10,11,12-dodecahydrotriphenylene	386
1,1'-Binaphthyl	388
Benzo[*ghi*]fluoranthene	390
Benzo[*c*]phenanthrene	391
9-Phenylanthracene	396
Cyclopenta[*cd*]pyrene	397
Benz[*a*]anthracene	398
Chrysene	*400*
Triphenylene	400
1,2'-Binaphthyl	405
7-Benz[*de*]anthrene	407
9-Phenylphenanthrene	407
Naphthacene	408
11-Methylbenz[*a*] anthracene	413
2-Methylbenz[*a*]anthracene	414
1-Methylbenz[*a*]anthracene	414
1-*n*-Butylpyrene	415
1-Methyltriphenylene	416
9-Methylbenz[*a*]anthracene	416
3-Methylbenz[*a*]anthracene	417
9-Methyl-10-phenylphenanthrene	417
8-Methylbenz[*a*]anthracene	418
6-Methylbenz[*a*]anthracene	418
3-Methylchrysene	418
5-Methylbenz[*a*]anthracene	419
2-Methylchrysene	419
12-Methylbenz[*a*]anthracene	419
4-Methylbenz[*a*]anthracene	420
5-Methylchrysene	420
6-Methylchrysene	421
4-Methylchrysene	421
1-Phenylphenanthrene	422
1-Methylchrysene	423
7-Methylbenz[*a*]anthracene	423
o-Quaterphenyl	424
2,2'-Binaphthyl	424
1,3-Dimethyltriphenylene	432
1,12-Dimethylbenz[*a*]anthracene	437
Benzo[j]fluoranthene	441
Benzo[b]fluoranthene	442
Benzo[*k*]fluoranthene	443
7,12-Dimethylbenz[*a*]anthracene	443
1,6,11-Trimethyltriphenylene	446
Benzo[*e*]pyrene	451
Benzo[*a*]pyrene	453
Perylene	456
1,3,6,11-Tetramethyltriphenylene	462
3-Methylcholanthrene	468
m-Quaterphenyl	473
Indeno[1,2,3-cd]pyrene	482
Pentacene	487
p-Quaterphenyl	488

Table 43 (continued)
POLYCYCLIC AROMATIC HYDROCARBONS
(2—6 RINGS)

Compound	I[a]
Dibenz[*a,c*]anthracene	495
Dibenz[*a,h*]anthracene	495
Benzo[*b*]chrysene	498
Picene	*500*
Benzo[*ghi*]perylene	501
Dibenzo[*def,mno*]chrysene	504
2,3-Dihydrodibenzo[*def,mno*]chrysene	504

Column Packing: P1 = 12-m by 0.28—0.30-mm I.D. glass capillary column coated with SE-52 methylphenylsilicone run at a temperature program of 2°/min from 50—250°C.

[a] Retention indices (I) are based on a special retention index scale described in the reference in which the I values for naphthalene, phenanthrene, chrysene, and picene are predefined as 200, 300, 400, and 500 retention index units, respectively.

REFERENCE

1. **Lee, M. L., Vassilaros, D. L., White, C. M., and Novotny, M.,** *Anal. Chem.,* 51, 768, 1979.

Table 44
POLYCYCLIC AROMATIC HYDROCARBONS
(2—4 RINGS)

Column packing	P1	P1	P2	P2	P3	P3	P4	P4
Temperature (rate; °C/min)	1	2	1	2	1	2	1	2
Reference	1	1	1	1	1	1	1	1

Compound	Retention temperatures (°C)[a]							
Indene	361	370						
Naphthalene	371	383	381	399			375	391
Biphenyl	390	406						
Fluorene	413	431	439	459	421	437	425	447
Phenanthrene	435	454			445	465	448	469
Fluoranthene	463							
Pyrene	468		509	533			481	505
Chrysene			549	573				

Column Packing: P1 = 25-m by 0.25-mm I.D. glass capillary column coated with SE-52; initial temperature, 80°C. P2 = 3-m stainless-steel column packed with 6% OV-101 on Chromosorb G initial temperature, 70°C. P3 = 13-m PLOT glass capillary column coated with Apiezon L; initial temperature, 100°C. P4 = 23-m PLOT glass capillary column coated with OV-101; initial temperature, 70°C.

[a] Retention temperatures calculated from relationships derived in reference.

REFERENCE

1. **Grant, D. W. and Hollis, M. G.,** *J. Chromatogr.,* 158, 3, 1978.

Table 45
ALKYLATED NAPHTHALENES

Column packing	P1	P1
Temperature (°C)	280	280
Reference	1	1

Compound	$\log_{10} U_a$[a]	Relative retention
Naphthalene	0.296	1.00
1-Methylnaphthalene	0.735	2.75
2-Methylnaphthalene	0.781	3.06
1,8-Dimethylnaphthalene	1.263	9.27
1,5-Dimethylnaphthalene	1.271	9.44
1,4-Dimethylnaphthalene	1.281	9.66
1,3-Dimethylnaphthalene	1.298	10.05
1,6-Dimethylnaphthalene	1.327	10.74
1,2-Dimethylnaphthalene	1.344	11.17
2,6-Dimethylnaphthalene	1.367	11.78
2,3-Dimethylnaphthalene	1.386	12.30
1,3,7-Trimethylnaphthalene	1.780	30.48
2,3,5-Trimethylnaphthalene	1.815	33.04
2,3,6-Trimethylnaphthalene	1.848	35.65

Column Packing: P1 = 10-m by 0.4-mm I.D. glass capillary column coated with Sterling® MTG graphitized carbon black.

[a] The $\log_{10} U_a$ values are the retention volumes of the compounds in cm^3/m^2 of the stationary phase.

REFERENCE

1. **Gonnord, M.-F., Vidal-Madjar, C., and Guiochon, G.,** *J. Chromatogr. Sci.,* 12, 839, 1974.

Table 46
TERPENE HYDROCARBONS

Column packing	P1	P1	P1	P1	P2	P2	P2	P2
Temperature (°C)	150	175	190	200	150	175	190	200
Reference	1	1	1	1	1	1	1	1
Compound	I							
Apopinene	924	935	940		998	1007	1010	
Tricyclene	963	971	981		1018	1023	1032	
α-Pinene	967	978	983		1020	1034	1038	
Camphene	995	1007	1013		1069	1077	1083	
β-Pinene	1022	1034	1040		1102	1113	1121	
Pinane	1039	1052	1060		1080	1091	1098	
Isocamphane	1039	1052	1060		1083	1094	1103	
Farnesane		1366		1369		1357		1360
Farnesene		1432		1439		1579		1584
Clovene		1440		1467		1511		1538
Isocaryophyllene		1447		1467		1550		1572
Caryophyllene		1470		1492		1572		1596
Dihydrohumulene		1483		1503		1566		1591
Caryophyllane		1491		1512		1527		1550
Clovane		1498		1525		1542		1571
Dihydroisocaryophyllene		1503		1519		1572		1588
Tetrahydrohumulene		1504		1523		1567		1590
Humulene		1507		1531		1622		1655
Humulane		1521		1541		1551		1573

Column Packing	P3	P3	P3	P4	P4
Temperature (°C)	150	175	200	175	200
Reference	1	1	1	1	1
Compound	I				
Apopinene	1060				
Tricyclene	1070				
α-Pinene	1077				
Pinane	1129				
Camphene	1130				
Isocamphane	1134				
β-Pinene	1173				
Farnesane		1344	1350	1348	1340
Farnesene		1678	1680	1827	1864
Clovene		1601	1633	1774	1837
Isocaryophyllene		1648	1670	1833	1886
Caryophyllene		1672	1700	1877	1938
Dihydrohumulene		1655	1679	1845	1899
Caryophyllane		1589	1617	1706	1771
Clovane		1621	1657	1787	1851
Dihydroisocaryophyllene		1661	1682	1832	1882
Tetrahydrohumulene		1653	1668	1803	1857
Humulene		1736	1766	1950	2032
Humulane		1609	1637	1710	1773

Column Packing: P1 = Two serial-connected columns of 4-m by 0.25-in. O.D. A1 packed with 15% Apiezon L on 70—80 mesh Celite 545. P2 = Two serial-connected columns of 4-m by 0.25-in. O.D. A1 packed with 15% Emulphor 0 on 70—80 mesh Celite 545. P3 = Two serial-connected columns of 4-m by 0.25-in. O.D. Al packed with 15% polyethyleneglycol 20M on 70—80 mesh Celite 545. P4 = Two serial-connected columns of 4-m by 0.25-in. O.D. Al packed with 20% ethyleneglycol succinate on 80—100 mesh Chromosorb G.

REFERENCE

1. **Luisetti, R. U. and Yunes, R. A.,** *J. Chromatogr. Sci.,* 9, 624, 1971.

REFERENCES

1. **Ravey, M.,** *J. Chromatogr. Sci.,* 16, 79, 1978.
2. **Nabivach, V. M. and Kirilenko, A. V.,** *Khim. Tverd. Topl. (Moscow),* 3, 90, 1979. Chem. Abstr., 91, 116878a.
3. **Grant, D. W. and Hollis, M. G.,** *J. Chromatogr.,* 158, 3, 1978.
4. **Dracheva, S. I., Bryanskaya, E. K., Sabirova, G. V., and Zhurba, A. S.,** *Rasshir. Utochnenie Programmy Issled. Neftei, Sb. Kzbr. Dokl. Mater. Vses. Nauchno-Tekh. Konf. 1975,* 1976, 139; Chem. Abstr., 84, 186858k, 1978.
5. **Luskin, M. M. and Morris, W. E.,** Spec. Tech. Publ. 557, American Society for Testing and Materials, Philadelphia, 1975.
6. **Bird, W. L. and Kimball, J. L.,** Spec. Tech. Publ. 557, American Society for Testing and Materials, Philadelphia, 1975.
7. **Ullrich, D. and Dulson, W.,** *Chromatographia,* 10, 537, 1977.
8. **Mitra, G. D., Mohan, G., and Sinha, A.,** *J. Chromatogr.,* 99, 215, 1974.
9. **Vanheetum, R.,** *J. Chromatogr. Sci.,* 13, 150, 1975.
10. **Pacakova, V. and Ullmannova, H.,** *Chromatographia,* 7, 75, 1974.
11. **Yakimenko, L. V. and Chebonenko, N. D.,** *Neftepererab. Neftekhim. (Moscow),* 1972 (4), 13; Am. Pet. Inst. Abstr., 20, 2148, 1973.
12. **Soroka, J. M.,** *City Univ., N.Y., Diss.,* 1979.
13. **Hirsch, R. F., Gaydosh, R. J., and Chretien, J. R.,** *Anal. Chem.,* 52, 723, 1980.
14. **Brander, B.,** *Analusis,* 7, 505, 1979.
15. **Matsuoka, S., Takano, T., and Tamura, T.,** *Bunseki Kagaku,* 27, 777, 1978.
16. **Bender-Ogly, A. O., Arustamova, L. G., Sultanov, N. T., and Babaev, F. R.,** Zerb. Khim. Zh., 1, 129, 1977.
17. **Ryba, M.,** *J. Chromatogr.,* 123, 327, 1976.
18. **Nabivach, V. M.,** *Collect. Czech. Chem. Commun.,* 42 (1), 259, 1977.
19. **Voorkees, K. J., Hileman, F. D., and Einhorn, I. N.,** *Anal. Chem.,* 47, 2385, 1975.
20. **Parcher, J. F., Weiner, P. H., et al.,** *J. Chem. Eng. Data,* 20, 145, 1975.
21. **Kulikov, V. I. and Kuzyaeva, V. V.,** *Zh. Anal. Chem.,* 28, 1565, 1973.
22. **Howery, D. G.,** *Anal. Chem.,* 46, 829, 1974.
23. **Pacokova, V., Hoch, K., and Smolkova, E.,** *Chromatographia,* 6, 320, 1973.
24. **Guichard-Loudet, N.,** *Analusis,* 2, 247, 1973.
25. **Sojak, L., Druscova, A., and Janak, J.,** *Ropa Uhlie,* 14, 238, 1972; *A.P.I.A.* 19, 9198, 1972.
26. **Castello, G., Lunardelli, M., and Berg, M.,** *J. Chromatogr.,* 76, 31, 1973; **Castello, G., Berg, M., and Lunardelli, M.,** *J. Chromatogr.,* 79, 23, 1973.
27. **Dimov, N. and Shopov, D.,** *J. Chromatogr.,* 63, 223, 1971.
28. **Deans, D. R. and Scott, I.,** *Anal. Chem.,* 45, 1137, 1973.
29. **Bach, R. W., Doetsch, E., Friedrichs, H. A., and Marx, L.,** *Chromatographia,* 4, 561, 1971.
30. **Robinson, P. G. and Odell, A. L.,** *J. Chromatogr.,* 57, 1, 1971.
31. **Schomburg, G.,** *Chromatographia,* 4, 286, 1971.
32. **Weingaertner, E., Guer, T., and Bayunus, O.,** *Z. Anal. Chem.,* 254, 28, 1971.
33. **Dondi, F., Betti, A., and Bighi, C.,** *J. Chromatogr.,* 66, 191, 1972.
34. **Sokolowski, S., Leboda, R., Slonka, T., and Swewryniak, M.,** *J. Chromatogr.,* 139, 237, 1977.
35. **Tarjan, G., Kiss, A., Kocsis, G., Meszaros, S., and Takacs, J. M.,** *J. Chromatogr.,* 119, 327, 1976.
36. **Kalashnikova, E. V., Kiselev, A. V., Poshkus, D. P., and Shcherbakova, K. D.,** *J. Chromatogr.,* 119, 233, 1976.
37. **Vanheertum, R.,** *J. Chromatogr. Sci.,* 13, 150, 1975.
38. **Rijks, J. A. and Cramers, C. A.,** *Chromatographia,* 7, 99, 1974.
39. **Ladon, A. W. and Sandler, S.,** *Anal. Chem.,* 45, 921, 1973.
40. **Schomburg, G. and Dielmann, G.,** *J. Chromatogr. Sci.,* 11, 151, 1973.
41. **Fonkich, A. G., Ambartsumov, P. A., and Mamedova, V. M.,** *Neftepererab. Neftekhim. (Moscow),* 8, 36, 1973; *A.P.I.A.* 21, 2198, 1974.
42. **Sojak, L., Hrivnak, J., Majer, P., and Janak, J.,** *Anal. Chem.,* 45, 293, 1973.
43. **Gerasimov, M. and Badinska, K.,** *Khim. Ind. (Sofia),* 1972, 207.
44. **Ettre, L. S.,** *Chromatographia,* 6, 489, 1973.
45. **Ettre, L. S.,** *Chromatographia,* 6, 525, 1973.
46. **Ettre, L. S.,** *Chromatographia,* 7, 39, 1974.
47. **Goedert, M. and Guiochon, G.,** *Anal. Chem.,* 45, 1188, 1973.
48. **Goedert, M. and Guiochon, G.,** *Chromatographia,* 6, 39, 1973.
49. **Mitra, G. D., Mohan, G., and Sinha, A.,** *J. Chromatogr.,* 91, 633, 1974.

50. **Martin, F. D.,** U.S. Patent 3,714,813, February 6, 1973.
51. **Caesar, F.,** *Chromatographia,* 5, 173, 1972.
52. **Louis, R.,** *Erdol Kohle,* 24, 88, 1971.
53. **Preston, S. T., Jr.,** A Guide to the Analysis of Hydrocarbons by Gas Chromatography, 2nd ed., Polyscience Corp., Evanston, Il., 1969.
54. **Ebing, W.,** *Chromatographia,* 2, 442, 1969.
55. **Guermouche, M., Fatscher, M., and Vergnaud, J. M.,** *J. Chromatogr.,* 52, 9, 1970.
56. **Loewenguth, J. C.,** *Column Chromatography,* Kovats, E. sz., Ed., Swiss Chemists' Association 1970, 182.
57. **Kaiser, R.,** *Chromatographia,* 2, 383, 1969.
58. **Takacs, J. and Kralik, D.,** *J. Chromatogr.,* 50, 379, 1970.
59. **Cramers, C. A., Rijks, J. A.,** Pacakova, V., and De Andrade, R., *J. Chromatogr.,* 51, 13, 1970.
60. **Nabivach, V. M. and Kirilenko, A. V.,** *Chromatographia,* 13, 29, 1980.
61. **Nabivach, V. M. and Kirilenko, A. V.,** *Chromotographia,* 13, 93, 1980.
62. **Nabivach, V. M. and Kirilenko, A. V.,** *Vopr. Khim. Khim. Tekhnol.,* 52, 139, 1978.
63. **Spivakovskii, G. I., Tishchenko, A. I., Zaslavskii, I. I., and Wulfson, N. S.,** *J. Chromatogr.,* 144, 1, 1977.
64. **Nabivach, V. M. and Krivoruchko, I. S.,** *Vopr. Khimii Khim. Tekhnol. Rep. Mezhdved. Nauch. Tekhn. Sb.,* 47, 100, 1977.
65. **Nabivach, V. M.,** *Vopr. Khim. Khim. Tekhnol.,* 46, 113, 1977.
66. **Shlyakhov, A. F., Anvaer, B. I., Zolotareva, O. V., Romina, N. N., Novikova, N. V., and Koreshkova, R. I.,** *Zh. Anal. Khim.,* 30 (4), 788, 1975.
67. **Mitra, G. D., Mohan, G., and Sinha, A.,** *J. Chromatogr.,* 99, 215, 1974.
68. **Mitooka, M.,** *Bunseki, Kagaku,* 21 (11), 1447, 1972.
69. **Schomburg, G.,** *Chromatographia,* 4, 286, 1971.
70. **Robinson, P. G. and Odell, A. L.,** *J. Chromatogr.,* 57, 11, 1971.
71. **Louis, R.,** *Erdoel Kohle, Erdgas, Petrochem.,* 24 (2), 88, 1971.
72. **Dimov, N., Shopov, D., and Petkova, T.,** *Proc. 3rd Anal. Chem. Conf.,* 1, 299, 1970.
73. **Lekova, K.,** *God. Nauchnoizsled. Inst. Khim. Prom.,* 11, 123, 1973.
74. **Poshkus, D. P. and Afreimovitch, A.,** *J. Chromatogr.,* 58, 55, 1971.
75. **Felscher, D.,** *Chem. Tech. (Leipzig),* 30, 147, 1978.
76. **Smolkova-Keulemansova, E.,** *Chromatographia,* 11, 70, 1978.
77. **Nabivach, V. M., Bur'yan, P., and Macak, J.,** *Zh. Anal. Khim.,* 33, 1416, 1978.
78. **Nabivach, V. M. and Kirilenko, A. V.,** *Vopr. Khim. Khim. Tekhnol.,* 52, 139, 1978; *Chem. Abstr.,* 91, 109859s.
79. **Nabivach, V. M. and Kirilenko, A. V.,** *Khim. Tverd. Topl. (Moscow),* 3, 90, 1979; Chem. Abstr., 91, 116878a.
80. **Dahlmann, G., Koeser, H. J. K., and Oelert, H. H.,** *Chromatographia,* 12, 665, 1979.
81. **Soroka, J. M.,** *City Univ., N.Y., Diss.,* 1979.
82. **Kalashnikova, E. V., Kiselev, A. V., Poshkus, D. P., and Sh'cherbakova, K. D.,** *J. Chromatogr.,* 119, 233, 1976.
83. **Nilsson, O.,** *Chromatographia,* 10, 5, 1977.
84. **Caesar, F.,** *Chromatographia,* 5, 173, 1972.
85. **Terauchi, O. and Kiwata, Y.,** *Aromatikkusu,* 25, 48, 1973.
86. **Sidorov, R. I., Khvostikova, A. A., and Bakhrusheva, G. I.,** *Zh. Anal. Khim.,* 32, 1650, 1977.
87. **Shlyakhov, A. F. and Koreshkova, R. I.,** *Tr., Vses. Nauchno Issled. Geologorazved. Neft. Inst.,* 196, 18, 1976.
88. **Dimov, N. and Papazova, D.,** *Chromatographia,* 12, 443, 1979.
89. **Dimov, M.,** *J. Chromatogr.,* 119, 109, 1976.
90. **Krawiec, Z., Gonnord, M. -F., Guiochon, G., and Chretien, J. R.,** *Anal. Chem.,* 51, 1655, 1979.
91. **Zhukhovitskii, A. A., Yanovskii, S. M., and Shvartsman, V. P.,** *Zh. Fiz. Khim.,* 52, 1442, 1978; *Chem. Abstr.,* 89, 139965z, 1978.
92. **Castello, G. and D'Amato, G.,** *J. Chromatogr.,* 116, 249, 1976.
93. **Hussey, C. L. and Parcher, J. F.,** *Anal. Chem.,* 46, 2237, 1974.
94. **Rappoport, S. and Gaeumann, T.,** *Helv. Chm. Acta,* 56, 1145, 1973.
95. **Hoshikawa, Y., Koike, K., and Kuriyama, T.,** *Jpn. Anal.,* 21, 307, 1972.
96. **Shlyakhov, A. F. and Telkova, M. S.,** *Zavod. Lab.,* 39, 9, 1973; Chem. Abstr., 78, 118957e, 1973.
97. **Jaeschke, A. and Rohrschneider, L.,** *Chromatographia,* 5, 333, 1972.
98. **Vigalok, R. V. and Vigdergauz, M. S.,** *Izv. Akad. Nauk S.S.R. Ser. Khim.,* 3, 718, 1972.
99. **Gerasimov, M. and Badinska, K.,** *Khim. Ind. (Sofia),* 1972, 207.
100. **Evans, M. P.,** *Chromatographia,* 6, 301, 1973.

101. Takacs, J., Szita, C., and Tarjan, G., *J. Chromatogr.*, 56, 1, 1971.
102. Takacs, E. C., Voros, J., and Takacs, J. M., *J. Chromatogr.*, 159, 297, 1978.
103. Sojak, L., Zahradnik, P., Leska, J., and Janak, J., *J. Chromatogr.*, 174, 97, 1979.
104. Raverdino, V. and Sassetti, P., *J. Chromatogr.*, 169, 223, 1979.
105. Lombosi, T. S., Lombosi, E. R., Bernat, I., Bernat, Zs. Sz., Takacs, E. C., and Takacs, J. M., *J. Chromatogr.*, 119, 307, 1976.
106. Dimov, N., *J. Chromatogr.*, 119, 109, 1976.
107. Sidorov, R. I. and Khovostikova, A. A., *Zh. Anal. Khim.*, 30, 340, 1975.
108. Sultanov, N. T. and Arustamova, L. G., *J. Chromatogr.*, 115, 553, 1975.
109. Castello, G. and D'Amato, G., *J. Chromatogr.*, 107, 1, 1975.
110. Wicar, S. and Novak, J., *J. Chromatogr.*, 95, 13, 1974.
111. Hartkopf, A., *J. Chromatogr. Sci.*, 10, 145, 1972.
112. Rohrschneider, L., *Anal. Chem.*, 45, 1241, 1973.
113. Castello, G., Lunardelli, M., and Berg, M., *J. Chromatogr.*, 76, 31, 1973.
114. Feltl, L., Smolkova, E., and Skurovcova, M., *Coll. Czech. Chem. Commun.*, 44, 1116, 1979.
115. Smolkova-Keulemansova, E., *Chromatographia*, 11, 70, 1978.
116. Jaworski, M. and Herzog, G., *Chem. Anal. (Warsaw)*, 24, 475, 1979.
117. Sojak, L., Zahradnik, P., Leska, J., and Janak, J., *J. Chromatogr.*, 174, 97, 1979.
118. Lukac, S. and Nguyen, T. Q., *Chromatographia*, 12, 231, 1979.
119. Dahlmann, G., Koeser, H. J. K., and Oelert, H. H., *Chromatographia*, 12, 665, 1979.
120. Soroka, J. M., *City Univ., N.Y., Diss.*, 1979.
121. Eisen, O., Orav, A., and Rang, S., *Chromatographia*, 5, 229, 1972.
122. Gerasimov, M. and Badinka, K., *Khim. Ind. (Sofia)*, 1972, 207.
123. Aime, P., Rang, S., and Eisen, O., *Isv. Akad. Nauk Est. S.S.R. Ser. Geol.*, 21 (1), 20, 1972.
124. Choubey, U. D. and Mitra, G. D., *Fert. Technol.*, 13 (4), 217, 1976.
125. Chre'tein, J. R. and Dubois, J. E., *Anal. Chem.*, 49, 747, 1977.
126. Choubey, V. D. and Mitra, G. D., *Fert. Technol.*, 13 (4), 217, 1976.
127. Orav, A., Kuningas, K., Rang, S., and Eisen, O., Eesti NSV Tead. Akad. Toim. Keem., 29, 18, 1980.
128. Janak, J. and Sojak, L., *Ber. Bunsenges. Phys. Chem.*, 77, 205, 1973.
129. Welsch, Th., Engewald, W., and Berger, P., *Chromatographia*, 11, 5, 1978.
130. Sidorov, R. I., Khvostikova, A. A., and Bakhrusheva, G. I., *Zh. Anal. Khim.*, 32, 1650, 1977.
131. Rang, S., Kuningas, K., Orav, A., and Eisen, O., *Chromatographia*, 10, 55, 1977.
132. Papazova, D. and Dimov, N., *J. Chromatogr.*, 137, 259, 1977.
133. Tarjan, G., *J. Chromatogr. Sci.*, 14, 309, 1976.
134. Takacs, J., Talas, Z., Bernath, I., Czako, G., and Fischer, A., *J. Chromatogr.*, 67, 203, 1972.
135. Eisen, O. G., Orav, A., and Rang, S., *Chromatographia*, 5, 229, 1972.
136. Sojak, L., Majer, P., Skalak, P., and Janak, J., *J. Chromatogr.*, 65, 137, 1972.
137. Ladon, A. W. and Sandler, S., *Column Chromatography*, Kovats, E. sz., Ed., Swiss Chemists' Association, 1970, 160.
138. Rang, S. A., Eisen, O. G., Kiselev, A. V., Meister, A. E., and Schcherbakova, K. D., *Chromatographia*, 8, 327, 1975.
139. Aime, P., Rang, S., and Eisen, O., *Isv. Akad. Nauk Est. S.S.R. Ser. Geol.*, 21 (1), 20, 1972.
140. Rang, S., Kuningas, K., Orav, A., and Eisen, O., *J. Chromatogr.*, 119, 451, 1976; 128, 59, 1976.
141. Rang, S., Kuningas, K., Orav, A., and Eisen, O., *J. Chromatogr.*, 128, 53, 1976.
142. Jaworski, M. and Herzog, G., *Chem. Anal. (Warsaw)*, 24, 475, 1979.
143. Engelward, W., Poerschmann, J., et al., *Z. Chem.*, 17 (10), 375, 1977.
144. Engelward, W., Epsh, K., et al., *J. Chromatogr.*, 119, 119, 1976.
145. Grenier-Loustalot, M. F., Bonastre, J., et al., *J. Chromatogr.*, 138, 63, 1977.
146. Gerasimov, M. and Badinska, K., *Khim. Ind. (Sofia)*, 1972, 207.
147. Dielmann, G., Schwengers, D., and Schomburg, G., *Chromatographia*, 7, 215, 1974.
148. Pascal, J. C., Heintz, M., Druilhe, A., and and Lefort, D., *Chromatographia*, 7, 236, 1974.
149. Besson, R. and Gaeumann, T., *Helv. Chim. Acta*, 56, 1159, 1973.
150. Howard, H. E., *Am. Chem. Soc. Div. Petrol. Chem. Prepr.*, 17, F54, 1972.
151. Lindeman, L. P., *Am. Chem. Soc. Div. Petrol. Chem. Prepr.*, 16, A85, 1971.
152. Vanek, J., Podrouzkova, B., and Landa, S., *J. Chromatogr.*, 52, 77, 1970.
153. Hala, S., Eyem, J., Burkhard, J., and Lanela, S., *J. Chromatogr. Sci.*, 8, 203, 1970.
154. Stopp, I., Engewald, W., Kuhn, H., and Welsch, T., *J. Chromatogr.*, 147, 21, 1978.
155. Dimov, N., *J. Chromatogr.*, 148, 11, 1978.
156. Rang, S., Kuningas, K., Orav, A., and Eisen, O., *Chromatographia*, 10, 115, 1977.
157. Sidorov, R. I. and Khovostikova, A. A., *Zh. Anal. Khim.*, 30, 340, 1975.
158. Takacs, J., Talas, Z., Bernath, I., Czako, G., and Fischer, A., *J. Chromatogr.*, 67, 203, 1972.

159. **Engewald, W., Epsch, K., Welsch, Th., and Graefe, J.,** *J. Chromatogr.,* 119, 119, 1976.
160. **Schomburg, G. and Dielmann, G.,** *Anal. Chem.,* 45, 1647, 1973.
161. **Engewald, W., Epsch, K., Graefe, J., and Weisch, T.,** *J. Chromatogr.,* 91, 623, 1974.
162. **Dielmann, G., Schwengers, D., and Schomburg, G.,** *Chromatographia,* 7, 215, 1974.
163. **Eisen, O. G., Orav, A., and Rang, S.,** *Chromatographia,* 5, 229, 1972.
164. **Shlyakhov, A. F. and Koreshkova, R. I.,** *Tr. Vses. N.-i. Geologorazved. Neft. In-t,* 193, 18, 1976.
165. **Vanek, J., Podrouzkova, B., and Landa, S.,** *J. Chromatogr.,* 52, 77, 1970.
166. **Luisetti, R. U. and Yunes, R. A.,** *J. Chromatogr. Sci.,* 9, 624, 1971.
167. **Smolkova-Keulemansova, E.,** *Chromatographia,* 11, 70, 1978.
168. **Nabivach, V. M., Bur'yan, P., and Macak, J.,** *Zh. Anal. Khim.,* 33, 1416, 1978.
169. **Nabivach, V. M. and Kirilenko, A. V.,** *Vopr. Khim. Khim. Tekhnol.,* 52, 139, 1978; *Chem. Abstr.,* 91, 109859s.
170. **Nabivach, V. M. and Kirilenko, A. V.,** *Khim. Tverd. Topl. (Moscow),* 1979 (3); *Chem. Abstr.,* 91, 116878a.
171. **Beernaert, H.,** *J. Chromatogr.,* 173, 109, 1979.
172. **Dahlmann, G., Koeser, H. J. K., and Oelert, H. H.,** *Chromatographia,* 12, 665, 1979.
173. **Schroeder, H.,** *J. High Resolut. Chromatogr. Chromatogr. Commun.,* 3, 38, 1980.
174. **Laub, R. J., Ramamurthy, V., and Pecsok, R. L.,** *Anal. Chem.,* 46, 1659, 1974.
175. **Saha, N. C. and Mitra, G. D.,** *J. Chromatogr.,* 71, 171, 1972.
176. **Louis, R.,** *Erdoel Kohle, Erdgas, Petrochem. Brennst.-Chem.,* 25, 582, 1972.
177. **Deans, D. R. and Scott, I.,** *Anal. Chem.,* 45, 1137, 1973.
178. **Gerasimov, M. and Badinska, K.,** *Khim. Ind. (Sofia),* 1972, 207.
179. **Weingaertner, E., Guer, T., and Bayunus, O.,** *Z. Anal. Chem.,* 254, 28, 1971.
180. **Roseira, A. N.,** *Rev. Brasil. Tecnol.,* 2 (1), 1, 1971.
181. **Engewald, W., Wennrich, L., and Poerschmann, J.,** *Chromatographia,* 11, 434, 1978.
182. **Vigdergauz, M. S., Gabitova, R. K., Vigalok, R. V., and Novikova, I. R.,** *Zavod. Lab.,* 45, 894, 1979.
183. **Vigalok, R. V. and Vigdergauz, M. S.,** *Izv. Akad. Nauk S.S.R., Ser. Khim.,* 1972 (3), 715.
184. **Tejedor, J. N.,** *J. Chromatogr.,* 177, 279, 1979.
185. **Engelward, W. and Wennrich, L.,** *Chromatographia,* 9, 540, 1976.
186. **Sojak, L. and Rijks, J. A.,** *J. Chromatogr.,* 119, 505, 1976.
187. **Kriz, J., Popl, M., and Mostecky, J.,** *J. Chromatogr.,* 97, 3, 1974.
188. **Gonnord, M. F., Vidal-Madjar, C., and Guiochon, G.,** *J. Chromatogr. Sci.,* 12, 839, 1974.
189. **Martire, D. E.,** *Anal. Chem.,* 46, 626, 1974.
190. **Bartle, K. D.,** *Anal. Chem.,* 45, 1831, 1973.
191. **Vigalok, R. V. and Vigdergauz, M. S.,** *Izv. Akad. Nauk S.S.R., Ser. Khim.,* 3, 715, 1972.
192. **Mitooka, M.,** *Bunseki Kagaku,* 21, 1437, 1972; *Chem. Abstr.,* 79, 26901a, 1973.
193. **Schomburg, G., Ziegler, E., et al.,** *Angew. Chem. Int. Ed. Engl.,* 11, 366, 1972.
194. **Mostecky, J., Popl, M., and Kriz, J.,** *Anal. Chem.,* 42, 1132, 1970.
195. **Duerbeck, H. W.,** *Z. Anal. Chem.,* 251, 108, 1970.
196. **Krupcik, J., Liska, O., and Sojak, L.,** *J. Chromatogr.,* 51, 119, 1970.
197. **Cook, L. E. and Raushel, F. M.,** *J. Chromatogr.,* 65, 556, 1972.
198. **Engewald, W., Wennrich, L., and Porschmann, J.,** *Chromatographia,* 11, 434, 1978.
199. **Sojak, L.,** *J. Chromatogr.,* 148, 159, 1978.
200. **Tejedor, J. N.,** *J. Chromatogr.,* 177, 279, 1979.
201. **Engewald, W. and Wennrich, L.,** *Chromatographia,* 9, 540, 1976.
202. **Sojak, L., Janak, J., and Rijks, J. A.,** *J. Chromatogr.,* 135, 71, 1977.
203. **Cno, A.,** *J. Chromatogr.,* 110, 233, 1975.
204. **Dimov, N.,** *J. Chromatogr.,* 137, 265, 1977.
205. **Sojak, L., Janak, J., and Rijks, J. A.,** *J. Chromatogr.,* 138, 119, 1977.
206. **Louis, R.,** *Erdoel Kohle, Erdgas, Petrochem. Brennst.-Chem.,* 25 (10), 582, 1972.
207. **Kriz, J., Popl, M., and Mostecky, J.,** *J. Chromatogr.,* 97, 3, 1974.
208. **Svob, V. and Deur-Siftar, D.,** *J. Chromatogr.,* 91, 677, 1974.
209. **West, S. D. and Hall, R. C.,** *J. Chromatogr. Sci.,* 13, 5, 1975.
210. **Wicar, S. and Novak, J.,** *J. Chromatogr.,* 95, 13, 1974.
211. **Rohrschneider, L.,** *Anal. Chem.,* 45, 1241, 1973.
212. **Matsumoto, H., Futami, H., Morita, F., and Morita, T.,** *Bull. Chem. Soc. Jpn.,* 44, 3170, 1971.
213. **Cook, L. and Raushel, F.,** *J. Chromatogr.,* 65, 556, 1972.
214. **Fike, W. W.,** *J. Chromatogr. Sci.,* 11, 25, 1973.
215. **Radecki, A., Lamparczyk, H., and Kaliszan, R.,** *Chromatographia,* 12, 595, 1979.
216. **Grimmer, G. and Boehnke, H.,** *Fresenius' Z. Anal. Chem.,* 261, 310, 1972.

217. **Popl, M., Dolansky, V., and Mostecky, J.,** *J. Chromatogr.,* 117, 117, 1976.
218. **Lee, M., Vassilaros, D. L., White, C. M., and Novotny, M.,** *Anal. Chem.,* 51, 768, 1979.
219. **Lee, M., Vassilaros, D. L., White, C. M., and Novotny, M.,** *Anal. Chem.,* 51, 768, 1979.
220. **Mitooka, M.,** *Bunseki Kagaku,* 21, 1437, 1972; *Chem. Abstr.,* 79, 26901a, 1973.
221. **Mostecky, J., Popl, M., and Kriz, J.,** *Anal. Chem.,* 42, 1132, 1970.
222. **Kaliszan, R. and Lamparczyk, H.,** *J. Chromatogr. Sci.,* 16, 246, 1978.
223. **Lee, M. L., Vassilaros, D. L., White, C. M., and Novotny, M.,** *Anal. Chem.,* 51, 768, 1979.
224. **Radecki, A., Lamparczyk, H., and Kaliszan, R.,** *Chromatographia,* 12, 595, 1979.
225. **Deaconeasa, V., Constantinesau, T., and Trestianu, S.,** *Rev. Chim.,* 28, 777, 1977.
226. **Ferapontov, V. A., Ostapenko, E. G., Gverdsiteli, D. D., and Litvinov, V. P.,** *Izv. Akad. Nauk S.S.S.R. Ser. Khim.,* p. 2417, 1970.
227. **Bartle, K. D., Lee, M. L., and Wise, S. A.,** *Chromatographia,* 14, 69, 1981.
228. **Lee, M. L., Vassilaros, D. L., White, C. M., and Novotny, M.,** *Anal. Chem.,* 51, 768, 1979.

INDEX